ZIRCONIA '88

Advances in Zirconia Science and Technology

Proceedings of the international conference Zirconia '88—Advances in Zirconia Science and Technology, held in Bologna, Italy, 16–17 December 1988, organized by the Italian Ceramic Center of Bologna with the sponsorship of ENEA and Agip.

ZIRCONIA '88

Advances in Zirconia Science and Technology

Edited by

S. MERIANI
University of Trieste, Italy

and

C. PALMONARI
University of Bologna, Italy

ELSEVIER APPLIED SCIENCE
LONDON and NEW YORK

ELSEVIER SCIENCE PUBLISHERS LTD
Crown House, Linton Road, Barking, Essex IG11 8JU, England

Sole Distributor in the USA and Canada
ELSEVIER SCIENCE PUBLISHING CO., INC.
655 Avenue of the Americas, New York, NY 10010, USA

WITH 43 TABLES AND 216 ILLUSTRATIONS

© 1989 ELSEVIER SCIENCE PUBLISHERS LTD
Softcover reprint of the hardcover 1st edition 1989
British Library Cataloguing in Publication Data

Zirconia '88. *Conference. (Bologna, Italy)*
   Zirconia '88: advances in zirconia science and
   technology
   1. Zirconium industries & trades
   I. Title   II. Meriani, S.   III. Palmonari, C.
   338.4'7669735

   ISBN-13: 978-94-010-7005-8   e-ISBN-13: 978-94-009-1139-0
   DOI: 10.1007/978-94-009-1139-0

Library of Congress CIP data applied for

FOREWORD

This meeting, **ZIRCONIA '88** – Advances in Zirconia Science and Technology, was held within the framework of the **7th SIMCER** – International Symposium on Ceramics (Bologna, December 14–17, 1988) organized by the Italian Ceramic Center of Bologna, with the sponsorship of ENEA and Agip and the endorsement of the American Ceramic Society, and under the auspices of the European Ceramic Society.

In the year 1988, the University of Bologna celebrated its 900th Anniversary. ZIRCONIA '88 was one of the celebration events which brought together academics and researchers from all over the world.

Under the chairmanship of Prof. C. Palmonari, Director of the Italian Ceramic Center of the University of Bologna, the Organizing Committee consisting of J. Castaing (C.N.R.S. Meudon, France), S. Meriani (University of Trieste, Italy), V. Prodi (University of Bologna, Italy) and J. Routbort (U.S. Dept. of Energy, Washington, USA) conducted a conference program of 47 contributions presented to the 220 enrolled Zirconia participants, out of the 775 enlisted within the main SIMCER framework.

The aim of ZIRCONIA '88 was to follow the stream of the well known International Conferences on the Science and Technology of Zirconia held in Cleveland, Ohio (1980), Stuttgart, Federal Republic of Germany (1983) and Tokyo, Japan (1986). SlMCER's goal was to bring together not only scientists and engineers directly involved with "advanced" ceramics but also a larger audience connected to the nearby Italian Ceramic District of Sassuolo.

The papers collected herein represent the majority of the contributions brought to the meeting. They have been edited in alphabetical order which does not reflect their "oral" or "poster" origin at the meeting. Others will be published in the new journal, entitled "Ceramica Acta" presented at the Conference with extended abstracts from all the SIMCER sessions. Editors look forward to receiving further contributions.

These proceedings have been published with the shortest delay possible thanks to the Authors who promptly sent in their contributions, and thanks also to the Editorial Staff, both in Bologna and Trieste, who prepared the manuscript.

Editors:
S. Meriani and C. Palmonari

CONTENTS

Page

THE ELECTRICAL PROPERTIES OF GRAIN AND GRAIN BOUNDARIES
IN Y-DOPED FLUORITE TYPE OXIDES.

P. ABELARD and J.F. BAUMARD
URA CNRS 899, ENSCI 47-73 Av. A. Thomas, 87065 Limoges-
Cedex FRANCE

ABSTRACT The complex impedance technique permit a
separate study of the grain and grain boundary
properties. Information on the dynamics of the oxygen
vacancy motion is contained in the bulk dielectric
function. The grain boundary properties are discussed
with respect to the microstructure.

INTRODUCTION

A ceramic material has a complex microstructure
which comprises grains, grain boundaries and quite
often second phases. Since the work of Bauerle /1/, it
has become customary to study the electrical properties
of ionic conductors such as doped fluorite type oxides
with the aid of the complex impedance technique. The
principle of the measurements is very simple. An
alternating voltage of frequency f (=w/2π) is applied
to the sample. The in phase and out of phase components
of the current with respect to the voltage are
determined separately. The complex impedance data
measured for a spectrum of frequencies are displayed in
the complex impedance plane (Re Z, -Im Z), see Fig.1.
Possible artefacts have been discussed in the
literature /2-4/.

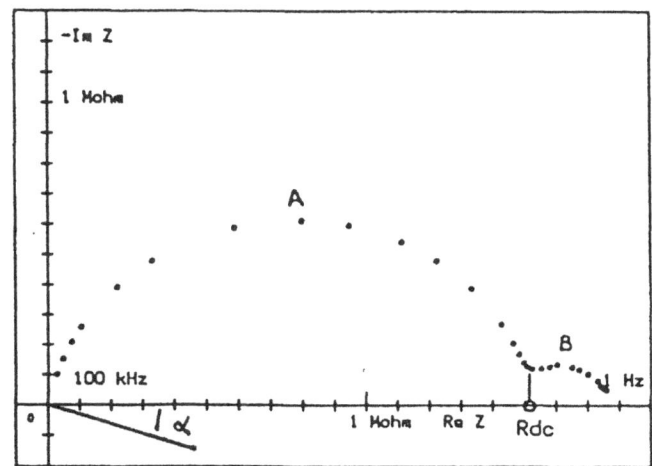

Figure 1: Example of a complex impedance spectra
obtained on a (CeO$_2$-Yb$_2$O$_3$) sample. A) grain- B) grain
boundary- arc of circle.

As a first approximation, each feature of the microstructure can be associated with an electrical circuit made of a resistor and a capacitor in parallel. The time constants $R_1 C_1$ of the grains and of the grain boundaries differ often by at least one order of magnitude and this explains why two distinct semicircles can be seen in the complex impedance plane. However this simple description does not permit a quantitative fit of the experimental data because the arcs of circle are depressed below the real axis by an angle α. Most often, a Cole-Cole expression /5/ is used to describe each contribution :

$$Z = \sum_i \frac{R_1}{1 + (jw\tau_1)^{1-n_i}} \qquad (1)$$

where n is simply related to the depressing angle, $\alpha = n.90°$, and the relaxation time τ_1 is equal to $R_1 C_1$. Going to lower frequencies or higher temperatures, other arcs of circle have been evidenced. They are associated with charge transfer phenomena at the electrodes /6/ and will not be discussed any further in this paper devoted to the electrical properties of the grain and grain boundaries

THE ELECTRICAL PROPERTIES OF THE BULK MATERIAL

From Eq.(1), it is sometimes inferred that the admittance Y of the bulk material obeys a power law as a function of frequency :

$$Y = 1/R_b + A (jw)^{1-n} \qquad (2)$$

where R_b is the bulk dc resistance of the sample. However, if such a dependence leads indeed to a depressed arc of circle, the inverse statement must be carefully scrutinized. This happens because a limited frequency range, $(-RC/10, 10\ RC)$ approximately, is sufficient to describe most part of the arc of circle while points outside this interval stack on both high and low frequency sides. Therefore the only way to ascertain Eq.(2) is to plot the data, and in the most efficient way the dielectric function $\varepsilon(w)$:

$$\varepsilon(w) = \frac{Y - 1/R_b}{j\ w} \qquad (3)$$

versus frequency on a logarithmic scale. On Figs.2 and 3 are depicted the experimental results obtained for the solid solution $(CeO_2 - Y_2 O_3)$ /7/. All the spectra can be satisfactorily fitted with a Cole-Cole expression !.

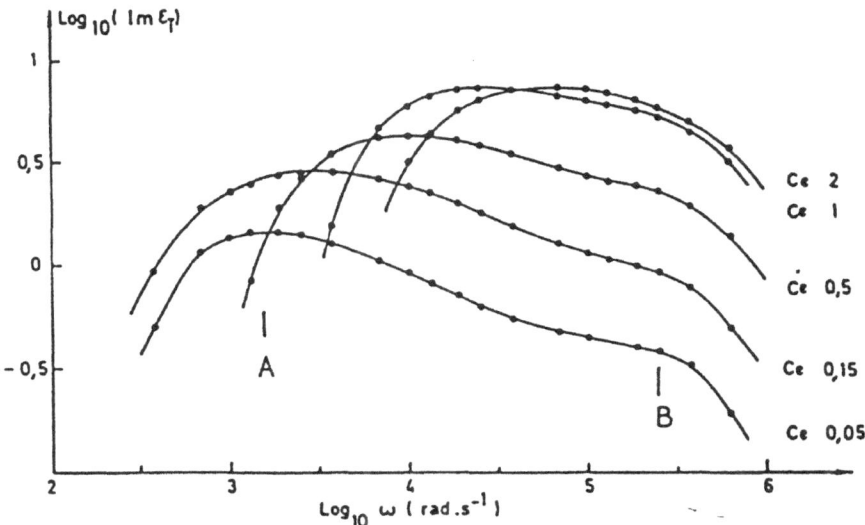

Figure 2 : Variations of ε(w) as a function of frequency for low doped samples. Ce x ≡ (1-x)CeO₂ +xY₂O₃ T=400K, from Ref.(7).

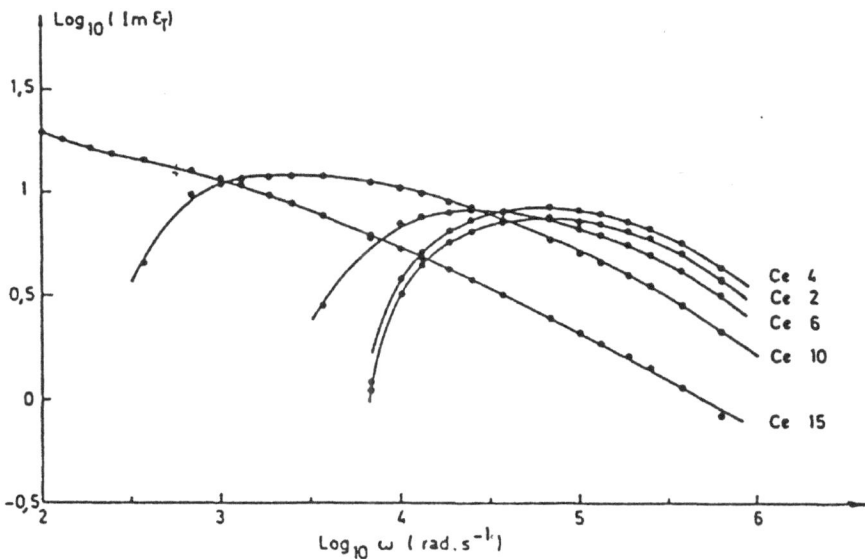

Figure 3 : Variations of ε(w) as a function of frequency for highly doped samples. Ce x ≡(1-x)CeO₂ +xY₂O₃ , T=400K, from Ref.(7).

In the case of ionic conductors such as the fluorite type oxides, two causes of dispersion of the electrical properties with frequency can be distinguished. The first one has been presented as early as in 1929 by Debye and is the reorientation of dipoles in the electric field with a relaxation time τ. Then ε(w) obeys the well known Debye law :

$$\varepsilon(w) = \varepsilon_\infty + \frac{\Delta\varepsilon}{1 + jw\tau} \quad (4)$$

The imaginary part of ε goes through a maximum for wτ=1. Peak B in Fig.2 is identified as a Debye relaxation process in agreement with internal friction measurements /8/. The relaxation time is thermally activated with an activation energy (0.6 eV) different from that of the dc conductivity (0.85 eV). The dipoles are formed of an immobile dopant ion and a mobile oxygen vacancy trapped in one of the neighboring sites, the two species carrying opposite charges. Similar results have been published in the case of Y-doped CSZ and TZP /9/.

However this mechanism contributes only very little to the depressing angle, even when the time constants τ and $R_b C_b$ are equal. The fact that the depressing angle does not vary much with temperature implies that another cause of dispersion exists which has something to do with long range transport. Indeed a diffusion process may lead to a frequency dependent conductivity if the jump frequency is not uniform /10/ which happens to be the case because of coulombic interactions between oxygen vacancies and dopant ions. The time evolution of the site occupancies is described by a Master equation /10/. The averaging procedure needed to derive the frequency dependent conductivity is non trivial and so far only approximate solutions have been published in the literature /11,12/. The advantage of the Continuous Time Random Walk theory developed by Scher and Lax /11/ is that all the information is contained in one function Q(t) which is the probability that a carrier remains a time t on its site before jumping. The frequency dependent conductivity is related to a statistical distribution of jump frequencies or equivalently of activation energies /13/. It has been stressed by several authors that the CTRW theory does not take into account percolation effects. In other words, it is particularly suitable if the energy of the jumping carrier at the saddle point does not depend on the environment, see Fig. 4a. Then the jump frequency is independent of the jump direction and is merely the inverse of a waiting time. Movaghar and Schirmacher have presented a selfconsistent theory of conductivity based on the Effective Medium Approximation which treats correctly the percolation paths at low frequency. However in its

simplest form /12/, it is valid for symmetrical jump
frequencies i.e. when the barrier height is the only
random quantity, see Fig. 4b. Clearly none of these
theories is adequate to fit quantitatively the
experimental data because the interactions between
defects are effective both at the saddle point and at
the equilibrium position. Moreover, they are not able
to include the Debye relaxation process as a special
case although it is described by a Master equation.
This is the consequence of another assumption, i.e. the
jump frequencies are spatially distributed at random.
If the dopant ions can be considered to be more or less
randomly dispersed, this is not true of the trapping
sites which are arranged in some ordered way around the
dopants.

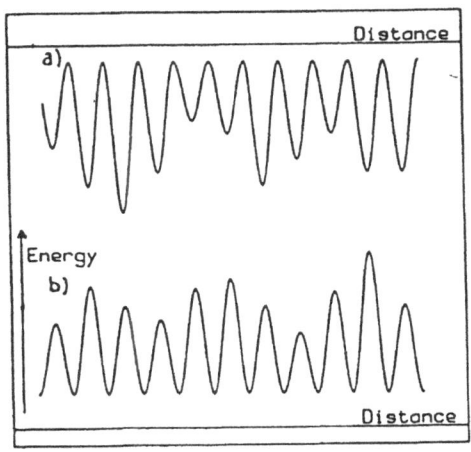

Figure 4 : Potential energy versus distance
experienced by an oxygen vacancy on its way. a),b) see
text.

Coming back to the experimental results, we may
give a qualitative interpretation. When the amount of
dopant is low, two maxima are discernable. The one at
high frequencies can be ascribed to a localized motion
of the oxygen vacancy and the other to long range
transport. They can be distinguished because the dopant
ions are far apart and because long range motion
necessitates the escape of the carriers out of the
traps in the vicinity of the dopant ions, see Fig.5a,
i.e. a large activation energy and a low jump
frequency. When the dopant concentration increases, the
spheres of influence of the dopant ions intersect, see
Fig.5b, and a continuous motion from trap to trap is
possible with a lower activation energy. The two peaks

merge into one. For highly doped samples, a mobile oxygen vacancy interacts with several dopant ions, randomly dispersed in the material, see Fig.5c. The spatial correlations between jumps, associated with a Debye relaxation process, are lost and the transport properties of these materials are not very different from that of alkali silicate or borate glasses /14/.

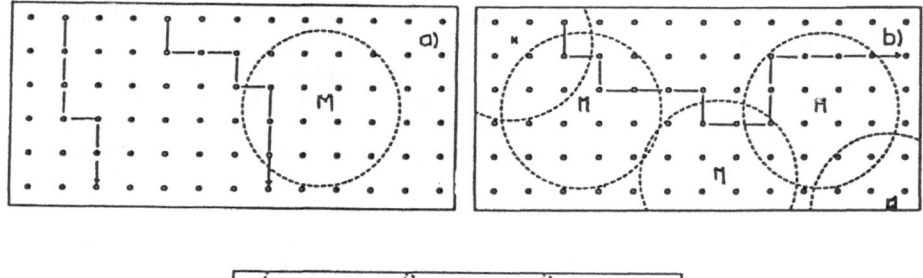

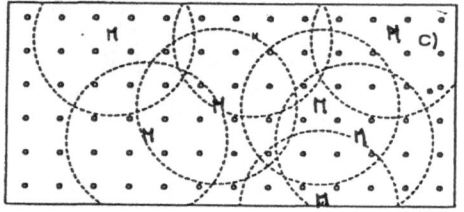

Figure 5 : Projection on the (001) plane of the oxygen sublattice. Only dopant cations (M) are depicted. a)isolated, b)percolating path c) multiple interactions.

THE GRAIN BOUNDARY ELECTRICAL PROPERTIES

A clear understanding of the electrical properties of the grain boundaries cannot be obtained without a detailed investigation of the microstructure of the ceramic samples. There is a general agreement that an intergranular second phase, rich in silica, calcia and in the dopant, is present at the grain boundaries of doped fluorite type oxides. Although the composition of this second phase varies from sample to sample, it seems to have most often a low melting temperature and to promote a liquid phase sintering. The way through which it is distributed at the interfaces depends on the nature of the oxide. In ceria, uniform films of variable thicknesses, up to 50 nm, have been observed /16,17/ while in zirconia, TEM studies show a more complex behavior. Uniform thin films prevail when the grain size is small enough, in TZP /18/ and CSZ /19/.

For large CSZ grains, the second phase forms small pockets /18,19/. However in the latter case, this does not mean that no intergranular layer exists between grains because of the limited resolution of the experimental technique .

Intergranular fracture surfaces can be examined by X-ray Photon Electron Spectroscopy and Auger Electron Spectroscopy /20,21/. Using the former technique, Hughes and coll. /19,20/ have analyzed the intergranular phase in several samples prepared from the same powder but sintered at different temperatures. Each step of the sintering process is characterized by a different behavior. Between 1280°C and 1400°C, densification is most effective. The atomic ratios O/Zr, Y/Zr and Si/Zr measured by XPS decrease at the grain boundaries and increase on the external surface of the disks indicating that part of the liquid phase is expelled from the bulk sample. At 1400°C, the final density (98%) is reached while grain growth proceeds. The decrease of the grain boundaries area and the observed increase of the XPS ratios suggest that the volume of the intergranular second phase is nearly constant. Above 1600°C, the XPS ratios decrease again, probably because of volatilization. This study clearly demonstrates that the average volume fraction and thickness of the grain boundary depends on sintering conditions.

Before drawing some conclusions about the influence of this intergranular layer on the electrical properties, we need to model the microstructure of the ceramic sample. In the following, we adopt the crude but simple brick layer pattern. It is visualized as an array of cubic grains of edge a, separated by flat grain boundaries of thickness d. Then the grain boundary resistance and capacitance are given by :

$$R_{gb} = \frac{1}{\sigma_{gb}} \frac{t}{A} \frac{d}{a} \quad \text{and} \quad C_{gb} = \varepsilon_0 \varepsilon_{gb} \frac{A}{t} \frac{a}{d} \quad (5)$$

where ε_{gb}, σ_{gb} are the dielectric constant and the conductivity of the second phase, while A, t are geometrical parameters of the disk sample. It may be noted that the critical frequency at the top of the arc of circle ($R_{gb}C_{gb}w = 1$) depends only on the specific properties of the grain boundary material. This appears to be the case in Ref. 19 where the only parameter is the sintering temperature.

Experimentally, it is usually found that the smaller is the average grain size, the larger is R_{gb} /22/. This is interpreted on the basis of Eq.(5) assuming a constant thickness of intergranular layer. We have seen that such an hypothesis is in most cases incorrect. The grain boundary resistance is indeed proportional to the volume of the intergranular layer,

$$V_{gb} = 3 \frac{d}{a} A.t \qquad (6)$$

Any process which decreases V_{gb} is going to lower the grain boundary resistance. We have alreday mentioned the densification stage of sintering /19/, and volatilization as possible mechanisms. Impurities may also segregate in the bulk as isolated inclusions. In this category can be ranged triple points, pockets observed in Y doped CSZ /19/, precipitates of definite compounds such as $Y_2Si_2O_7$ or Y_2SiO_5 /16/. An interesting case is provided by doping with alumina /23/. It is soluble to some extent in the bulk material (0.1mol.% at 1300°C) and this affects the bulk conductivity. It may dissolve into the liquid phase forming yttrium aluminosilicates on cooling /20/ with a corresponding increase of both V_{gb} and R_{gb}. It may also act as a scavenger /24/ for other impurities, decreasing V_{gb} and R_{gb}.

The grain boundary arc of circle is always depressed below the real axis by a large angle (10°-30°). The analysis is very difficult because the grain boundaries are numerous and constitute a complex network. The dispersion of the electrical properties arises both from an intrinsic behavior of the grain boundary and from a statistical distribution of the parameters of the grain boundaries. While the former is usually neglected, the latter is discussed on the basis of the brick layer model. The impedance is that of a serial assembly of resistors and capacitors in parallel:

$$Z = R_{bulk} + R_{gb} \int \frac{G(\log \tau)}{1 + jw\tau} d(\log \tau) \qquad (7)$$

where τ is the relaxation time (=RC) and $G(\log \tau)$ is a continuous distribution function. The advantage of such a description is that analytical expressions /5/ exist which depend on a small number of parameters, the Cole-Cole expression beeing the most popular, see Eq.(1). Then the functional G is given by :

$$G(\log \tau) = \frac{1}{2\pi} \frac{\sin (\pi n)}{\cosh\{(1-n)\log(\tau_0 / \tau)\} - \cos(\pi n)} \qquad (8)$$

which depends on two adjustable parameters τ_0 and n. As already mentioned, the quality of the fit cannot be really appreciated in the complex impedance plane but in any case, we do not expect to have more than an order of magnitude of the width of the distribution. Looking at published data, it is surprising to note that the depressing angle is rarely greater than 30°. As depicted in Fig.6, such an upper limit does not correspond to a very broad statistical distribution. This could be due to percolation effects. Let us consider the complex network associated with the real microstructure. We remove all impedances and replace them one by one in increasing order with respect to their values. For some critical impedance, which may depend on frequency, a path opens from one electrode to the other. Adding more elements, new paths are created but of higher impedances. It can be concluded that in Eq.(8), only a subset of impedances is really effective.

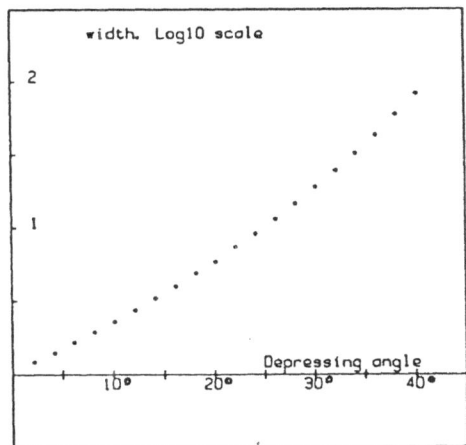

Figure 6: Width at half height of the statistical distribution of relaxation times , Eq.(8), versus the depressing angle.

It has been suggested several times that space charge could be present at the interfaces and contributes to the grain boundary resistance. If this

mechanism is operative, non linear current-voltage and
capacitance-voltage characteristics ought to be
detected. Unfortunately, it is not possible to use dc
techniques because most of the applied voltage drop
occurs at the electrodes. This can be avoided by using
an alternating voltage with an appropriate frequency
such that the corresponding image in the complex
impedance plane lies on the grain boundary arc of
circle. Then the contribution of the metal ceramic
contact is negligible (R_e C_e w >> 1) and that of the bulk
grains reduces to a resistance (R_b C_b << 1) which can be
substracted. The experimental set-up and the
theoretical analysis of the data have been described
elsewhere /17/. This technique has been used to study
the electrical properties of ceramic samples of
composition $((1-x)CeO_2 - xY_2O_3)$. The conductive and
capacitive components of the current versus the applied
voltage are depicted in Fig.7. The latter is a linear
function which means that the capacitance is voltage
independent. The I(V) characteristic is well
represented by the Poole-Frenkel law, usually observed

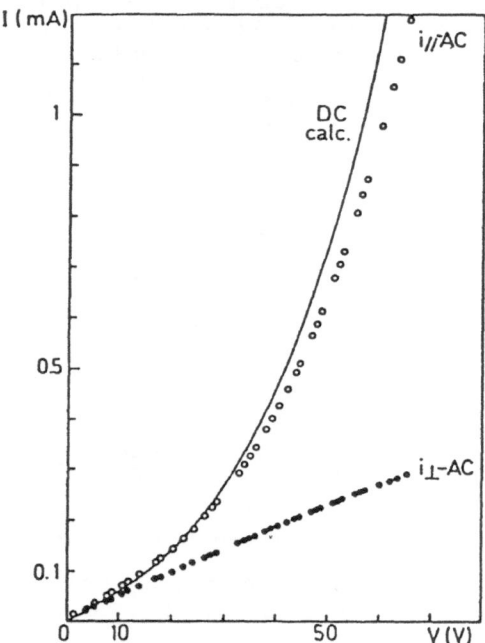

Figure 7: Conductive ($\parallel$) and capacitice ($\perp$)
components of the current with respect to the applied
voltage. From Ref.(17).

for insulators. The effective thickness of the grain boundary layer, deduced from a fit of the data is in good agreement with TEM observations. Space charge effects are of no importance in this case.

REFERENCES

1-J.E. BAUERLE, J. Phys. Chem. Solids, 30, 2657-2669 (1969).
2-P. LUPOTTO, M. VILLA and G CHIODELLI, J. Phys. E Sci. Instrum., 20, 634-636 (1987).
3-T. DICKINSON and R. WHITFIELD, Electrochim. Acta, 22, 385-389 (1977).
4-J.J. BENTZEN, N.H. ANDERSEN, F.W. POULSEN and O.T. SORENSEN, in *Proceedings of the 6th Int. Conf. on Solid State Ionics* (1987) Eds. W. Weppner and H. Schulz, North Holland Publ. Co., 550-559 (1988).
5-C.J.F. BÖTTCHER and P. BORDEWIJK, in *Theory of Electric Polarization*, Elsevier Publ., Vol II (1978).
6-M. KLEITZ, H. BERNARD, E. FERNANDEZ and E. SCHOULER, in *Advances in Ceramics,* Eds. A.H. Heuer and L.W. Hobbs, The Am. Ceram. Soc., Columbus Ohio, 310-335 (1981).
7-P. ABELARD, A. SYLVESTRE and J.F. BAUMARD, in *Proceedings of the 6th Riso Int. Symp. On Metallurgy and Materials Science*, Eds. F.W POULSEN, N.H. ANDERSEN, K. CLAUSEN, S. SKAARUP and O.T. SORENSEN, Riso-Denmark, 273-278 (1985).
8-A.S. NOWICK, J. Phys.Paris,46-C10, 507-511 (1986).
9- M. WELLER, H. SCHUBERT, J. DIEHL, and G. PETZOW, to be published in Zirconia III.
10- H.R. ZELLER, H.U. BEYELER, P. BRÜESCH, L. PIETRONERO and S. STRÄSSLER, Electrochim. Acta,24, 793-797 (1979).
11-H. SCHER and M. LAX, Phys. Rev., B7, 4491-4501 (1973).
12-W. SCHIRMACHER, ibid Ref.4, 129-133 (1988)
13-P. ABELARD and J.F. BAUMARD, Phys. Rev., B26, 1005-1017 (1982).
14-P. ABELARD and J.F. BAUMARD, Solid State Ionics,14, 61-65 (1984).
15-R. GERHARDT and A.S. NOWICK, J. Am. Ceram. Soc.,69, 641-646 (1986).
16-R. GERHARDT, A.S. NOWICK,M.E. MOCHEL and I. DUMLER, J. Am. Ceram. Soc.,69, 647-651 (1986).
17-J. TANAKA, J.F. BAUMARD and P. ABELARD, J. Am. Ceram. Soc.,70-[9], 637-643 (1987).
18-N. BONANOS, J. DRENNAN, R.K. SLOTWINSKI, B.C.H. STEELE and E. P. BUTLER, Silicates Ind.,[9-10], 127-132 (1985).

19-S.P.S. BADWAL and J. DRENNAN, J. Mater. Sci.,22, 3231-3239 (1987).

20-A.E. HUGHES, in *Ceramic Developments* Eds. C.C. SORRELL and B. BEN-NISSAN, Materials Science Forum,Trans. Tech. Publ. Switzerland, Vol. 34-36, 243-247 (1988).

21-A.J.A. WINNUBST, P.J.M. KROOT and A.J. BURGGRAAF, J. Phys. Chem. Solids,44, 955-960 (1983).

22-A. EL BARHMI,Thesis, University of Grenoble (1987).

23-M. MIYAYAMA, H. YANAGIDA ans A. ASADA, Am. Ceram. Soc. Bull., 64-[4], 660-664 (1982).

24-E. P. BUTLER and J. DRENNAN, J. Am. Ceram. Soc.,65, 474-478 (1982).

CHARACTERIZATION OF NANOCRYSTALLINE ZIRCONIA POWDERS BY ELECTRON OPTICAL TECHNIQUES

L.A. Bursill[1,2], E. Bernstein[2] and M. G. Blanchin[2]

1 - School of Physics, University of Melbourne - 3052, Vic, Australia

2 - Département de Physique des Matériaux, Université Claude Bernard Lyon I - 43, Bvd. du 11 Novembre 1918, 69622 Villeurbanne Cédex, France

Abstract

Electron optical techniques are described for the characterization of the size distribution of agglomerates, aggregates and primary micro-and nanocristallyne of as-processed zirconia powders. These techniques allow for direct identification of individual crystallites as tetragonal or monoclinic, by optical transform of high-resolution electron micrographs. The latter also permit surface morphology to be examined with atomic resolution. Application to a range of pure and doped zirconia powders, of recent commercial interest, are presented, which enable the results of concurrent studies by sedimentation, surface specific area measurements, porosity and sinterability to be correctly interpreted.

1 - Introduction

The use of microcrystalline (submicron) and nanocrystalline (10 nm) particles as precursor materials for fabrication of advanced ceramics, for example high temperature superconductors as well as oxide-based ceramics, requires new techniques for characterization. Thus classical optical microscopy (resolution $\simeq$ 1 μm) or even scanning electron microscopy (resolution $\simeq$ 5-10 nm) are no longer effective for examination of 1-10 nm particles. At the Département de Physique des Matériaux, Université Claude Bernard, we have developed techniques appropriate for the direct determination of the morphology, size, homogeneity and states of aggregation, agglomeration or flocculation of ceramic powders. The onset of sintering may be sensitively detected and porosity may be visualized down to the manometre scale. In addition the atomic structure of vicinal surfaces and facets, as well as intergranular interfaces may be explored with respect to variation of processing route (chemical and physical parameters). Observations of zirconia powders of recent industrial interest are summarized below. These emphasize the potential contribution of transmission electron microscopy (T.E.M.) at medium resolution, as well as transmission high-resolution electron microscopy (H.R.E.M.), for the routine analysis of ultrafine ceramic powders. These techniques act as both complementary to existing characterization techniques, as well as providing quite new types of information.

2 - Size distribution of particles

Two zirconia powders were prepared [1] by calcination of crystallized acetates (C) and amorphous acetates (A). Preliminary characterization included determination of the size distribution of weakly-bonded agglomerates by sedimentation, measurement of the specific surface areas by B.E.T. and of the mean crystallite sizes by X-ray diffractometry. The powders were then examined by TEM [2]. For example, Fig. 1a,b compares these two powders at low magnification, following dispersion of dry powder onto a transparent carbon support film. Individual particles range in size from 0.05 μm $\leq$ D $\leq$ 0.7 μm for specimen A and 0.05 μm $\leq$ D $\leq$ 1.5 μm for specimen C.

Sedimentation measurements using a centrifugal particle size distribution analyzer and "optimal dispersion conditions" yielded agglomerate dimensions 0.5 μm $\leq$ D(A) $\leq$ 8 μm for A and 0.5 μm $\leq$ D(C) $\leq$ 3 μm for C. The surface area results were 6.2 m^2/g for A and 12.2 m^2/g for C, corresponding to equivalent spherical diameters Ds = 160 nm for A and 80 nm for C respectively [1]. Conventional TEM, using both bright and dark-field imaging modes, as well as electron diffraction [3] showed clearly that powder A forms weakly-bonded agglomerates of primary crystalline particles (see Fig. 1c) whereas powder C formed both strongly-bonded aggregates of primary crystallites as well as agglomerates of the latter (see Fig. 1d). A HREM image [4] of powder C (Fig. 2) clearly reveals the polycrystalline nature of such an aggregate. The lattice fringes [4] having different spacings (0.02-0.5 nm) and orientations, reveal the extent of the primary crystallites as well as the crystallite boundaries. Obviously the unusual aggregates of this powder tend to have a platelet habit, with some preferred alignment of primary crystallites. These should be compared to the loosely-bonded agglomerates of powder A, composed of distinct polyhedra (see Figs. 1c and d). The difference may be attributed to the different chemical processing routes, via crystalline and amorphous acetates respectively, since the calcination temperatures were identical.

Note that the distribution of secondary particles (aggregates or agglomerates), as determined by TEM was much narrower than that measured by sedimentation techniques which proves that even if the dispersive conditions are

Fig.1 - TEM images showing typical distribution of particles.

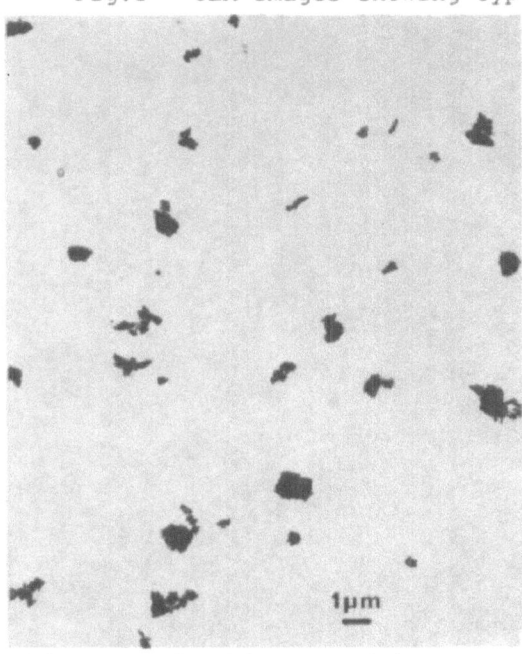

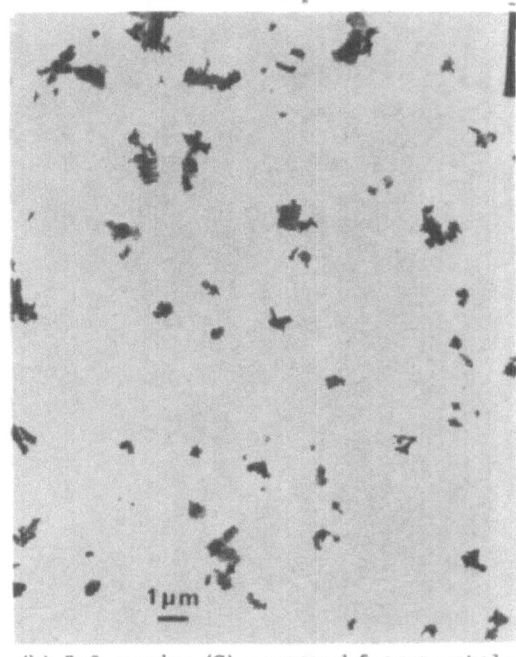

(a) ZrO$_2$ powder (A) prepared from amorphous acetate calcined at 850°C.

(b) ZrO$_2$ powder (C) prepared from crystallized acetate calcined at 950°C.

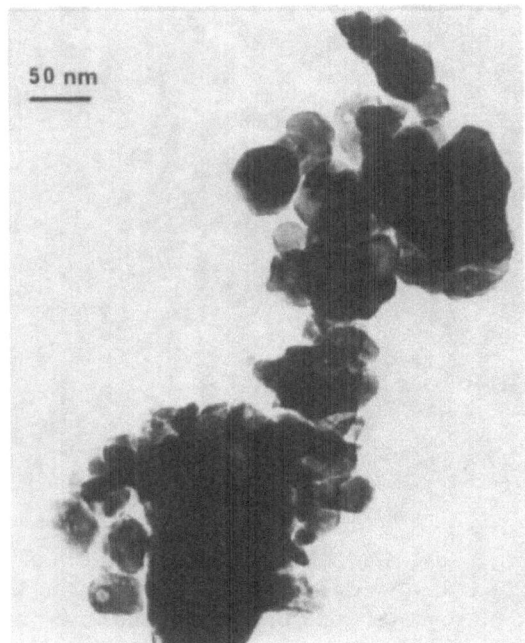

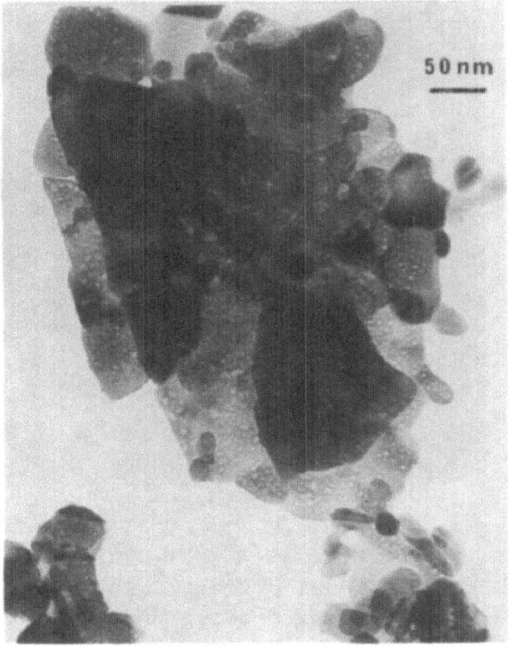

(c) shows details of a weakly-bonded agglomerate of primary crystallites for powder A.

(d) gived details of a strongly-bonded aggregate for powder C.

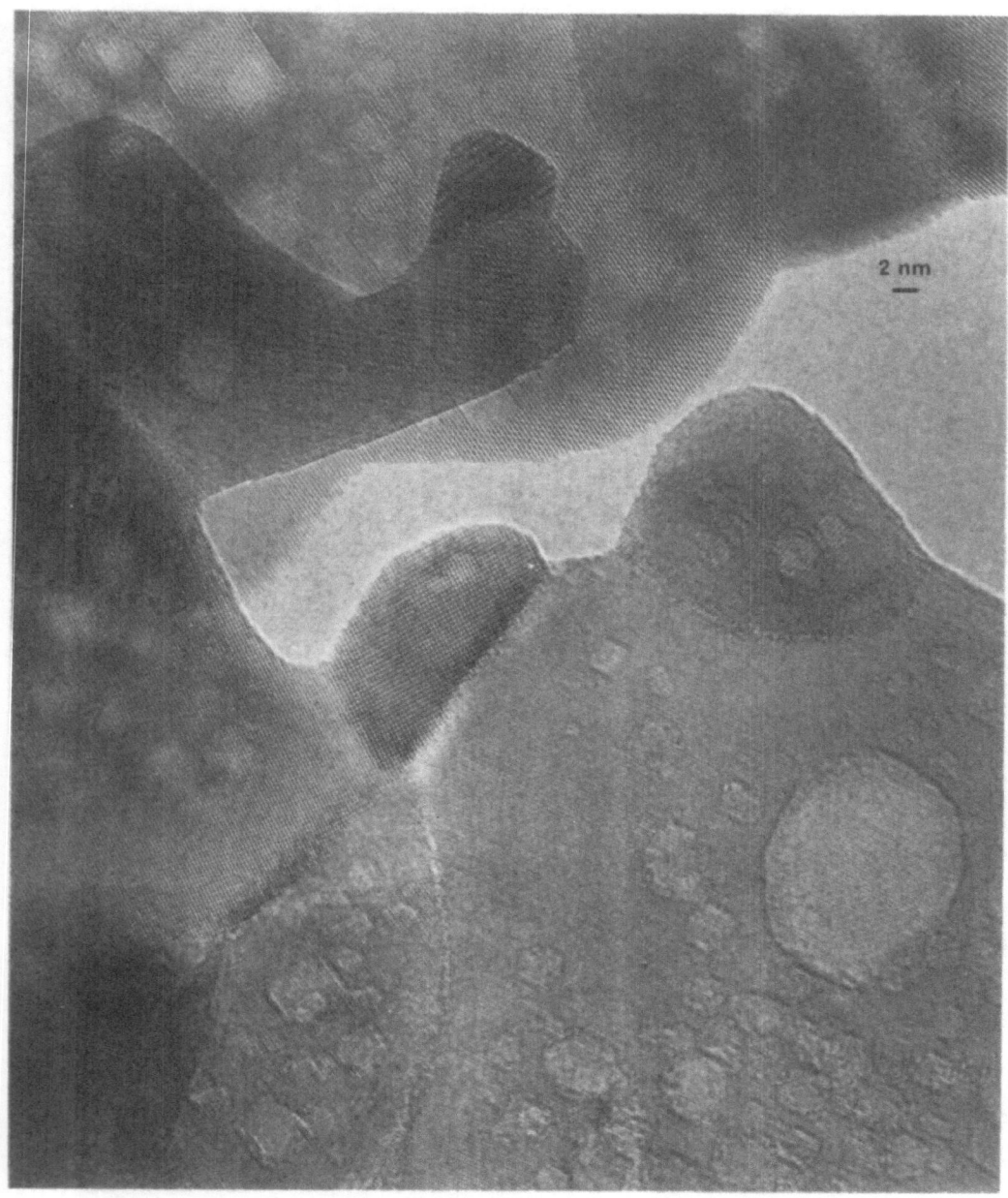

2 nm

Fig.2 – HREM IMAGE OF POWDER C (cf Fig. 1d). Lattice fringes show extent of individual primary crystallites in the aggregate, as well as details of the surface profile. Note presence of vicinal surfaces and steps having height equal to one interatomic spacing.

optimized (by controlling the pH of the dispersing liquid), zirconia remains heavily agglomerated in solution. Consequently, as described above, the particle size analyzer does not necessarily give either the size of the agglomerates or aggregates existing in the original dry powder preparations. Equivalent spherical diameters, calculated from specific surface area measurements have also to be examined with caution. In the case of powder A, the primary particles may be considered as approximately spherical, but they present a wide size distribution, with diameters determined by TEM ranging from 22 nm-250 nm. Powder C is formed rather of plate-shaped aggregates, approx. 30 nm thick, as determined by tilting experiments in the TEM. For the latter case the equivalent spherical diameter, obtained from B.E.T. measurements, has little physical relevance. Furthermore the high resolution image of powder C (Fig. 2) shows that the surace area is greatly increased by extensive pitting of the surface, as well as by the appearance of vicinal surface steps in the surface profile. Thus the effects of aggregation of primary crystallites is partially counteracted by surface roughening of individual crystallites; again implying caution with respect to interpretation of the B.E.T. measurements.

3 - Effect of calcination temperature

Ultrafine zirconia powders were processed by calcination of an amorphous hydrated zirconia oxide at temperatures between 300 °C and 1100 °C (see Table I for values of typical ceramic parameters).
X-ray examination showed the sequence:

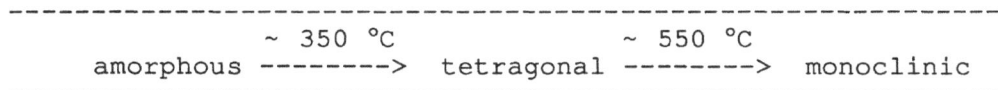

```
------------------------------------------------------------
                  ~ 350 °C              ~ 550 °C
      amorphous -------> tetragonal -------> monoclinic
------------------------------------------------------------
```

with increasing temperature. Figs. 3a-c reproduce three HREM images obtained for samples calcined for 1 hour at 400°C, 700 °C and 800 °C respectively. Thus the crystal lattices of whole primary grains have been imaged at 0.22 nm resolution. As well the surface profiles have been obtained, providing a new scale of morphological information. The set of images reveals correlated changes in crystallite sizes and development of surface facetting. With increasing temperature the crystalline change from roughly spherical to strongly polyhedral in shape, with a preference for relatively extensive facetting along $(111)_M$ and $(010)_M$ planes (Fig. 3b). As well the crystal structure changes from predominantly tetragonal to predominantly monoclinic.

TABLE I

Temperature of calcination (°C)	Specific surface area BET (m² /g)	Phases detected by X-ray diffraction
300	141	Traces of tetragonal (T)
400	112	100% T
500	63	100% T
600	25	T + traces of monoclinic (M)
700	14	60% T, 40% M
800	7.7	25% T, 75 M
900	6.3	M + traces of T
1000	5.4	100% M
1100	4.5	100% M

The proportion of phases represents the ratio T/M and not the total percentage, since there is also an amorphous phase present at low temperatures.

Differential thermal analysis places crystallization temperature at about 430 °C.

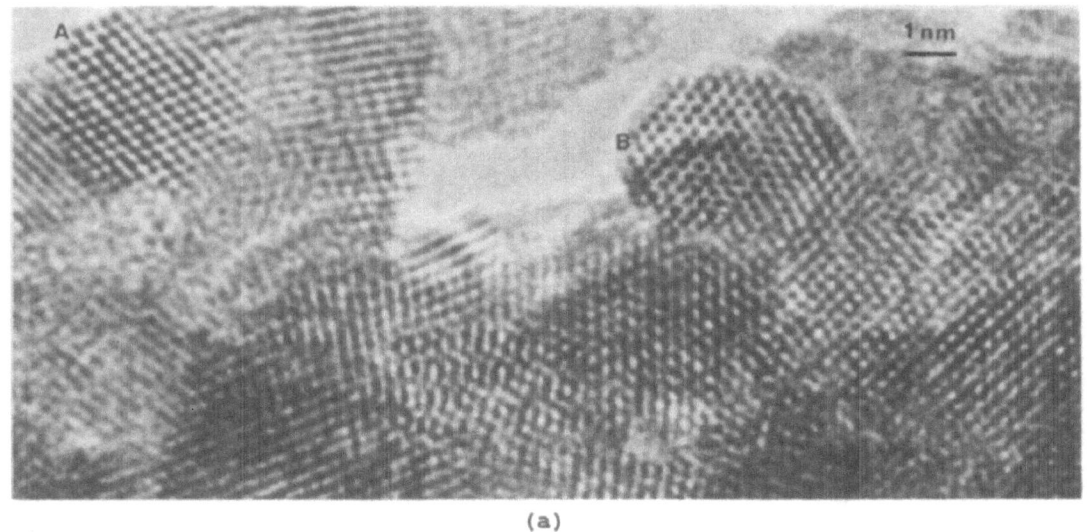

(a)

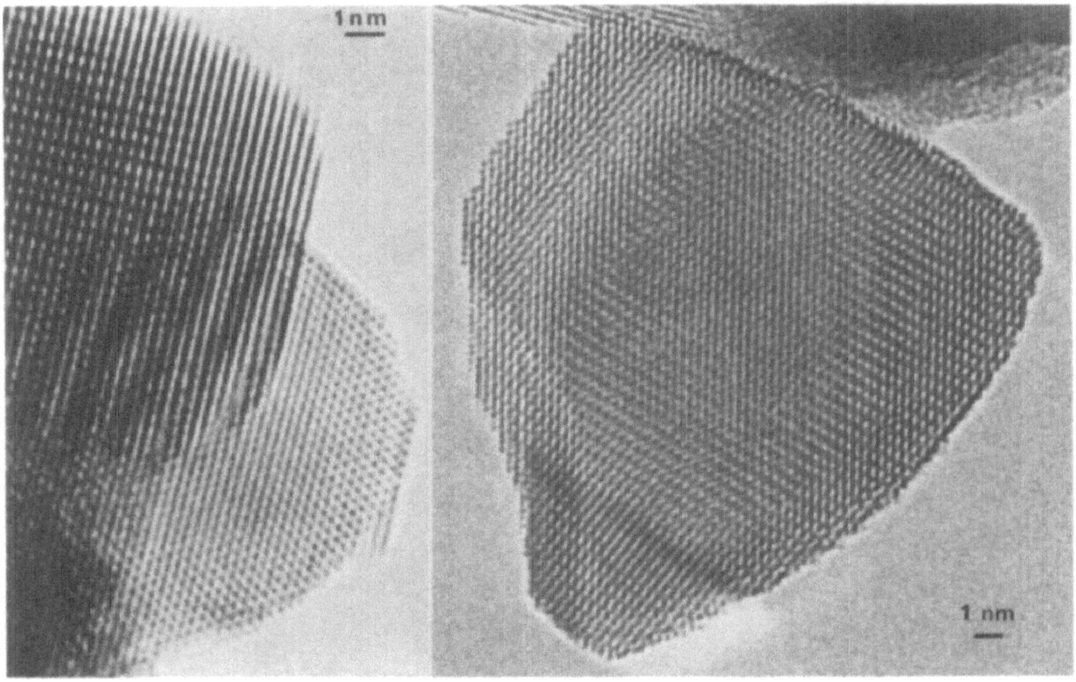

(b)   (c)

Fig.3 - HREM images of zirconia powders calcined at 400°C (a), 700°C (b)
and 800°C (c), showing increase in crystallite size with increasing
temperature. Note also development of $(111)_M$ and (010) facets in (b) and
(c).

4 – Identification of crystalline modification

Optical diffractograms, obtained by forming the Franuhofer diffraction pattern of the images using laser illumination on an optical bench [4], allow the symmetry and reciprocal lattice geometry of the crystallites to be determined, even for single nanocrystalline grains of diameter $\leq$ 10 nm like in Fig. 3a. This proved to be a powerful tool for identifying the tetragonal and/or monoclinic phases of zirconia. Thus diffractograms of Figs. 4a,b show $[100]_T$ and $[111]_T$ orientations of tetragonal zirconia for the areas labelled A and B in Fig. 3a, whereas Figs. 4c,d identify $[\bar{1}01]_M$ and $[\bar{1}10]_M$ orientations of the monoclinic phase for areas C and D of Figs. 3b,c respectively.

5 – Atomic surface structural information

It is clear from Figs. 3a,b,c that much detailed information concerning the atomic surface topology is available from the HREM images. Thus, whereas the tetragonal phase (from a specimen calcined at 400 °C) contained roughly circular nanocrystals, with an uniformly high distribution of vicinal steps, the monoclinic phase at 700 and 800 °C exhibited marked development of $(111)_M$ and $(010)_M$ facets, imposing a more polyhedral habit for the crystalline grains.

Even within the monoclinic phase details of the surface morphology vary significantly, depending on the preparation route. For example, two powders were examined [5], exhibiting specific surface areas of 6.5 m^2/g and 20.6 m^2/g respectively after calcination at identical temperatures. The first showed relatively large, facetted microcrystals, whereas the second had extensive surface roughening associated with the vicinal surfaces of essentially spherical crystallites.

Such differences are expected to be important for understanding surface reactivity and sintering rates, as well as understanding and interpreting the classical B.E.T. results.

6 – Twin boundaries

Structural defects are readily apparent in HREM images. Thus Fig. 5 shows a twin boundary in a monoclinic grain of a powder processed as described in section 3 but from a different precursor calcined at 1000 °C for 1 hour. Such twin boundaries occurred relatively frequently at that temperature but were absent from powders processed from acetate calcined at 950 °C for 12 hours.

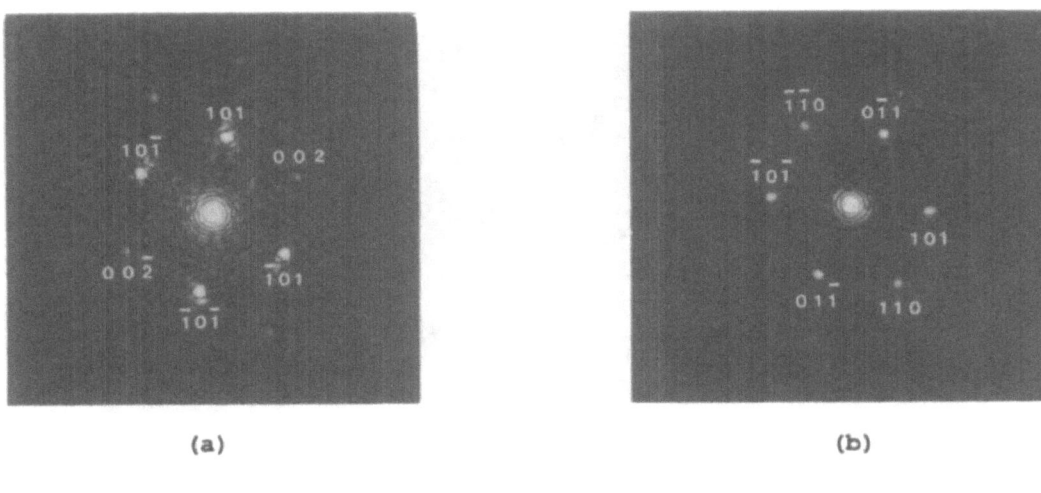

(a)   (b)

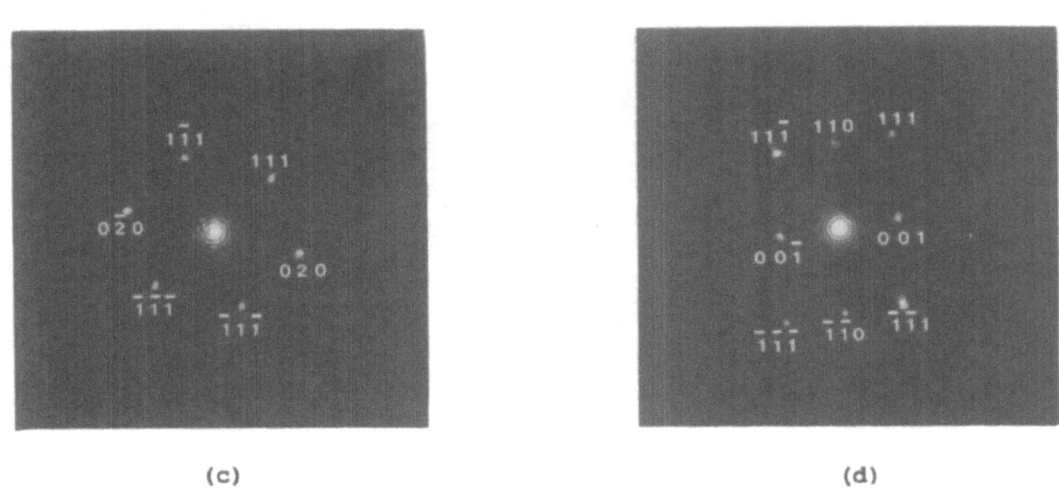

(c)   (d)

Fig.4 - (a) and (b) show optical transforms obtained from areas A and B of Fig. 3a, allowing the $[100]_T$ and $[111]_T$ projections of the tetragonal phase to be identified; (c) and (d) show $[\bar{1}01]_M$ and $[\bar{1}10]_M$ projections of the monoclinic phase to be identified for areas C and D of Figs. 3b and c respectively.

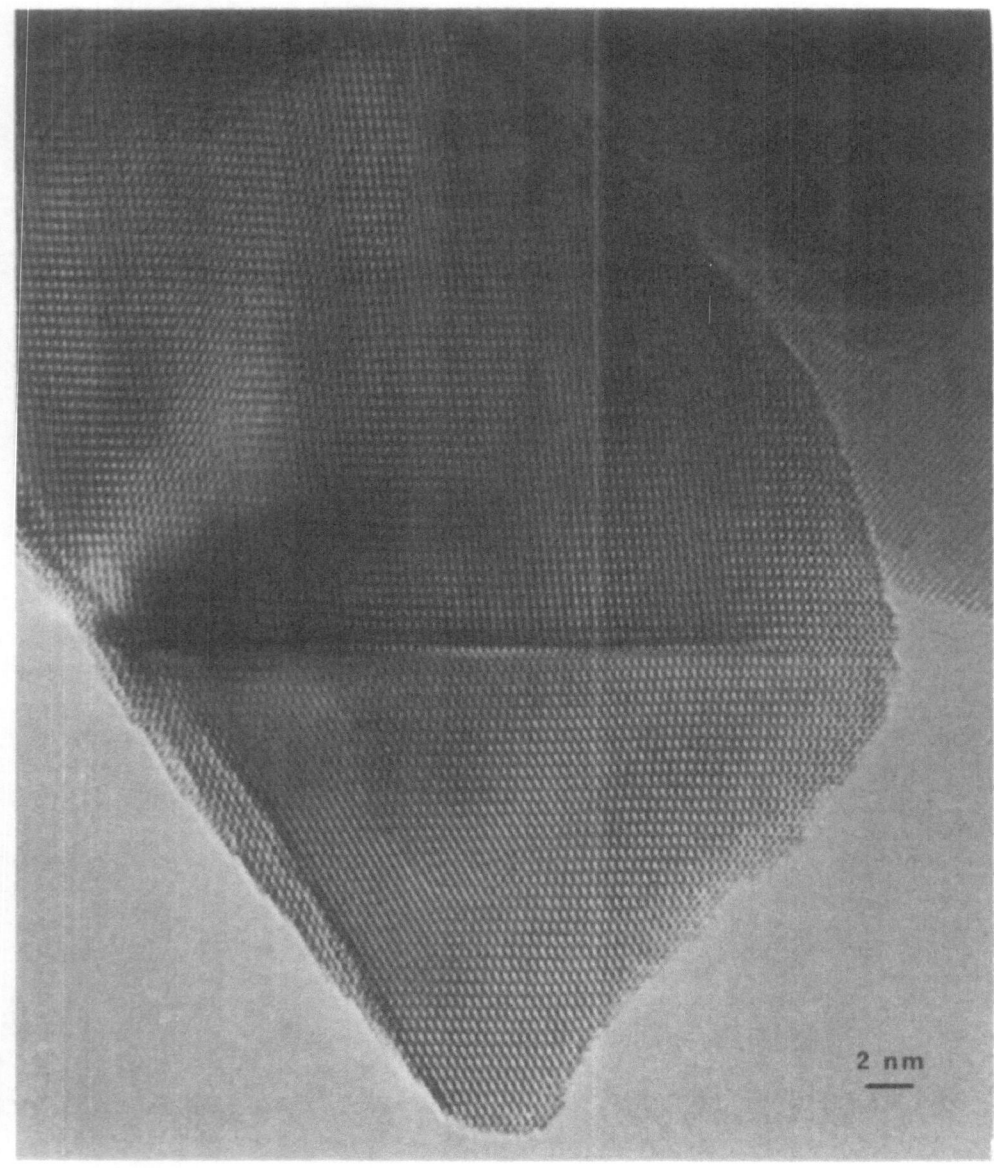

Fig.5 – HREM image of a zirconia powder calcined at 1000°C for 1 hour, showing detail of a $(100)_M$ twin boundary.

Crystallographic studies of the microscopic twinning elements, whereby screw axes and/or glide twins may be distinguished from rotation or mirror twins are proceeding using a series of HREM images obtained for different axial projections.

The atomic structure of the twins is interesting from the point of view of understanding the mechanism of tetragonal -----> monoclinic structural transformation.

7 - Intergranular interfaces and porosity due to sintering

Fig. 6 shows a TEM image of a zirconia powder processed from crystallized acetate after calcination at 950 °C for 12 hours. Note the bubble like nature of the aggregate, with development of planar minimal surfaces between many of the grains, and curved bubble shapes at the external surfaces. The grains increase in diameter at the higher temperature and longer calcination times. Note also the appearance of void-space, which basically gives rise to porosity and the less than theoretical densities measured typically for ceramics. It is clear that it will be difficult to fill those spaces, without extensive long-range ionic mobility (~ 0.5 μm). This can only be imposed by the application of elevated temperatures and pressures and/or repeated grinding and sintering.

Clearly, the observational technique depicted here (at magnifications in the range 20-100,000 X) has the potential to provide a routine analysis in view of understanding details of the mechanism and progress of sintering in individual cases.

8 - Yttria-stabilized tetragonal zirconia

Fig. 7a shows a TEM image exhibiting the texture of a tetragonal zirconia obtained by thermal decomposition at 950°C of mixed zirconium and yttrium acetates [6]. Comparison with Fig. 6 shows a similar development of grain size and polyhedral facetting, together with a degree of sintering. However the proportion of void-space is relatively low. Thus this preparation appears to yield a good precursor from the point of view of homogeneity of grain size and yttria content; factors which lead to sinterability of commercial interest [6].

Fig. 7b gives a HREM image of this preparation. It shows a bicrystal formed by sintering of two grains. Note the formation of a reentrant angle. The atomic surface profile exhibites once more the combination of facetting of $\{101\}_T$ planes and increasing frequency of vicinal surface steps

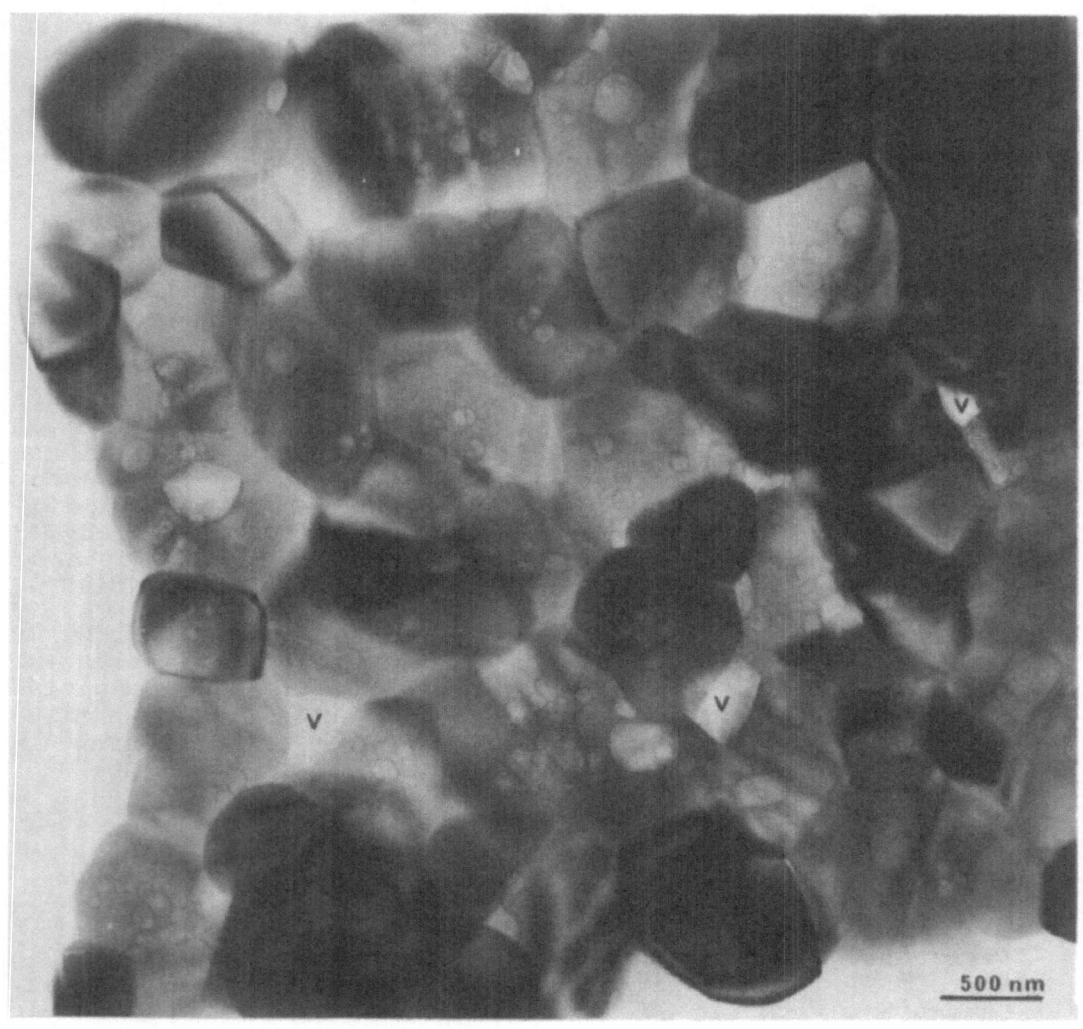

Fig.6 = TEM image of a zirconia powder calcined at 950°C for 12 hours, showing bubble-like nature of a strongly-bonded aggregate of crystallites (monoclinic phase). Note initial stages of sintering, indicated by planar interfaces between grains. Voids are indicated which will contribute to porosity.

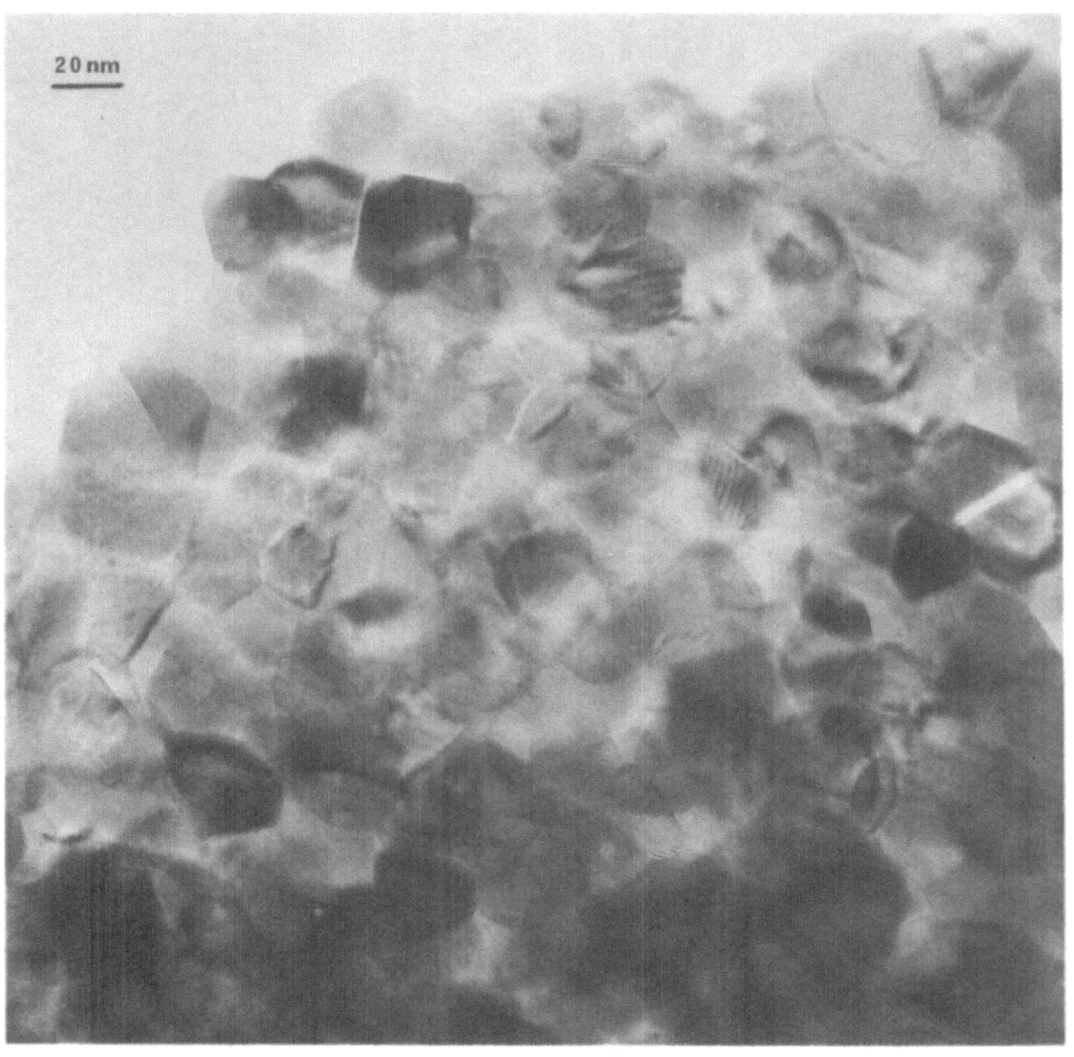

Fig.7 (a) - TEM image of yttria-doped zircon (stabilized tetragonal) after finally calcining at 950°C. Note homogeneous crystallite size distribution and relative absence of pores (cf Fig.6).

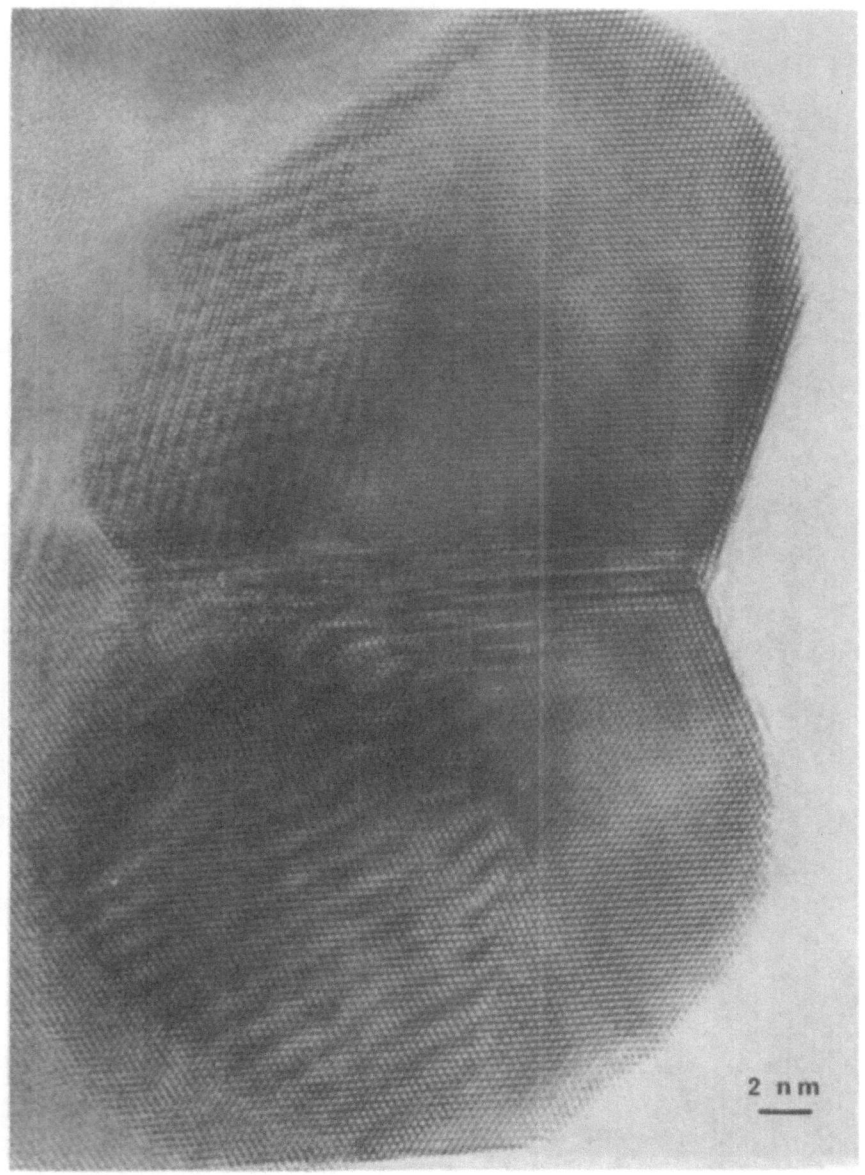

Fig.7 (b) – shows HREM image of a bicrystal formed by sintering of two crystallites symmetrically oriented. Note detail of vicinal surfaces with development of $\{101\}_T$ facets. A reentrant angle is formed naturally.

required to construct the rounded portions of the grains. Regarding the grain boundary structure it is again apparent that it is timely to begin systematic analysis of the frequency of occurrence of relatively low surface energy coincidence site lattices as well as more complex multinodal intergranular structures.

9 - Effects of different processing routes

After structural and textural studies have been made for a number of preparation conditions, and the image contrast mechanism understood, then it is reasonable to proceed to use TEM and HREM as tools for the evaluation and optimization of different preparative routes for chemical and physical processing of ceramic precursor and ceramic products. Thus the structural characteristics of both pure and doped zirconia powders, prepared using hydroxychlor, sulphate, acetate and molten nitrate routes, as well as the effects of washing with isopropanol before sintering, have ben investigated in this laboratory [3,5-8]. The effects of dewatering by calcination of amorphous hydrous zirconia on tetragonal phase content of zirconia powder have also been examined using TEM techniques in another laboratory [9].

10 - Conclusion

The techniques of TEM and HREM are capable of routine application in research and industrial laboratories. Thus specimen preparation methods are relatively trivial for powders. One hour of HREM time provides images of many thousands of particles, allowing a reasonable statistical analysis to be achieved. On-line image processing techniques have been developed for particle-size and statistical analysis of the distributions, a technique widely used by biologists.

Image digitization using the VT camera usually attached to a HREM together with Fast-Fourier algorithms and an array processor allows rapid (on-line) collection of the equivalent of the optical transform of HREM images of individual grains. Thus readily interpretable nanodiffraction data, by comparison with standards stored in the computer, may be obtained along with the morphological and surface structure data.

It is worth noting that use of a high-resolution video cassette recorder (VCR) or a video disc greatly reduces the need to use expensive, time-consuming photographic processing and storage techniques. Thus the images and Fourier tranforms may be conveniently stored and reviewed.

Clearly, the above techniques are capable of further development in the field of ceramic processing. Careful attention to specimen preparation is likely to reveal new combinations of particle size distribution and surface morphologies, of interest for ceramic, as well as catalytic, applications.

REFERENCES

[1]   SAMDI, A. These Lyon I, Ordre N. 92/87;
      SAMDI, A. GROLLIER-BARON, Th; DURAND, B., and ROUBIN M.
      Ann.Chim. Fr. 13 (1988) 171 and 188.

[2]   HIRSCH, P.B., HOWIE, A., NICHOLSON, R.B., PASHLEY, D.W.
      and WHELAN M.J., "Electron Microscopy of Thin Crystals"
      Butterworths, London (1965).

[3]   BERNESTEIN, E., BLANCHIN, M.G., and SAMDI, A., Ceramics
      Intern. in press (1989).

[4]   SPENCE, J.C.H., "Experimental High-Resolution Electron
      Microscopy", Clarendon Press Oxford (1981) pp. 121,278.

[5]   BLANCHIN, M.G., BURSILL, L.A., and BERNSTEIN, E.,
      "L'industrie Ceramique", in press (1989).

[6]   SAMDI A., BERNSTEIN, E., GROLLIER-BARON Th., DURAND, B.
      ROUBIN, M. and BLANCHIN M.G., Memoire Scientif. de la
      Revue de Metallurgie, in press (1988).

[7]   BERNSTEIN, E., Thèse LYON I, in preparation (1989).

[8]   RAVELLE-CHAPUIS, R. and BLANCHIN M.G., in preparation
      (1989)

[9]   KOSMAC, T., GOPALKRISHNAN, R., KRASEVEC, V. and KOMAC,
      M., J. de Physique, Colloq. CI, Suppl. no. 2, 47 (1986)
      CI-43.

EFFECT OF PHYSICOCHEMICAL CHARACTERISTICS OF ZrO_2-Y_2O_3 POWDERS ON THE COMPACTION BEHAVIOR AND MICROSTRUCTURE DEVELOPMENT.

J.Y. CHANE-CHING, A. M. LE GOVIC and D. BROUSSAUD.
Rhône-Poulenc Recherches, Aubervilliers, France.

ABSTRACT

Inter-relations between compaction ability, density homogeneity, green strength and particle morphology were explored using submicronic, ZrO_2-Y_2O_3 (3% molar Y_2O_3) commercial powders. The effect of chemical characteristics such as impurities content and dopants additions on the densification behaviour was illustrated. Within the high sintering temperatures domain, Al_2O_3 addition or yttrium inhomogeneity in the powders, were shown to enhance dedensification, by altering phase relations or promoting the formation of cubic nuclei . A mechanism , based on pore growth along a bimodal grain growth accompanying partitioning reaction into low Y_2O_3 tetragonal and high Y_2O_3 cubic phases, was proposed for the observed dedensification.

INTRODUCTION

Difficulties in selecting good powder characteristics in ceramic fabrication arise from several considerations: On the one hand, powders having low surface area and large particle size distribution were found to provide processing ease [1] . On the other hand, fine and uniform microstructures, desirable for high performance ceramics were produced from fine, non agglomerated, monosized and equiaxed powders [2]. Besides, the chemical characteristics of powders and dopants addition [3] are also shown to greatly affect the microstructure development during sintering.

In the present paper , the role of morphology of submicronic , commercial ZrO_2-Y_2O_3 (3% molar Y_2O_3) powders on compaction behavior will be illustrated. The effect of chemical characteristics or dopants additions will also be discussed with special focus on high sintering temperatures domain and a mechanism for the observed dedensification proposed.

1- EFFECT OF ZrO2-Y2O3 POWDER CHARACTERISTICS ON COMPACTION BEHAVIOUR.

High green densities and minimal density gradients are usually preferred characteristics to achieve the required properties of the fired ceramics. Adequate green strength to survive die ejection and for handling or post machining prior to sintering might also be required.
Although the use of lubricants and binders are known to be quite effective to control compacts properties, the powder characteristics were shown to greatly contribute to determine the properties of green pieces shaped by compaction.
High green densification has already been reported on $ZrO_2-Y_2O_3$ powders, processed by electrorefining [4] or chemical routes [5]. For three powders A, B and C as illustrated in fig 1 , synthesized by different chemical routes , we demonstrated that the highest green density (56% dth at 250 MPa) was obtained with powder A (fig 2), showing the highest tapped density and a morphology producing a lower interparticle porous volume as determined by mercury porosimetry. A lower green density (49% dth at 250 MPa) was observed for monosized spherical particles in powder C [5]compared to powder A.

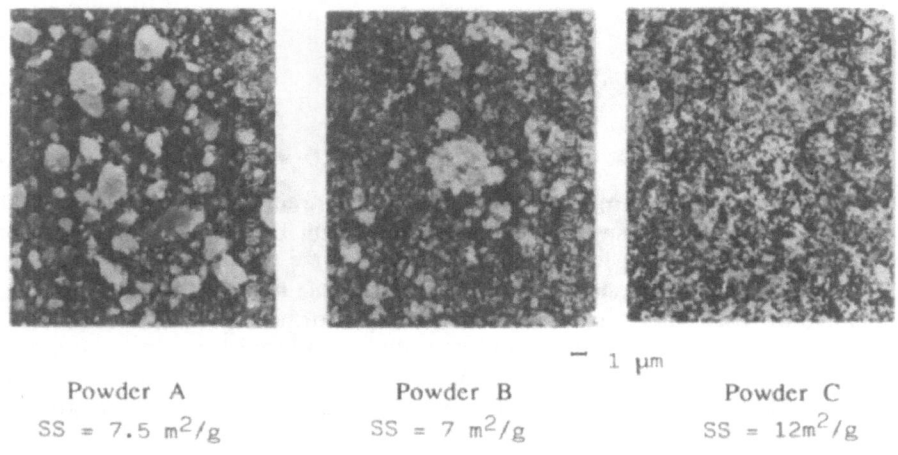

Powder A	Powder B	Powder C
SS = 7.5 m^2/g	SS = 7 m^2/g	SS = 12m^2/g

— 1 µm

Fig 1 : Scanning Electron Micrograph of powders A, B and C

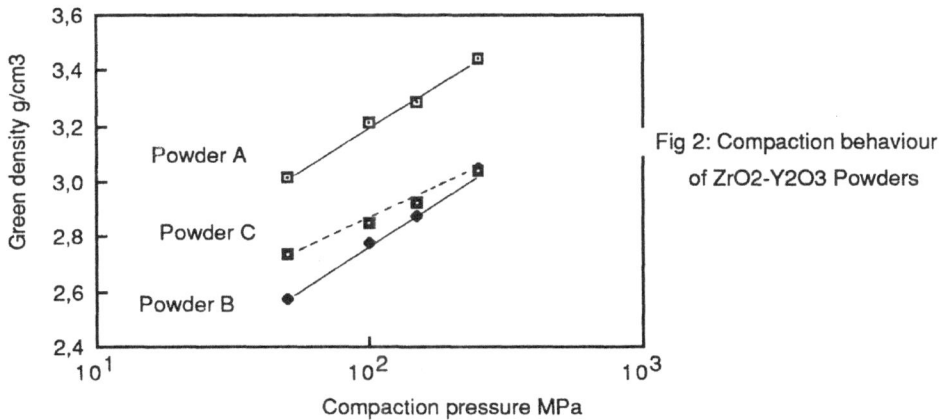

Fig 2: Compaction behaviour
of ZrO2-Y2O3 Powders

Powder characteristics will also affect the transmission of stresses during powder compaction. For an uniaxial compression in a cylindrical die , the measurement of stress and refering to applied stress in the vertical direction and radial stress induced in the horizontal direction respectively , demonstrated a higher stress ratio for powder C as reported in a previous study [6] . Thus, a more homogeneous density and the possibility of producing more complex shapes were expected for green compacts , prepared from powder C.

The production of high strength green compacts make the green machining easier and the use of binders was known to improve the green strength . However, the comparison of the yield locus of various powders without binder outlined a large effect of powder morphology on the strength of green pieces [6]. A lower tensile strength was determined on green compacts prepared from spherical, monodisperse particles of powder C.

These various observations illustrated the important effect of morphology , a compromise must be worked out for powder selection to obtaining the required properties of green compacts depending on the complexity of shapes.

2-EFFECTS OF CHEMICAL CHARACTERISTICS ON THE SINTERING BEHAVIOUR AND THE MICROSTRUCTURAL DEVELOPMENT OF ZrO2-Y2O3 CERAMICS.

Two ZrO_2-Y_2O_3 (3% molar Y_2O_3) powders , D and E, resulting from different processing routes exhibiting differences in impurities content as presented in table 1 , and in yttrium distribution , as shown by Electron Probe Microanalysis, were examined. The role of inhomogeneity of yttrium distribution and dopant addition on the sintering behaviour and microstructure development of $ZrO_2 - Y_2O_3$ ceramics has been investigated.

	Powder D	Powder E	
SiO_2	< 0,05%	0.2 %	
MgO	< 150 ppm	<0.02%	
Al_2O_3	< 50 ppm	< 0.01%	Table 1:
Fe_2O_3	< 20 ppm	< 0.02%	Impurities content
TiO_2		0.12%	in powders D and E

2-1-Sintering behaviour and microstructural development of $ZrO_2 - Y_2O_3$ ceramics prepared from highly homogeneous Y distribution and low impurities content powders.

Powder D was prepared from a processing route yielding a homogeneous distribution of yttrium and low impurities content.

The evolution of sintered densities of powder D , versus firing temperatures showed a large sintering temperatures range (1400°C to 1600°C) for obtention of nearly complete densification (99% dth). A fine regular microstructure, with 0,2 μm grain size was observed on samples sintered at 1400°C. After sintering at 1600°C , a uniform grain growth could be noted , with an average grain size of 1 μm (fig 3).

— 10 μm

Fig 3: Microstructure of sintered ZrO2-Y203 (powder D) showing invariance of grain size distribution

1400°C   1600°C

Alumina addition up to 0,3 % molar ratio to ZrO_2 -Y_2O_3 , incorporated homogeneously as $Al(NO3)_3$ during powder processing was investigated on the sintering behavior. As already described by several authors [7-11], lower densification temperatures were observed : density values of 97,5 % dth were obtained at 1300°C. Al_2O_3 addition was also shown to impede complete densification in association with an enhancement of grain growth . A maximum value of 97,5 % dth was measured with 0.3 molar alumina , in comparison with a value of 98,8 % dth without doping(fig 4). Microstructural observation indicated a bimodal distribution of grain size as illustrated on fig 5.

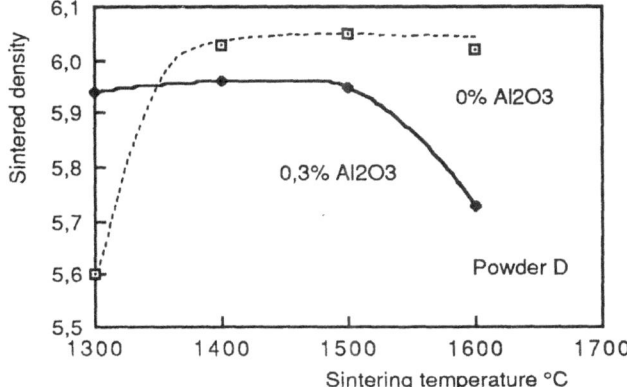

Fig 4: Effect of alumina on densification behaviour of ZrO2-Y2O3 , powder D

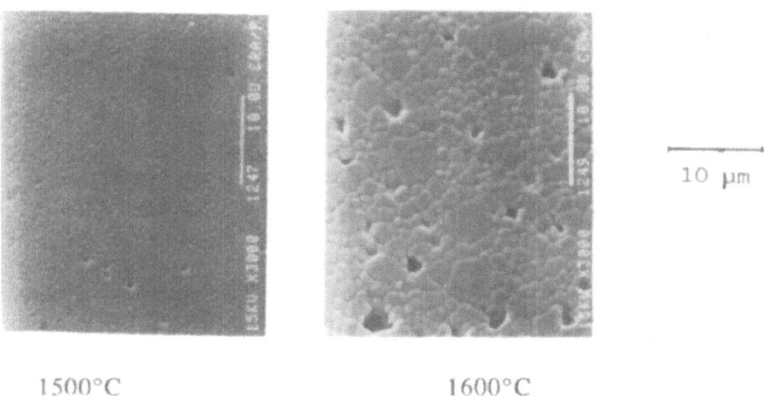

1500°C   1600°C

Fig 5 : Bimodal distribution of grain size of ZrO2-Y2O3 Ceramics (powder D, Al2O3 additions)

Larger grains above 5 /μm are found surrounded by smaller grains from 1 /μm to 2 /μm after sintering at 1600°C. Porosity remained intergranular , samples fired at 1600°C underwent no abnormal grain growth. Another observation was the decrease of density at these high sintering temperatures: a drop to 94% dth was found after firing at 1600°C . This dedensification was associated with pore growth as seen from microstructural examination and pores were found to be located either close to larger grains or in regions of smaller grains (fig 5). A slight structural evolution was outlined by X-ray diffraction, showing some cubic phase development with increased sintering temperatures (fig 6).

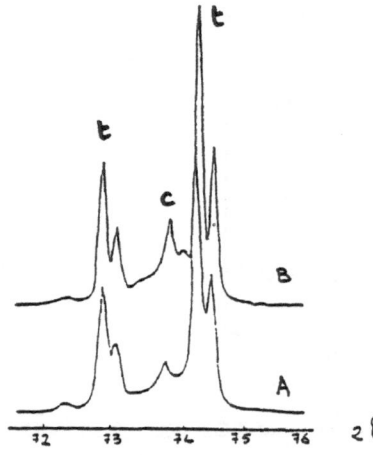

Fig 6 : X ray diffraction patterns of ZrO2-Y2O3 ceramics which were A(1400°C) and B (1600°C), showing a slight cubic phase development (powder D)

2-2-Sintering behaviour of ZrO2-Y2O3 powders having high SiO2 content and less homogeneous yttrium distribution.

Powder E exhibited a higher silica content than powder D as shown in table 1 and a lower yttrium homogeneous distribution. The high content of silica was due to the nature of raw materials.
   A narrow plateau of sintering temperatures (100°C) leading to high density values was observed for this powder (fig 7).
A slight decrease of densities at high sintering temperatures was observed. A typical microstructure of a sample sintered at 1600°C (fig 8) showed the development of a bimodal grain size distribution with isolated large grains of 8 /μm and smaller grains of 0,8 /μm. The decrease of sintered density values was associated with pore growth. Different types of pores could be sorted out : a first type was found either in the vicinity of the grains (intergranular type) or within the large grains (intragranular type) with convex or concave curvature ; a second type, surrounded by smaller grains was characterized by a high coordination number and convex curvature . X-ray diffraction
evidenced a strong phase transformation associated with the dedensification phenomenon. Tetragonal and cubic phases were shown to be transformed into monoclinic, cubic and t' which is a non transformable yttrium-rich tetragonal phase (fig 9).

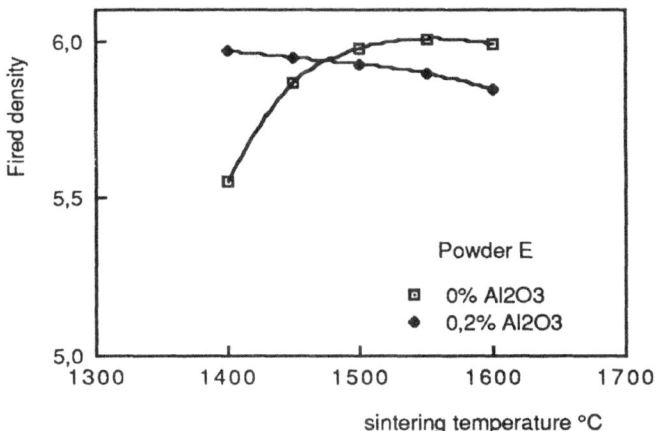

Fig 7: Effect of Al_2O_3 addition on sintering behavior of powder E

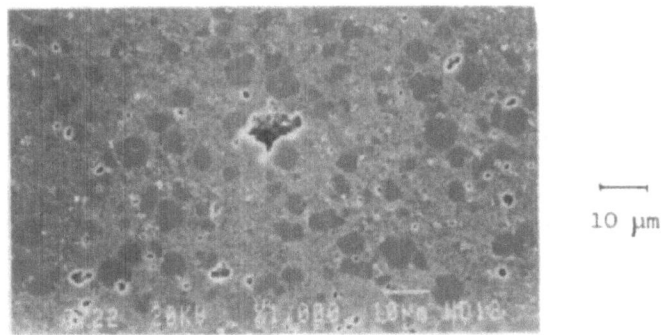

10 μm

Fig 8 : Bimodal grain size distribution of ZrO2-Y2O3 ceramics showing high coordination number of pores (powder E)

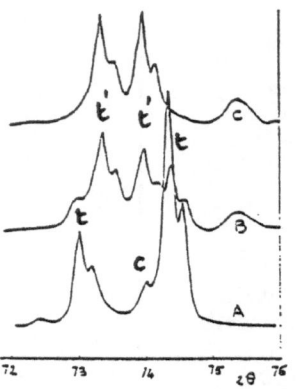

Fig 9: X ray diffraction pattern of ZrO2-Y2O3 ceramics (powder E) showing large structural transformation of tetragonal phase (sintering temperature 1600°C A: 10 mn, B: 360 mn, C: 600 mn)

Addition of alumina enhanced sinterability of the powder, lowering the sintering temperature down to 1400°C (fig 7). Pronounced dedensification was also shown to occur at firing temperatures lower than observed without alumina . Once again , the development of a bimodal grain size distribution with pore growth is to be noted for the microstructure evolution with temperature.

2-3-Discussion

The development of a bimodal grain growth in the two phases , low Y_2O_3 tetragonal and high Y_2O_3 cubic [12] ZrO_2-Y_2O_3 ceramics, during sintering was previously described by Ruhle and al [3] . They suggested that the presence of large cubic grains , with higher Y_2O_3 content than in matrix grains, was due to a more rapid grain growth by grain boundary phase diffusion rather than nucleation of new cubic grains during sintering in the matrix. By T.E.M. examination of several samples , they also emphasized a liquid phase sintering.
Dedensification has been earlier observed on ZrO_2-Y_2O_3 ceramics by several authors. Production of high pressure gaz pockets such as CO_2 [13], SO_2 [14] were reported. Exaggerated grain growth and development of pores on ZrO_2-Y_2O_3 ceramics during high temperature aging was also associated with the highly strained crystal lattice [15]. Lange [16] has ascribed dedensification to oxygen release during cubic and low yttrium content phases partitioning. The role of inhomogeneity of yttrium distribution on the sintering behaviour of ZrO_2-Y_2O_3 ceramics has been previously described [14] to possibly play a role in the high temperature dedensification.

2-3-1-Role of inhomogeneity of Yttrium distribution on microstructure development of ZrO_2-Y_2O_3 ceramics.

The dedensification observed at high sintering temperatures was responsible for the smaller sintering plateau with maximal sintered densities . As reported earlier (14) , we confirmed by dilatometry measurements that the dedensification phenomenon was associated with a volume expansion , occurring at high sintering temperatures, typically above 1550°C.

Microstructural observations made on samples prepared from powder E showed pores development either close to large grains or close to small grains. The examination of the morphology of these large pores showed high coordination number for pores with convex curvatures suggesting pore growth as elsewhere described by Kingery and Francois[17]. They reported conditions of pores developpement, giving special attention to details of pore grain boundary configuration. This pore development phenomenom is usually inhibited by concurrent grain growth during densification as described by Lange [18] who evidenced grain growth and rearrangement processes leading to the decrease of the coordination number of remaining pores and allowing them to disappear during last stages of sintering. In our $ZrO_2 - Y_2O_3$ ceramics grain growth at high sintering temperatures occurred in a bimodal mode , and did not allow optimal rearrangement of pores morphology with the decrease of pore coordination number , and thus make pores development easier . The inhomogeneity of yttrium distribution on powders might play a major role on the nucleation of cubic phase , then allowing grain growth by diffusion along grain boundaries and dedensification.

2-3-2-Role of Al_2O_3 addition on sintering behaviour and microstructure development.

Al_2O_3 was shown to be an effective sintering aid that would enhance the low temperature sintering of cubic $ZrO_2 - Y_2O_3$ (12% wt Y_2O_3)[7]. K. Radford and R. Bratton [8] described that densification occured via a liquid phase mechanism and suggested the formation of a liquid phase at 1355°C in the $Al_2O_3 - SiO_2 - MgO$ system or at 1546 °C in the $Al_2O_3 - SiO_2$ system. This mechanism was rejected by Bernard [9] who demonstrated no evidence of liquid phase formation by microstructural examination on $ZrO_2 - Y_2O_3$ (9%) samples containing 0.44% mole of Al_2O_3 . A limited solubility of 0.1 mole% at 1300°C was detected suggesting that a partial reason for the role of Al_2O_3 might be that some Al^{3+} ions substitutes for the Zr^{4+} ions in the lattice[9]. By TEM observations of several Y-TZP ceramics obtained from commercial sources , A. Ruhle and al. [3] showed that a continuous grain boundary yttrium silicate phase containing Al_2O_3 was always detected with variation of the width of the glassy layer according to purity of raw materials. The role of yttria in conjonction with alumina on the composition of the grain boundary glass was confirmed by Montross[10] who reported the effect of an yttria-alumina-silica ternary eutectic (1.45 moles Y_2O_3 to 1 mole Al_2O_3 to 2.465 moles SiO_2 with a eutectic temperature of 1395°C[19]).

In our experiments, the effect of Al_2O_3 addition on lowering the sintering temperature of ZrO_2-Y_2O_3 (3 molar %) was observed either with powders with low impurities content (powder D) or with powders with high SiO_2 content (powder E). Enhancement of low temperature sintering could reasonnably be attributed to liquid phase sintering for powder E . As for powder D, addition of Al^{3+} might play the same role as Fe^{3+}[20] or other transition metals[21]: a partial solubility of Al^{3+} in the Zr^{3+} sites will create oxygen vacancies in the crystal and enhancement of lattice diffusion (D_l)with improved densification and grain growth (11).

Concerning the dedensification phenomenon observed with alumina addition , Lange (16) reported the effect of the presence of Al_2O_3 inclusions and glass phase boundary on microstructure developments. The bloating or decrease of density values observed at high sintering temperatures was attributed to the fact that the cubic phase formation during the partitioning reaction is accompanied by rejection of oxygen. For Al_2O_3 addition , a similar effect to that described for the role of yttrium homogeneity could be proposed: limited solid solubility of Al^{3+} might have similar effect to that Y^{3+} in the ZrO_2 structures and could promote the formation of high vacancy domain . Alumina addition in pure cubic ZrO_2-Y_2O_3 (> 5% molar Y_2O_3) was shown to promote grain growth , with invariance of the shape of the grain size distribution[11]. For our material (3% molar Y_2O_3) the bimodal grain growth observed should involved same mechanism as previously described for role of inhomogeneity of yttrium: Growth of existing cubic nuclei occurred more rapidly than nucleation of new cubic ZrO_2-Y_2O_3 nuclei, and enhanced, as previously described, pore growth.

2-3-3- Densification , grain growth in monoclinic plus cubic system: Grain boundary diffusion and role of grain boundary composition.

Table 2 summarized the observations made on powders D and E at high sintering temperatures where simultaneous densification , grain growth and partitioning reactions occurred.

1/ homogeneous Y distribution distribution.	3/ Inhomogeneous Y distribution
No or little glassy phase boundary	Presence of glassy phase boundary
Powder D	Powder E

Slight uniform grain growth

Partitioning and bimodal
grain growth
with liquid phase boundary

2/ homogeneous Y distribution

4/ Inhomogeneous Y
distribution

No or low glassy phase boundary

Presence of glassy phase
boundary

Alumina addition
Powder D

Alumina addition
Powder E

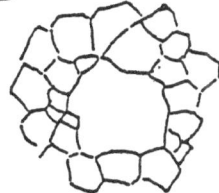

Partitioning and
bimodal grain growth

Partitioning and bimodal
grain growth with liquid
boundary phase

Table 2

Addition of alumina , or the use of a powder with inhomogeneous yttrium distribution was shown to promote dedensification. In these two cases , dedensification was associated with bimodal grain growth and pore development. Experiments performed without alumina addition , on powder D showing monodisperse particule size distribution, resulted in an uniform grain growth, i.e. with invariance of shape of grain size distribution. The driving force for the exaggerated bimodal grain growth observed for the same powder D with alumina addition , could not thus be attributed to the classical effect of boundary curvature .
The highest dedensification rate was observed for powder E having high SiO_2 content and Al_2O_3 addition. This, added to Lange's results showing the importance of glassy phase residu from milling balls on the bloating phenomenon [16] demonstrates that mass transport mechanism is promoted by the presence of a glass boundary phase suggesting large effect of grain boundary diffusion on the bimodal grain growth mechanism. Boundaries should thus be yttrium-rich

regions to allow diffusion of yttrium from a high yttrium concentration domain towards a low yttrium concentration domain.

Lattice diffusion and grain boundary diffusion are then expected to be dominating respectively in the densification process and grain growth mechanism of ZrO_2-Y_2O_3 ceramics.

From these observations, the role of grain boundary diffusion on grain growth and bloating was emphasized. Grain boundary composition defined by the impurities content in the powders must largely affect grain boundary diffusion.

The high rate of grain growth is also likely to decrease the ratio : rate of densification / rate of grain growth and thus impedes obtention of complete densification [22]. This is consistent with the afore mentioned effect of alumina addition which was shown to decrease the maximum value of the fired density (fig 4).

CONCLUSIONS

Monomodal spherical particles were shown to promote homogeneity of stress distribution, but to decrease strength and densification of green compacts, then a compromise is to be found for practical use of ZrO_2-Y_2O_3 powders.
Impurities content or dopants addition such as Al_2O_3 , largely affect the densification behaviour. Al^{3+} was shown to enhance densification by lattice diffusion independently of the grain boundary composition.
A dedensification phenomenon was observed in the high sintering temperatures domain, enhanced by Al_2O_3 additions, altering phase relations. It was shown to be also related to yttrium inhomogeneity which promotes the formation of cubic nuclei. This dedensification was associated in both cases with a bimodal grain growth and pore development . A mechanism describing dedensification based on pore growth along a bimodal grain growth associated with partitioning reaction was proposed.

BIBLIOGRAPHY

1- The Role of Powder Properties in Ceramic Processing, J. A. Mangels, Ceram. Eng. Sc. Proc. , 1986, 7, (9-10), 1112-1121.

2- Formation , Packing and Sintering of Monosized TiO2 Powders, E.A. Barringer and H.K. Bowen, J. of Am. Ceram. Soc. 65, 12- C 199 (1982)

3- Microstructural Studies of Y2O3 Containing Tetragonal ZrO2 Polycrystals ZrO2 Y-TZP , M. Ruhle , N. Claussen and A. H.Heuer, Science and Technology of Zirconia 2 , Advances in Ceramics , Vol 12, 1984.

4- Toughened Zirconia Ceramics from Electrorefined PSZ Powders, S. Blacburn , C.R. Kerridge , P. G. Senhenm, 89 th Am. Ceram. Soc. Meeting, Pittsburgh , USA, 1987.

5- Effect of ZrO2-Y2O3 Powder characteristics on compaction Behavior, J. Y. Chanc-Ching, N. Paraud, D. Bortzmeyer and M. Abouaf, Proc. of 2nd Int. Conf. in Powder Processing, Berchtesgaden, FRG, 1988, To be published.

6- Morphology Characteristics and Yield Locus of ZrO2-Y2O3 Powders, D; Bortzmeyer, M. Abouaf, J.Y. Chane-Ching and N . Paraud, In preparation.

7 - US patent 3 573 107 (1971),R.A. Paris and G. Paris.

8 - Zirconia Electrolytes Cells , Part 1- Sintering Studies, K. C. Radford and R.J. Bratton, J. Materials Science, 14, 1, 59-65 (1979).

9- Sintered Stabilized Zirconia Microstructure and Conductivity, H. Bernard, Rep CEA, R 5090, Commissariat à l'energie atomique, CEN Saclay, France , 1981.

10- Sintering Aids for the Yttria Partially Stabilized Zirconia, C.S. Montross, Thesis 1987, University of Washington.

11- Effects of Al2O3 Additions on Resistivity and Microstructure of Yttria Stabilized Zirconia, M. Miyayama, H. Yanagida , A. Asada, Am. Ceram. Soc. Bull., 64, (4), 660-664, 1985.

12- H. G. Scott, J. of Mater. Science , 10 (1975) 1527.

13- Influence of Precipitating Atmosphere on Sintering of ZrO2+ 12 mol % Y2O3 , M.A. Thompson, D.R. Young and E.R. McCartney, J. of the Am. Ceram. Soc. , Vol 56, N° 12, 1973, 648-654.

14-Sinterability of Tetragonal ZrO2 Powders, A. Smith and J.F. Baumard, Am. Ceram. Soc. Bull. 66, (7), 1144-1148, 1987.

15- Effect of Densification Conditions on the stabilization of the tetragonal phase in ZrO2 Polycrystals, V. K. Pujari and I. Jawed, J. Am. Ceram. Soc. 68 (9) C 242-243.

16- Effects of Attrition Milling and Post Sintering Heat Treatment on Fabrication , Microstructure and Properties of Transformation

17- The Sintering of Cristalline Oxides . 1- Interactions Between grains Boundaries and Pores, W.D. Kingery and B. Francois, G.C. Kuzynke , N.A. Hooten and G.F. Gibbon , Ed Gordon Breach N.Y. (1967).

18- Sinterability of Agglomerated Powders , F.F. Lange , J. Am. Ceram. Soc. , Vol 67-2 , 83-89 , 1984.

19- Al2O3-Y2O3-SiO2 Phase Diagram, I.A. Bondar, F.Y. Galakhov, Ivz. Akad. Nauk, SSSR, Scr. Khim, Vol 7, 1963, 1325.

20- Iron oxide Doped Yttria Stabilized Zirconia Ceramics , R.V. Wilhelm , J. and D; Howarth, Ceramic Bulletin, Vol 58, N°2, (1979) 228-232.

21- Transition Metals Oxides Doped YTZ , H. Okamura, N. Kimura, 2nd Int. Conf. Powder Processing, Berchtesgaden, FRA, 1988.

22- Fabrication Principles for the Production of Ceramics with Superior Mechanical Properties, R. J. Brook, Proc. Brit. Ceram. Soc. 32,7-24, 1982.

SINTERING BEHAVIOR OF 3 MOLE % YTTRIUM DOPED CHEMICALLY SYNTHESIZED ZIRCONIA

M.DESCEMOND, C.BRODHAG, F.THEVENOT
Ecole Nationale Supérieure des Mines, Saint-Etienne, France
P.HOMERIN, E.ROTHMAN
CRICERAM, Division CERALTECH, PECHINEY Group, France

ABSTRACT

Sintering behavior of Yttria stabilized ZIRCONIA (SYGMA+3Y™) was examined for chemically synthesized powders produced by CRICERAM's sublimation hydrolysis method. Results were obtained for different surface areas : from 6 to 40 m^2/g. These studies indicated densification at a relatively low temperature (<1500°C) with good mechanical properties.

INTRODUCTION

Sintering studies of a series of 3 mole % Yttria doped Zirconia powders (CRICERAM: SYGMA+3Y™) of various surface areas were performed at the ECOLE NATIONALE SUPERIEURE DES MINES DE SAINT-ETIENNE. The powders were chemically synthesized by a vapor phase reaction [1], generating naturally fine, pure zirconia powder. Surface areas from 6 to 40 m^2/g were examined (only surface areas of 6, 12, and 18 are commercially available). Dilatometry was used to optimize sintering conditions. Based on the resulting physical properties, the optimal sintering temperature for surface areas of 6, 12, and 18 m^2/g was determined. Parameters optimized included : X-ray analysis for surface monoclinic phase content [2], fired density and mechanical strength.

EXPERIMENTAL

Powders : Surface area measurements were made using a BET single point nitrogen adsorption (Areameter II of Ströhlein), median agglomerate size determined by sedimentation (Sedigraph D5000) and monoclinic phase content by XRD (Siemens D500).

Pressing : As recieved powders were pressed first uniaxially followed by isostatic pressing at 400MPa.

Sintering : Dilatometry was performed (Adamel DI 24) in a static atmosphere. Heating and cooling rates were constant 2.5°C/min. Maximum temperature level was held for 1 hour.

Sintered Properties : Archimedes method was used for density measurements and XRD analysis for determination of the fraction of the monoclinic phase. Mechanical properties were determined using the 3 point bend test on bars cut from uniaxially pressed pellets or from isostatically pressed blocks.

RESULTS

For surface areas varying from 6 to 40 m^2/g the median agglomerate size decreases from 0.7 to 0.2 micron and the fraction of monoclinic phase from 0.83 to 0.09 (Table 1). Green density decreases from 3.53 to 3.18 which is 58% to 52% of theoretical fired density (6.09), respectively, as surface area increases.

Table 1- Characteristics of the zirconia powders and of the sintered bodies obtained with these powders.

	POWDER				SINTERED BODY		
Name	Surface area (m2/g)	Median agglom. size(μm)	Monoclinic phase content (%)	Green density (400MPa)	Sintering temperature (°C)	Fired density	Monoclinic phase content (%)
3Y6	8,5	0,70	83	3,52±0,02	1420	5,82	47
					1440	5,84	47
3Y12	12,7	0,48	67	3,53±0,015	1360	6,06	9
				3,44*	1370	6,01	8,5
				3,53±0,015	1450	6,07	13
3Y18	17,3	0,27	33	3,41	1315	6,08	0
				3,37*	1315	6,02	0,1
				3,41	1450	6,07	6
3Y20	19,7	0,28	27	3,31*	1400	6,02	0,2
3Y30	28,9	0,19	16	3,27*	1400	6,02	0,5
3Y40	37,2	0,17	9	3,18±0,04*	1400	6,03	0

* pressing in a 10x8 mm^2 die (the others : in a 25x8 mm^2 die)

Illustrated in Fig.1 (a,b) are the shrinkage and shrinkage rate dilatometer curves for powders of 6 and 18 m^2/g. Four characteristics of the dilatometric curves are noted: (A) the beginning of shrinkage (the point for 1% of shrinkage is considered), (B) the maximum rate of shrinkage, (C) the overall shrinkage and (D) the volumetric expansion upon cooling due to martensitic transformation.

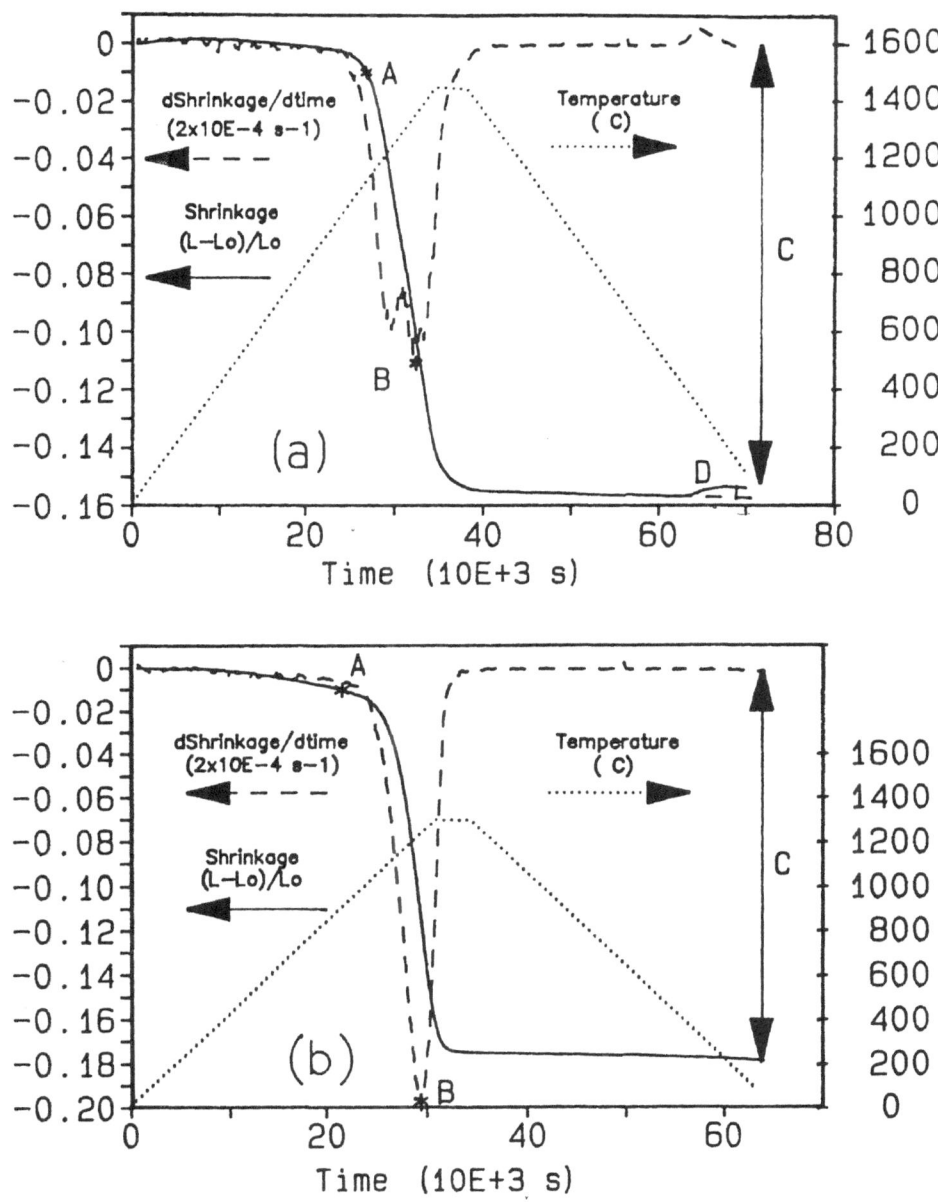

Fig.1- Shrinkage and shrinkage rate curves for :
    (a) powder of 6 m^2/g sintered to 1440°C,
    (b) powder of 18 m^2/g sintered to 1315°C.

For the powder of low surface area (6 m^2/g), the shrinkage rate curve presents two peaks. The first one, observed at approximatively 1200°C, does not appear for powder of higher surface area. It could be related to the monoclinic to tetragonal transformation during the heating of the compact. This transformation, associated with a diminution of volume, results in creation of porosity in the compact and in a fall-off of the shrinkage rate. When the fraction of monoclinic phase in the powder decreases, the peak of transformation does not appear on the shrinkage rate curve. For the second peak observed at higher temperature (1350°C), the shrinkage rate of the tetragonal phase is maximum.

For increasing surface area, the temperature (A) at which the shrinkage begins decreases (Fig.2) as well as the temperature (B) at which the shrinkage rate is at a maximum (Fig.3).

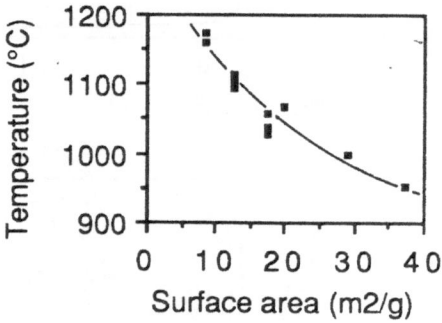

Fig.2 - Temperature (A) at which the shrinkage begins against the surface area of the powder.

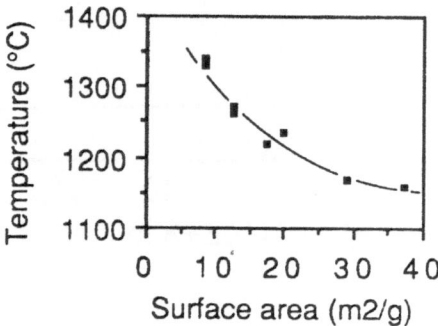

Fig.3 - Temperature (B) at which the shrinkage rate is at a maximum against the surface area of the powder.

The relative shrinkage (C) increases with surface area (Fig.4). This agrees with low green densities observed with high surface area powders.

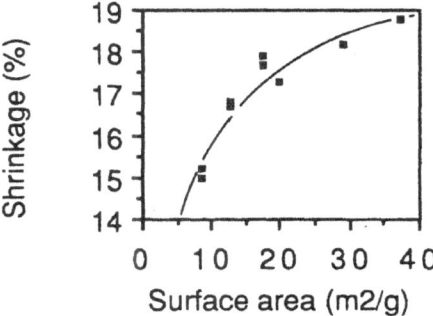

Fig.4 - Overall shrinkage (C) against the surface area of the powder.

The volumetric expansion (D) during cooling due to the retransformation of tetragonal to monoclinic phase is almost non existant for surface areas above 18 m^2/g but increases sharply below 12 m^2/g (Fig.5).

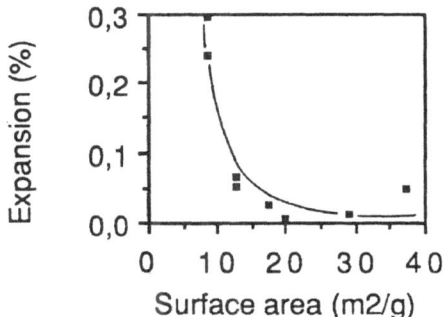

Fig.5 - Volumetric expansion (D) during cooling against the surface area of the powder.

A correlation of this expansion with the surface monoclinic phase content appears clearly. For surface areas $\geq$18 m^2/g, this content is in the range of 0 to 6 % and for $\leq$12 m^2/g it is in the range of 8.5 to 47 % (Table 1). The percentage of monoclinic phase varies with the maximum sintering temperature.

All sintered densities are greater than 6 except for the SYGMA+3Y6™ (Table 1). In this table, variations in fired densities within a surface area are due largely to differences in pressing techniques. Powders of all surface areas sinter well below normal sintering temperatures; less than 1500°C as opposed to 1600°C, standard firing temperature.

For optimal sintering conditions of SYGMA+3Y12™ and SYGMA+3Y18™, isostatically pressed samples, tested in 3 point bend, mechanical strength is 960±120 MPa. For SYGMA+3Y6™, uniaxially pressed, the strength is 825 MPa.

Under the stated experimental conditions, the performance of 'ready to use' powders, SYGMA+3Y12B™ and SYGMA+3Y18B™, does not show significant variations when compared to binderless powders.

DISCUSSION

Using optimized sintering conditions, yielded good mechanical properties, high density and low monoclinic phase content at temperatures below 1500°C. These characteristics were obtained, using pressing as the forming method, for high surface area powders. On the contrary, higher surface areas led to lower green densities and important shrinkage, creating difficulties in controlling near net shape forming. Moreover, very high surface area powders present handling difficulties. Surface areas of 12 and 18 m^2/g appear to be optimum. However, lower surface areas (6 m^2/g) when pressed, do not produce sintered bodies with comparable mechanical properties (this powder is best suited for slip casting or injection molding).

Increasing surface area is beneficial up to a certain level. For example, for powder of surface area >18 m^2/g there is no significant decrease in volumetric expansion upon cooling or in the temperatures corresponding to the beginning and fastest rate of densification. However, the green density continues to decrease linearly as surface area increases.

The study clearly shows the importance of sintering temperature optimization: higher temperatures lead to increase shrinkage, however, upon cooling there is some reconversion to the monoclinic phase weakening the structure. This results from the grain growth which occurs during the last stage of sintering. Some of the grains exceed a critical size and then transform spontaneously from the tetragonal to the monoclinic phases [3,4].

CONCLUSION

This study shows that the CRICERAM's Yttria Stabilized Zirconia powders of middle surface area (12 and 18 m^2/g), sintered to an optimal temperature (<1500°C), lead to materials with a high density, a low monoclinic phase content and good mechanical properties.

REFERENCES

[1] - R.BARRAL, M.TITEUX, *Synthèse industrielle de poudres céramiques ultra-fines de haute pureté : cas de l'alumine et de la zircone,* L'industrie céramique, 790 [1], 49-50, 1985.

[2] - P.HOMERIN, *Préparation et caractérisation de matériaux à base d'alumine renforcés par une dispersion de zircone,* Thèse, Institut National des Sciences Appliquées de Lyon, 29 septembre 1987.
R.FILLIT, P.HOMERIN, J.SCHAFER, H.BRUYAS, F.THEVENOT, *Quantitative XRD analysis of zirconia-toughened alumina ceramics ,* J.Mater. Sci., 22, 3566-3570, 1987.

[3] - F.F.LANGE, *Transformation Toughening : Part 1,* J. Mater. Sci., 17 [1], 225-234, 1982.

[4] - F.F.LANGE, *Transformation Toughening : Part 3,* J. Mater. Sci., 17 [1], 240-246, 1982.

Reaction-Sintered Mullite-Zirconia Composites: Mechanism and Properties

J.V. Emiliano* and A.M. Segadães
Dep.º de Eng.ª Cerâmica e do Vidro, Universidade de Aveiro
3800 AVEIRO - PORTUGAL

It is well known that zirconia additions to mullite-based refractories, either fusioncast or conventionally sintered, improve their ability to withstand corrosion by molten slags and glasses. Special attention has been directed to reaction-sintering prepared mullite-zirconia composites from zircon sand and alumina, but the simplicity of the process can be somewhat overcast by the difficulty in establishing the mechanism through which reaction+sintering occur.

In the present work, zircon sand and two different aluminas were used to prepare mullite-zirconia composites. Dry pressed test pieces with stoichiometric composition were fired between 1300 and 1650°C and the decomposition of zircon and the formation of mullite were followed by X-ray diffraction. Changes in density and microstructure were observed and the cold bending strength and Young's modulus were determined and correlated with the observations made.

I. Introduction

Mullite is an alumino-silicate extensively used in traditional refractory applications and one of most studied crystaline phases in the SiO_2-Al_2O_3 system [1-3]. Its singular properties, namely low thermal expansion, high creep resistence and high chemical stability, make it most suitable to high temperature technological use.

However, mullite seems to have mechanical characteristics not as good as those of other competitive ceramics (for sintered mullite, $\sigma_f \simeq 150$ MPa and $K_{Ic} \simeq 2$ MPa m$^{1/2}$ [4]) and it is difficult to produce silica + alumina homogeneous mixtures [5,6].

---

* Dep.º de Eng.ª de Materiais, Universidade Federal de São Carlos, Cx.P. 676, 13560 São Carlos-SP, BRASIL

In the past few years, several authors [7-19] have tried to improve the properties of mullite and conceive alternative synthesis methods. Among these, the production of zirconia-added mullite through reaction-sintering of $Al_2O_3+ZrSiO_4$ mixtures has atracted most attention due to its potentialities. The relevant equation is:

$$(3+x)Al_2O_3 + 2ZrSiO_4 \xrightarrow{\frac{CaO, MgO}{TiO_2}} 3Al_2O_3.2SiO_2 + 2ZrO_2 + xAl_2O_3 \qquad (1)$$

where CaO, MgO and TiO_2 act as sintering aids and lower the temperature of decomposition of $ZrSiO_4$ [13-15].

The work done showed that the process, although conceptually simple, is difficult to control due to the competition between densification and reaction and to its extreme sensibility to the characteristics of the starting powders [20].

Also, other authors [19,21-24] showed that the powder processing technique used in the reaction-sintering of alumina-zircon powders can produce microstructural deffects which can greatly hinder the benefits of zirconia additions (mechanical mixing of powders usually favours agglomerate formation and other inhomogeneities. Agglomerates tend to densify before the ceramic body as a whole and pore traping, or even pore growth (if there is a broad enough grain size distribution inside the agglomerate), can occur, hindering the densification process).

In the present work, the reaction-sintering process of two mixtures of alumina and zircon, prepared in the same proportions but with different aluminas, is studied by following the changes in microstructure and mechanical characteristics of test pieces subjected to thermal treatment between 1350 and 1650°C for 1 to 4 h.

II. Experimental Procedure

(1) Preparation of samples

Two different aluminas and one zircon, as described in Table 1, were used to prepare two mixtures with a slight excess (≈1%) alumina as referred to an alumina:silica ratio of 3:2, in HDPE atrition mill with alumina balls as grinding media and isopropyl alcohol as grinding aid (better mixing and lower contamination [12]). It was found that after 8h grinding the wear of the griding media was <0.1% in weight for a 1000g batch. Also, 3% PVAL in weight was added to assist subsequent pressing of the pellets.

Table I. Characteristics of the raw materials used

Aluminas:
ALCOA A16 (referred to in the text as A16), *Aluminum Company of America, Pa.*
    average grain size, $d_{50\%} \simeq 0.5$ μm, 100% < 1μm
    specific surface area $\simeq 7$ m^2/g

ALCOA APC3000 (referred to in the text as APC), *ALCOA do Brasil*
    average grain size, $d_{50\%} \simeq 0.6$ μm
    specific surface area $\simeq 3$ m^2/g
    chemical analysis: Al$_2$O$_3$ TiO$_2$ CaO   Na$_2$O (wt %)
                       99.7   0.08   0.03   0.025

Zircon:
ZrSiO$_4$ ALW (referred to in the text as ALW), *MAGNESITA S.A., Brazil*
    average grain size, $d_{50\%} \simeq 0.5$ μm
    chemical analysis: SiO$_2$ ZrO$_2$ Al$_2$O$_3$ TiO$_2$ (wt %)
                       33.1   65.5   0.9   0.08

After drying at 110°C for 12h and sieving through 100 mesh, 10% water, in weight, was spray-added to induce the plasticizer action, the mixtures were sieved again for homogenization and test pieces 10mm in diameter were prepared by uniaxial pressing followed by isostatic pressing at 200 MPa. Pellets were then fired in air in an electric furnace with MoSi$_2$ heating elements for 4h (heating rate of 5°C/min and 2h soak at 600°C to eliminate PVAL), at temperatures ranging from 1350 to 1650°C.

Test bars (50×5×5 mm) were also prepared and heat treated as described above, to determine the mechanical characteristics.

(2) Characterization of samples after heat treatment

Relative quantitative analysis of crystalline phases present in the samples (α-alumina, mullite, zircon, tetragonal- and monoclinic-zirconia) was carried out by X-ray diffraction of the surface *in natura*.

The relative amounts of zirconia and mullite formed were evaluated using Boch and Giry [20] equations:

$$\% \ ZrO_2(m+t) = \frac{I\{ZrO_2-m[11\bar{1}] + ZrO_2-t[111]\}}{I\{ZrO_2-m[11\bar{1}] + ZrO_2-t[111]\} + I\{ZrSiO_4[200]\}} \qquad (2)$$

$$\% \ 3Al_2O_3.2SiO_2 = \frac{I\{A_3S_2[210]\}}{I\{A_3S_2[210]\} + I\{ZrSiO_4[200]\}} \qquad (3)$$

where I{compound[ijk]} is the integrated intensities of the corresponding X-ray line. The relative amount of tetragonal-zirconia retained after each heat treatment was evaluated according to Garvie-Nicholson method [25], using the equation:

$$\% \ ZrO_2-t = \frac{I\{ZrO_2-t[111]\}}{I\{ZrO_2-t[111] + ZrO_2-m[111] + ZrO_2-m[11\bar{1}]\}} \qquad (4)$$

The apparent density (D_{ap}) was determined by Archimedes method in Hg and the theoretical density (D_o) was calculated assuming that the reaction was carried out to completion and 30% tetragonal zirconia was retained ($D_o=3.79\times10^3$ kg/m³).

Microstructures were observed under reflected light optical microscope and scanning electron microscope.

Mechanical properties were measured by three-point bending test, with 30 mm span and 0.5 mm/min cross-head speed. Only totally reacted test pieces (100% mullite formed and zircon dissociated) were used to calculate σ_f and E.

III. Results and discussion

(1) Microstructural features

Figures 1 and 2 are representative electron-micrographs of mixtures ALW-APC and ALW-A16, respectively, after sintering at 1300°C for 240 min. In both cases, alumina (black spots) and zircon (white spots) agglomerates of various sizes can be seen, inbeded in a light-grey matrix undergoing reaction.

Since the tendency to form aglomerates increases with specific surface area, larger agglomerates were to be expected in mixture ALW-A16, because of the higher specific surface area of the A16 alumina.

However, while in ALW-A16, the alumina agglomerates are about 10μm in size and zircon agglomerates are about 20μm, in mixture ALW-APC much bigger agglomerates can be seen ($\approx$80μm for alumina and $\approx$50μm for zircon).

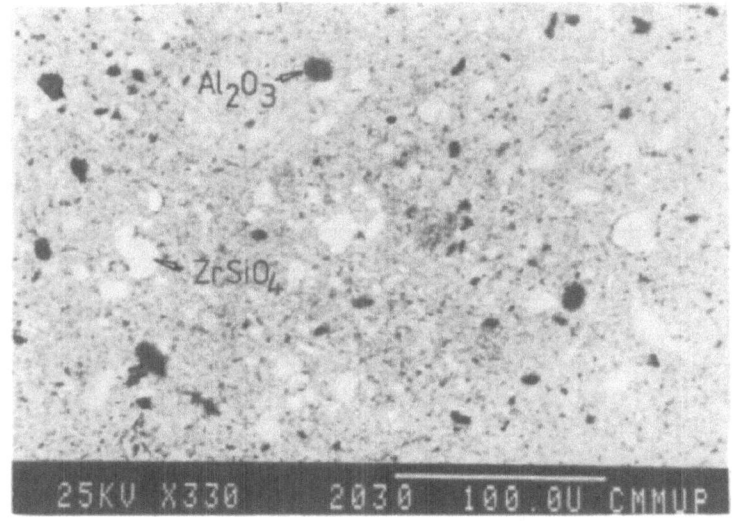

Fig.1. Microstructure of mixture ALW-APC after sintering at
1300°C for 4h, showing alumina and zircon agglomerates

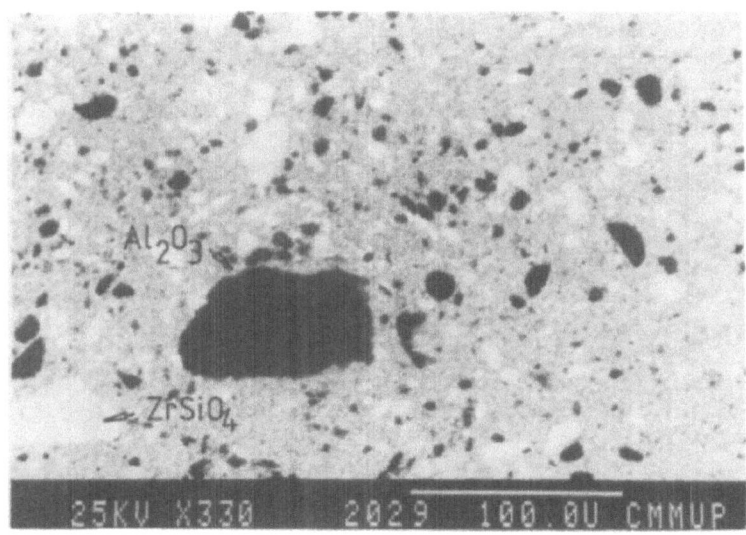

Fig.2. Microstruture of mixture ALW-A16 after sintering at
1300°C for 4h, showing alumina and zircon agglomerates

Figure 3 shows needle-like mullite grains in mixture
ALW-APC after sintering at 1600°C for 4h, evidence of a
liquid phase present (probably localized and transient)

as well as areas of amorphous silica. No mullite needles were found in mixture ALW-A16.

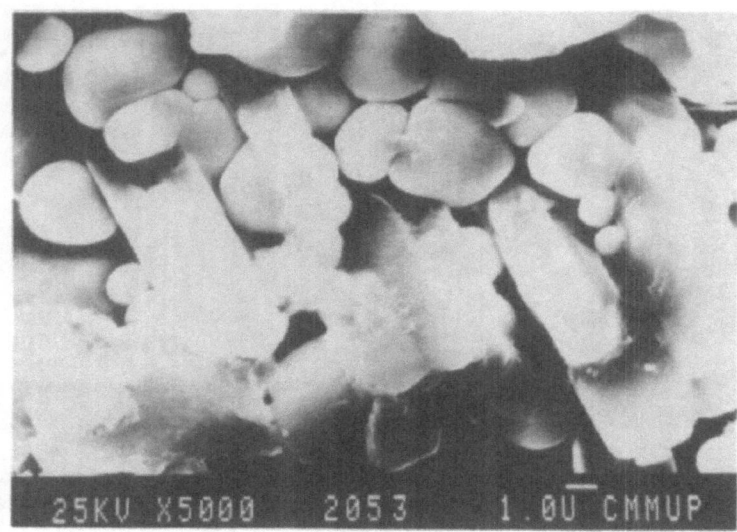

Fig.3. Needle-like mullite grains and amorphous silica in mixture ALW-APC sintered at 1600°C for 4h

In spite of the larger agglomerates, the presence of this liquid phase, presumably developped from the TiO_2 and Na_2O existing in the raw materials, can lead to a higher densification of mixture ALW-APC.

Figure 4 shows that the agglomerates and larger zircon grains (lower left corner) decompose from the outside to the centre producing zirconia (bright white dots) and amorphous silica (continuous medium-grey phase). This in turn, transforms into ß-crystobalite and reacts with alumina to produce mullite (light-grey phase), which is deposited on the surface of the grain or agglomerate.

The so-formed mullite layer acts as a diffusion barrier, and the silica produced by the decomposition of zircon has now to migrate to the outside, before it can react, leaving a pore behind. The resulting small pores migrate and grow by Kirkendall effect [26] and the zirconia grains produced in the same reaction form hollow agglomerates.

These features can be seen in Figure 5 which shows a zirconia-lined cavity surrounded by a mullite layer.

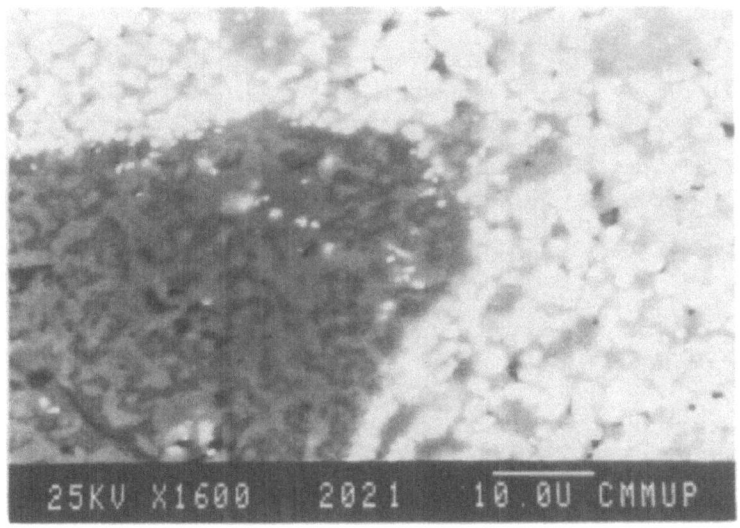

Fig.4. Large agglomerate of zircon grains
undergoing decomposition

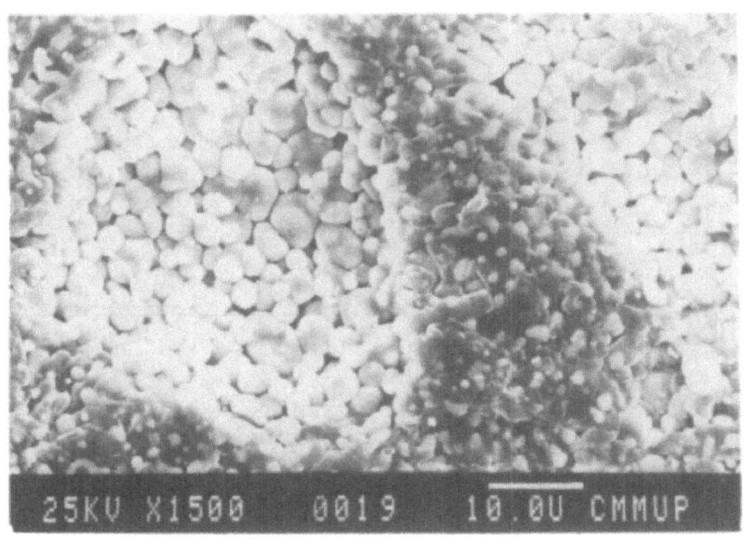

Fig.5. Zirconia agglomerates formed during the
dissociation of zircon (1600°C, 4h)

58

(2) The reaction - densification process

Mullite is a low density phase ($\simeq 3.16\times10^3$ kg/m^3), as compared to zircon ($\simeq 4.63\times10^3$ kg/m^3) or alumina ($\simeq 3.99\times10^3$ kg/m^3), and although the zirconia also produced in the reaction has higher density ($\simeq 5.68\times10^3$ kg/m^3 for the monoclinic phase and $\simeq 6.09\times10^3$ kg/m^3 for the tetragonal phase), the overall reaction, as expressed by equation (1), has a de-sintering effect on the composite.

Figure 6 shows the relative density (D_{ap}/D_o) of samples sintered for 240 min, as a function of sintering temperature (ranging from 1350 to 1650°C).

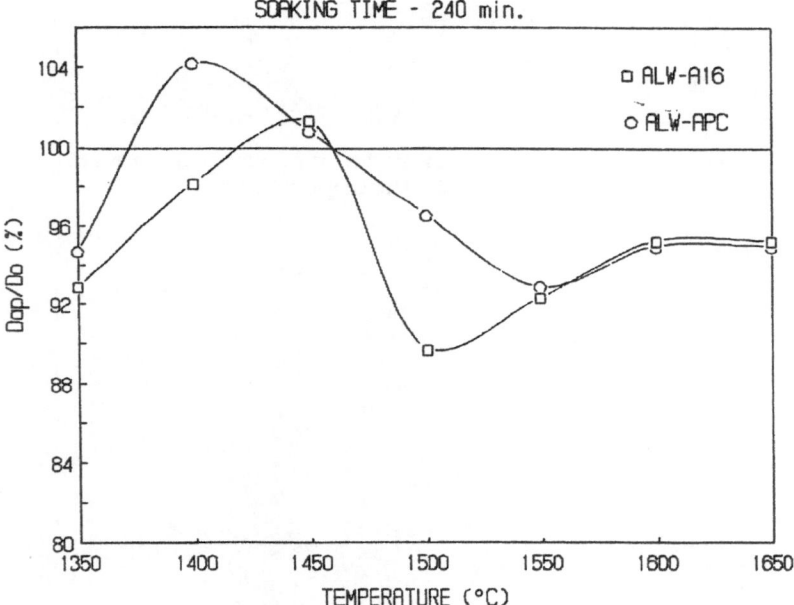

Fig.6. Relative densities of samples as a function of sintering temperature

Relative density values above 100%, based on the D_o value for complete reaction, are reached in the early stages of the heat treatments. This clearly shows that the sintering process of the reactants starts before reaction takes place and, in fact, the maxima observed can be related to the on-set of the reaction to form mullite (reffer to Figures 7 and 8). To obtain the usual relative density plots, a more sophisticated method for calculating the theoretical density, namely a step-by-step calculation as a function of sintering temperature and time, has to be used [27].

Maximum density is reached at 1400°C for sample ALW-APC
and at 1450°C for sample ALW-A16.

Figures 7 and 8 compare, for each mixture, the effect
of temperature on densification and the amounts of
mullite and zirconia formed.

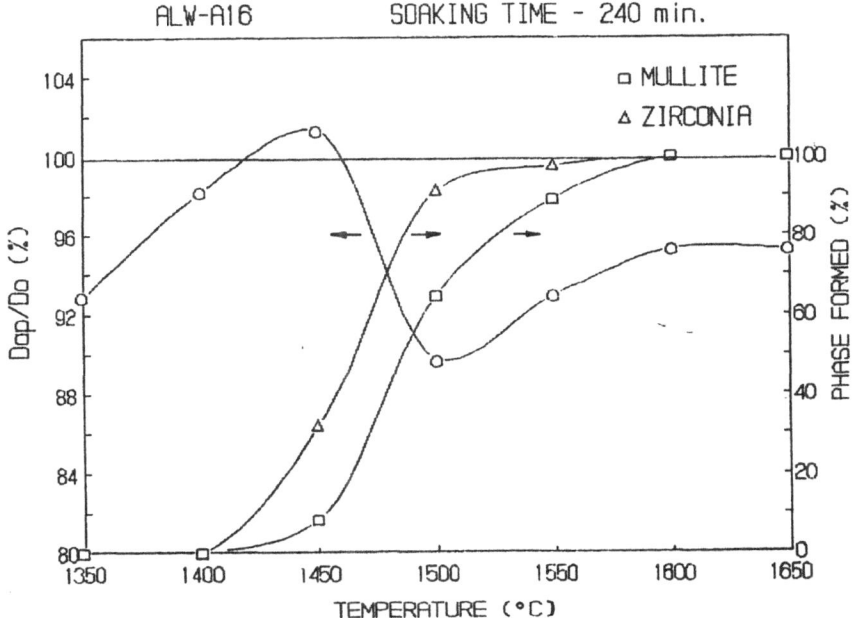

Fig.7. Relative density and amounts of mullite and zirconia
formed versus sintering temperature, in mixture ALW-A16

As mentioned above, sintering starts earlier than the
reaction and, since theoretical density was calculated
on the basis of complete reaction and the reactants are
denser than the products of the reaction, relative
densities higher than 100% are reached in the early
stages of the heat treatment of the mixtures.

Above 1400°C, de-densification occurs and a minimum
density is reached at 1500°C for mixture ALW-A16
(Figure 7) and at 1550°C for ALW-APC (Figure 8). The
decline of densification rate coincides with the start
of zircon decomposition (equation 2) and the on-set of
the reaction to form mullite (equation 3). This is in
agreement with Boch and Giry's results [20].

In mixture ALW-APC this stage of the process seems to
be slower than in mixture ALW-A16 (spread over a
broader range of temperature).

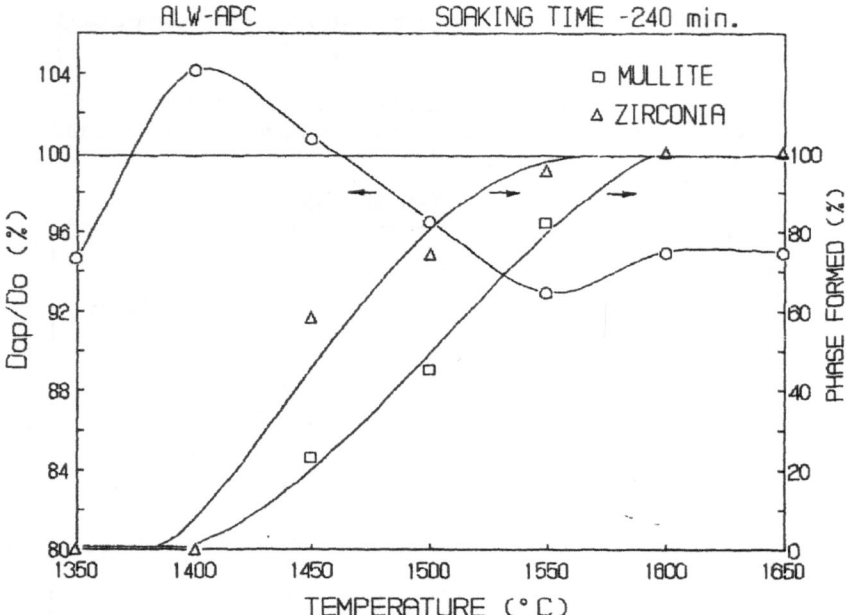

Fig.8. Relative density and amounts of mullite and zirconia
formed versus sintering temperature, in mixture ALW-APC

When the decomposition of zircon and the formation of
mullite approach completion, the relative densities of
the two mixtures start rising again, a sign that the
sintering process of the products of the reaction is
taking over.

In both mixtures, the final relative density reached,
calculated as described, is naturally lower than the
maximum occuring before the reaction starts.

Evidence such as shown in Figures 4 and 5 suggests that
there is a certain time delay between the start of
zircon decomposition (production of silica) and the
begining of mullite formation which is in agreement
with Di Rupo and Anseau's results [28].

In fact, the X-ray analysis of the sintered samples
(Figure 9-a and b) show that after 1h at 1600°C all
ZrSiO$_2$ is decomposed. However, due to the delayed
reaction to form mullite, even longer heat treatments,
at this temperature still shows a slow decline in
relative density in both samples (Figure 10).

Presumably, once the formation of mullite is complete,
sintering should definitely take over and relative

density should increase again. This effect should be noticeable, at 1600°C, for soaking periods longer than 4h.

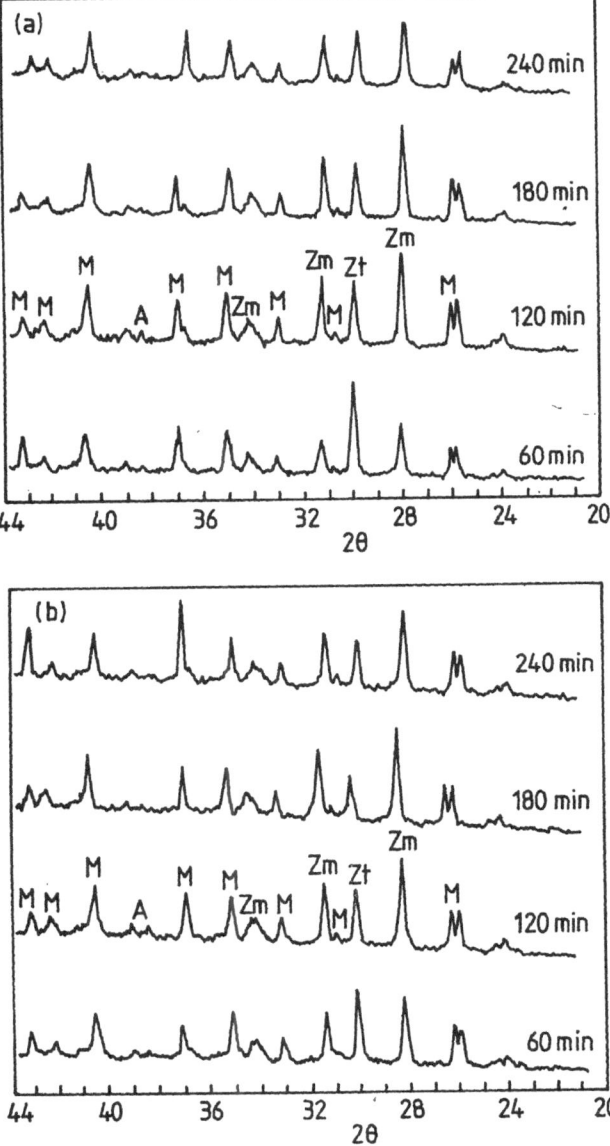

Fig.9. X-ray diffractograms of mixtures ALW-APC (a) and ALW-A16 (b) heat-treated at 1600°C for several soaking times

Nevertheless, the decrease in relative density can also be attributed to the changes observed in the

microstrutural   deffects   produced   during   the
decomposition of zircon (Figure 5).

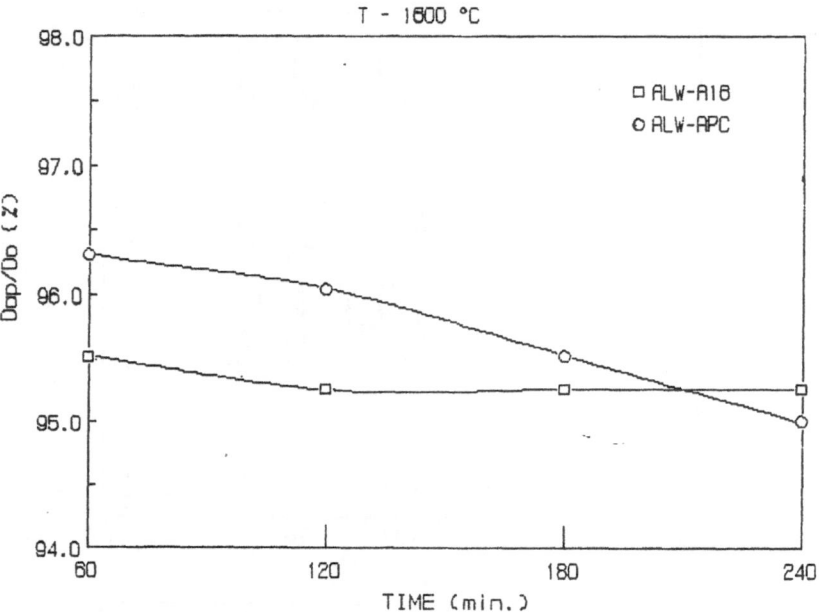

Fig.10. Relative densities of mixtures versus duration
of heat treatment at 1600°C

As the sintering process is being furthered, with its
associated grain growth, the zirconia fraction retained
as tetragonal (*i.e.* those grains below the critical
size, that do not transform martensitically to the
monoclinic phase) decreases when sintering time is
extended, as shown in Figure 11-a.

Figure 11-a also shows that zirconia grains grow faster
in mixture ALW-APC (larger decrease in tetragonal
zirconia retained up to 2h), probably due to Ostwald
ripening [29] in the presence of a liquid phase.

For a fixed soaking time, when the temperature is
raised the kinetics of formation of new zirconia nuclii
is initially favoured, to be overtaken, after a certain
temperature, by the kinetics of grain growth [30].
Therefore, the percent tetragonal zirconia retained
goes through a maximum and then decreases. This is
shown in Figure 11-b.

Figure 11-b also shows that grain growth, as compared
to nuclii formation, is favoured in mixture ALW-A16 in
which the maximum tetragonal zirconia (≈28%) is less

than in mixture ALW-APC (≈48%) and occurs later (at a
higher temperature).

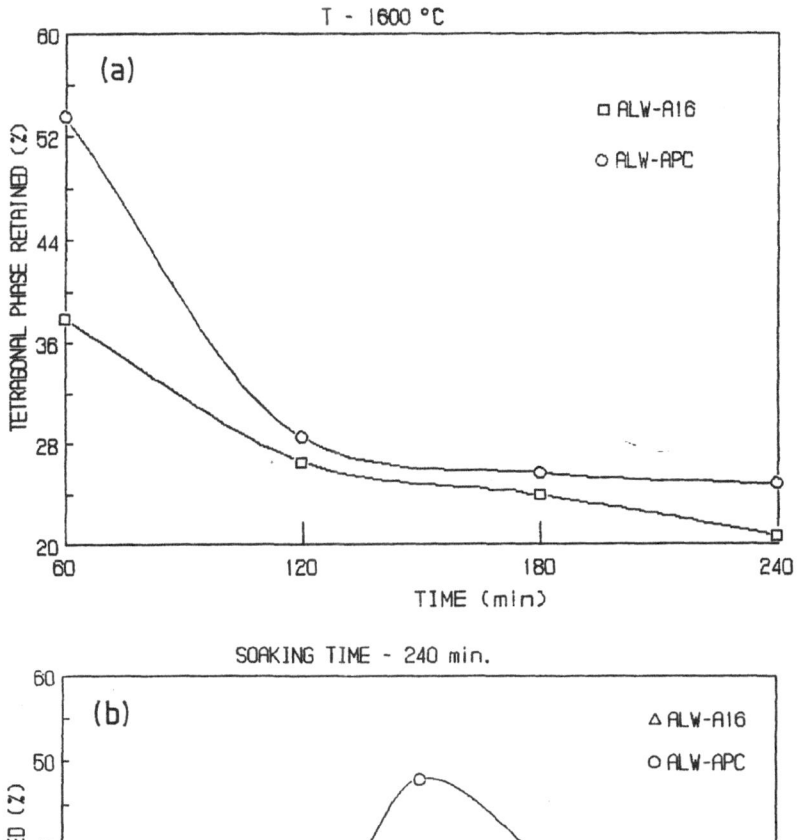

Fig.11. % tetragonal zirconia retained as a function of
soaking time (a) and temperature (b)

(3) Mechanical properties

Measured values of σ_f and E (after 4h at 1600°C) were 254 MPa and 146 GPa for mixture ALW-APC , and 264 MPa and 156 GPa for mixture ALW-A16, respectively.

The σ_f values show that the addition of zirconia to mullite, through reaction-sintering, is effective in increasing the flexural resistence as compared to conventionally processed mullite, but they are quite below the values reported in literature.

On the other hand, the elastic modulus in both mixtures is lower than sintered mullite modulus ($\approx$200 GPa).

In spite of the curves shown in Figure 11-a and b, one can not ascertain that the strengthening observed is due to the microcraking originated by the martensitic transformation of tetragonal zirconia and in what extent it is hindered by the microstrutural defects reported. It is likely that both factors play a determined role, which needs further investigation.

IV. Conclusions

The results obtained generally confirm the opinion of other authors: in reaction-sintering preparation of mullite-zirconia composites, the separation of densification stage from reaction stage is only possible with powders having specific characteristics, and this is detrimental to the possibilities of the process.

Although one lower-quality alumina (APC) was used, with roughly half reactivity and higher impurity contents, still comparable cold-strength values were obtained. Some dissimilar hot-strength behaviour is to be expected due to the localized liquid phase detected in the samples prepared with APC alumina. In any case, at high temperatures (>900°C) the zirconia present in both mixtures is stable in the tetragonal form and the strengthening mechanism through martensitic transformation no longer applies.

The microstrutures observed show the need to use more sophisticated techniques for mixing and powder processing, as variables like compaction pressure, heating rate, temperature and soaking time, strongly affect the changes in microstruture and, hence, the final properties of the composite. This also makes it difficult to compare results reported by different authors.

Finally, the mechanical properties measured show that microstrutural defects produced during the process hinder the benefits to be gained with the zirconia additions to mullite.

References

1. Mazdiyasni,K.S. and Brown,L.M., "Synthesis and Mechanical Properties of Stoichiometric Aluminum Silicate (Mullite)", J.Am.Ceram.Soc. 55[11]548-52 (1972)

2. Aramaki,S. and Roy,R., "Revised Phase Diagram for the System $Al_2O_3-SiO_2$", J.Am.Ceram.Soc. 45[5] 229-42 (1962)

3. Klug,F.J., Prochazka,S. and Doremus,R.H., "Alumina-Silica Phase Diagram in the Mullite Region", J.Am.Ceram.Soc. 70[10] 750-9 (1987)

4. Mah,T. and Mazdiyasni,K.S., "Mechanical Properties of Mullite", J.Am.Ceram.Soc. 66[10] 699-703 (1983)

5. Kanzani,S., Kabata,H., Kumazawa,T. and Ohta,S., "Sintering and Mechanical Properties of Stoichiometric Mullite", J.Am.Ceram.Soc. 68[1] C.6-7 (1985)

6. Ghate,B.B., Hasselman,D.P.H. and Spriggs,R.M., "Synthesis and Characterization of High Purity, Fine Grained Mullite", Am.Ceram.Soc.Bull. 52[9] 670-2 (1973)

7. Claussen,N. and Jahn,J., "Mechanical Properties of Sintered, In Situ Reacted Mullite-Zirconia Composites", J.Am.Ceram.Soc. 63[3-4] 228-9 (1980)

8. Di Rupo,E., Gilbart,E., Carruthers,T.H. and Brook,R.J., "Reaction Hot-Pressing of Zircon-Alumina Mixtures", J.Mat.Sci. 14 705-11 (1979)

9. Anseau,M.R., Leblud,C. and Cambier,F., "Reaction Sintering (RS) of Zircon-based Powders as a Route for Producing Ceramics Containing Zirconia with Enhanced Properties", J.Mat.Sci.Lett. 2 366-70 (1983)

10. Pena,P., Moya,J.S., de Aza,S., Cardinal,E., Cambier,F., Leblud,C. and Anseau,M.R., "Effect of Magnesia Additions on the Reaction Sintering of Zircon/Alumina Mixtures to Produce Zirconia Toughened Mullite", J.Mat.Sci.Lett. 2 772-4 (1983)

11. Wallace,J.S., Petzow,G. and Claussen,N., "Microstructure and Property Development of in situ-reacted Mullite-ZrO_2 Composites", Advances in Ceramics, vol.12: "Science and Technology of Zirconia II", The Am.Ceram.Soc. inc., Columbus, Ohio, 1983, p.436-43

12. Prochazka,J., Wallace,J.S. and Claussen,N., "Microstructure of Sintered Mullite-Zirconia Composites", J.Am.Ceram.Soc. 66 C-125 (1983)

13. Rincon,J.M. and Moya,J.S., "Microstructural Study of Toughened ZrO_2/Mullite Ceramic Composites Obtained by Reaction Sintering with TiO_2 Additions", Br.Ceram.Trans.J. 85 201-6 (1986)

14. Pena,P., Miranzo,P., Moya,J.S. and de Aza,S., "Multicomponent Toughened Ceramic Materials Obtained by Reaction Sintering,

part 1 - System $ZrO_2-Al_2O_3-SiO_2-CaO$", J.Mat.Sci. **20** 2011-22 (1985)

15. Miranzo,P., Pena,P., Moya,J.S. and de Aza,S., "Multicomponent Toughened Ceramic Materials Obtained by Reaction Sintering, part 2 - System $ZrO_2-Al_2O_3-SiO_2-MgO$", J.Mat.Sci. **20** 2702-10 (1985)

16. Melo,M.F., Moya,J.S., Pena,P. and de Aza,S., "Multicomponent Toughened Ceramic Materials Obtained by Reaction Sintering, part 3 - System $ZrO_2-Al_2O_3-SiO_2-TiO_2$", J.Mat.Sci. **20** 2711-8 (1985)

17. Orange,G., Fantozzi,G., Cambier,F., Leblud,C., Anseau,M. and Leriche,A., "High Temperature Mechanical Properties of Reaction Sintered Mullite-Zirconia and Mullite-Alumina-Zirconia Composites", J.Mat.Sci. **20** 2533-40 (1985)

18. de Portu,G. and Henney,J.W., "The Microstructure and Mechanical Properties of Mullite-Zirconia Composites", Br.Ceram.Trans.J. **83** 69-72 (1984)

19. Moya,J.S. and Osendi,M.I., "Microstruture and Mechanical Properties of Mullite-ZrO_2 Composites", J.Mat.Sci. **19** 2904-14 (1984)

20. Boch,P. and Giry,J.P., "Preparation and Properties of Reaction-Sintered Mullite-ZrO_2 Ceramics", Mat.Sci.Eng. **71** 39-48 (1985)

21. Paulus,M., Laher-Lacour,F., Dugleux,P. and Dubou,A., "Defects and Transitory Liquid Phase Formation During the Sintering of Mixed Powders", Br.Ceram.Trans.J. **82** 90-8 (1983)

22. Cambier,F., Baudin de la Lastra,C. and Pilate,P., "Formation of Microstructural Defects in Mullite-Zirconia and Mullite-Alumina-Zirconia Composites Obtained by Reaction Sintering of Mixed Powders", Br.Ceram.Trans.J. **83** 196-200 (1984)

23. Lange,L., "Transformation Toughening, part 1 - Size Effects Associated with the Thermodinamics of Constrained Transformations", J.Mat.Sci. **17** 225-34 (1982)

24. Heuer,A.H., "Transformation Toughening in ZrO_2 Containing Ceramics", J.Am.Ceram.Soc. **70**[10] 689-98 (1987)

25. Garvie,R.C. and Nicholson,P.S., "Phase Analysis in Zirconia Systems", J.Am.Ceram.Soc. **55**[6] 303-5 (1972)

26. Paulus,M., "Relationship between Densification, Crystal Growth and Mechanisms of Formation in Ceramics", Mat.Sci.Res. vol. 6: "Sintering and Related Phenomena", ed. G.C. Kuczynski, 1973, p.236-40

27. Emiliano,J.V. and Segadães,A.M., to be published

28. Di Rupo,E. and Anseau,M.R., "Solid State Reactions in the $ZrO_2.SiO_2-\alpha Al_2O_3$ System", J.Mat.Sci. **15** 114-8 (1980)

29. Fichmeister,H. and Grinvall,G., "Ostwald Ripening - A Survey", Mat.Sci.Res. vol. 6: "Sintering and Related Phenomena", ed. G.C. Kuczynski, 1973, p.119-49

30. Leriche,A., Cambier,F. and Brook,R.J., "Study of Some Factors Influencing the Microstructural Development of Mullite-Zirconia Composites Obtained by Reaction-Sintering", Br. Ceram. Proceedings, Special Ceramics 8, 1986, p.167-77

AGEING EFFECT ON MICROSTRUCTURAL AND MECHANICAL

PROPERTIES OF MULLITE – ZrO$_2$-TiO$_2$ COMPOSITES

M. Fátima Melo*   J. S. Moya**

*LNETI/DTM Materials Techn. Dept., Lumiar, 1699 Lisbon
(Codex)

**Instituto de Ceramica y Vidrio. C.S.I.C. Arganda del
Rey (Madrid)

ABSTRACT

The microstructural and mechanical properties of
mullite-zirconia composites with TiO$_2$ additions have
been studied, after ageing the samples over a wide
temperature range (1000°C-1500°C) for long periods of
time (100-200h).

The ageing behaviour was studied by X-ray diffraction,
by RLOM, SEM and changes of bend strength and toughness
during the ageing treatment are also reported.

In the sample with 0.25 mole of TiO$_2$ addition, changes
in solid state compatibility at temperatures below
1450°C were detected. In the sample containing 1 mol
TiO$_2$, decomposition of Al$_2$TiO$_5$ occurs at $\leq$1200°C. Both
compositions exhibit no increment in zirconia average
grain size during ageing and, concomitantly, there is no
strength degradation until higher temperatures (>1400°C)
are reached, which become more drastic when Al$_2$TiO$_5$ is
present.

1 - INTRODUCTION

Mullite-zirconia composites with TiO_2 additions have been intensively studied as a reaction sintered material, obtained from zircon-alumina-titania mixtures, with reasonable strength and a quite remarkable toughness (1-3).

Several authors had already given attention to the ageing behaviour of Y-TZP ceramic and partially stabilized zirconia (4-8), it is observed that the strength of these materials is greatly degraded at high temperature and that the changes developed in their microstructures are considered to be due to t->m transformation.

In this work we are studying microstructural changes on mullite-zirconia composites with an emphasis on the effect of zirconia average grain size and the addition of titania on the degradation of mechanical properties during high temperatures ($\geq$ 1000°C) ageing treatments.

2 - EXPERIMENTAL PROCEDURE

Taking into account the compatibility relationships in the $ZrO_2-SiO_2-Al_2O_3-TiO_2$ system (9), two different kinds of composition were formulated, according to the equations and the following molar proportions:

$$2ZrSiO_4+(3+X)Al_2O_3+XTiO_2 \rightarrow 2ZrO_{2(ss)}+Al_6Si_2O_{13(ss)} \qquad [1]$$

$$\text{with} \quad X = 0.25$$

$$2ZrSiO_4+(3+X)Al_2O_3+XTiO_2 \rightarrow 2ZrO_{2(ss)}+Al_6Si_2O_{13(ss)}+Al_2TiO_5 \qquad [2]$$

$$\text{with} \quad X = 1$$

Appropriate quantities of zircon*, Al_2O_3** and titania*** required to produce mullite-zirconia and mullite-zirconia-aluminium titanate composites as indicated by reactions [1] and [2], were used.

In both cases the powders were homogenized by attrition milling in a teflon coated attritor using alumina balls in an Isopropyl Alcohol suspension. After drying and sieving, the raw compositions were isostatically pressed at 200 MPa.

The resulting cylindrical green bars were sintered at 1500°C 2h (X=1 composition) and 1550°C 2h (X=0.25) in air. Sintered bars were subsequently aged at 1000, 1200, 1400, 1450 and 1500°C for times ranging from 0-200h.

Powdered samples (d < 35μm), after sintering and ageing, were then analysed by X-ray diffraction (XRD) using Cu Kα radiation. The $t-ZrO_2$ content was determined by using the Garvie expression (10). In this case the samples were tested as sintered.

The final density of the sintered and aged material was determined by the Archimeds method using destilled water.

The sample microstructure was observed by RLOM and by SEM-EDS after polishing and thermal etching. Zirconia average grain size was estimated in samples submitted to different heat treatments, by measuring lineal intercepts on SEM micrographs according to the Fullman and Richmon technique (11).

Bending strength ($σ_F$) was determined by the three point bend method on cylindrical bars with ≈4mm diameter and ≈80mm length. The toughness (K_{IC}) value of different samples was determined by the indentation method (12).

---

* – Zircon Opazir-S QUIMINSA S.A., Spain
** – Al_2O_3 CT 3000 SG ALCOA, USA
*** – Titania MERCK, West Germany

3 - RESULTS AND DISCUSSION

In Fig.1 the phase evolution vs. ageing time at different temperatures for X=0.25 and X=1 compositions are reported.

The RLOM micrographs corresponding to X=0.25 and X=1 samples are illustrated in Fig.2.

SEM micrographs corresponding to X=0.25 and X=1 are shown in Fig.3.

In Fig.4 the σ_F vs. time at different temperatures for X=0.25 and X=1 compositions are plotted.

The density and K_{IC} values for the different samples are reported in Table I. For X=1 composition (Fig.1) the only change observed in the initial starting phases, during ageing at different temperatures occurs at 1200°C as a consequence of Aluminium Titanate eutectoide decomposition into rutile plus alumina. At this particular temperature, the dissociation rate of Aluminium Titanate reaches its maximum as reported by P. Pena et. al., (13).

Conversely when the composition is located in the solid solution region (X=0.25), the appearance of zircon during the ageing treatment at 1000°C, could be related to the shift of mullite composition from the one as sintered (fired at 1550°C) towards others with a high alumina content. This fact very depends on temperature as reported by Okada et. al. (14).

The lower the ageing temperature treatment, the higher the alumina content in the corresponding mullite.

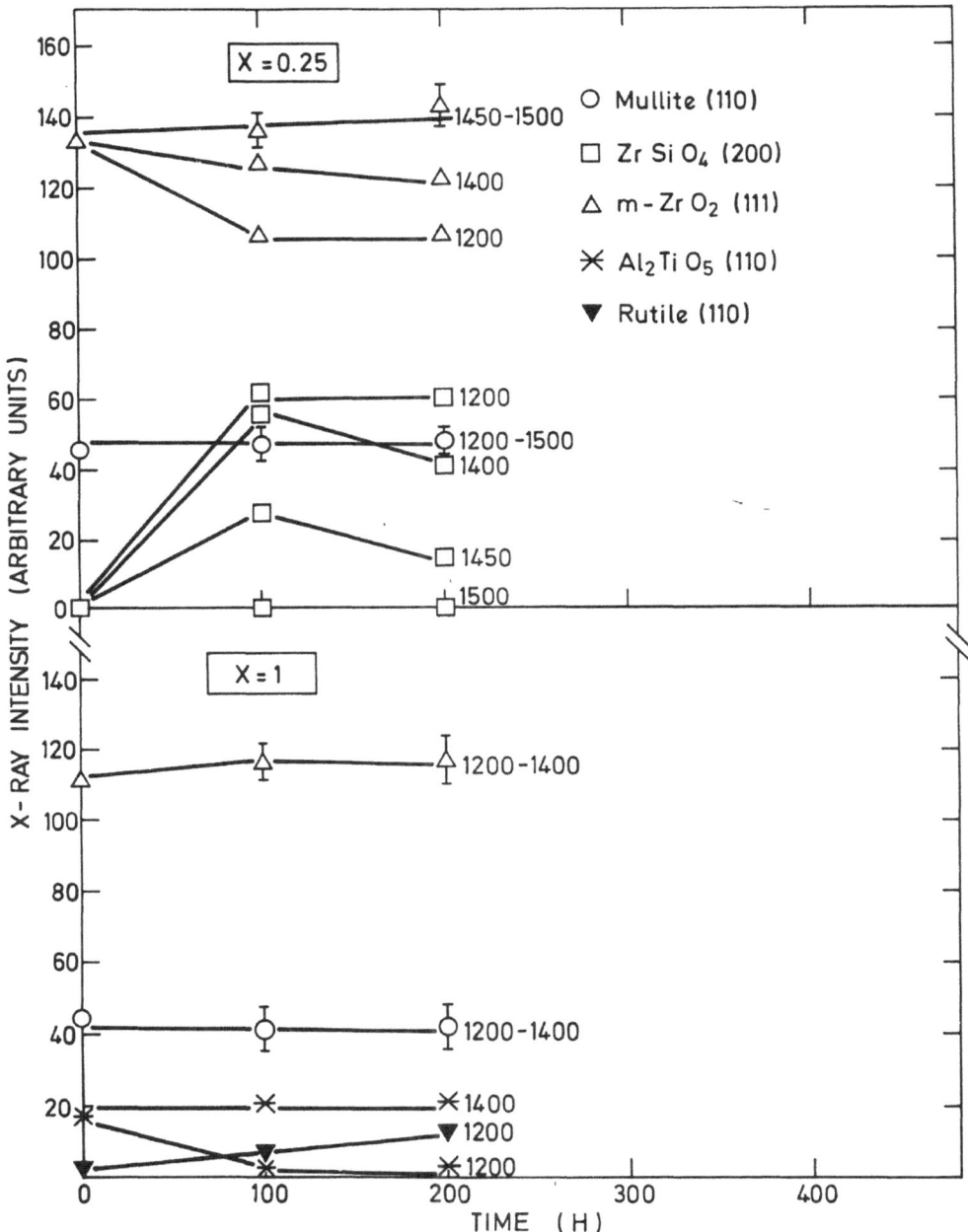

Fig. 1 — Phase evolution vs ageing time at differents temperatures for both compositions.

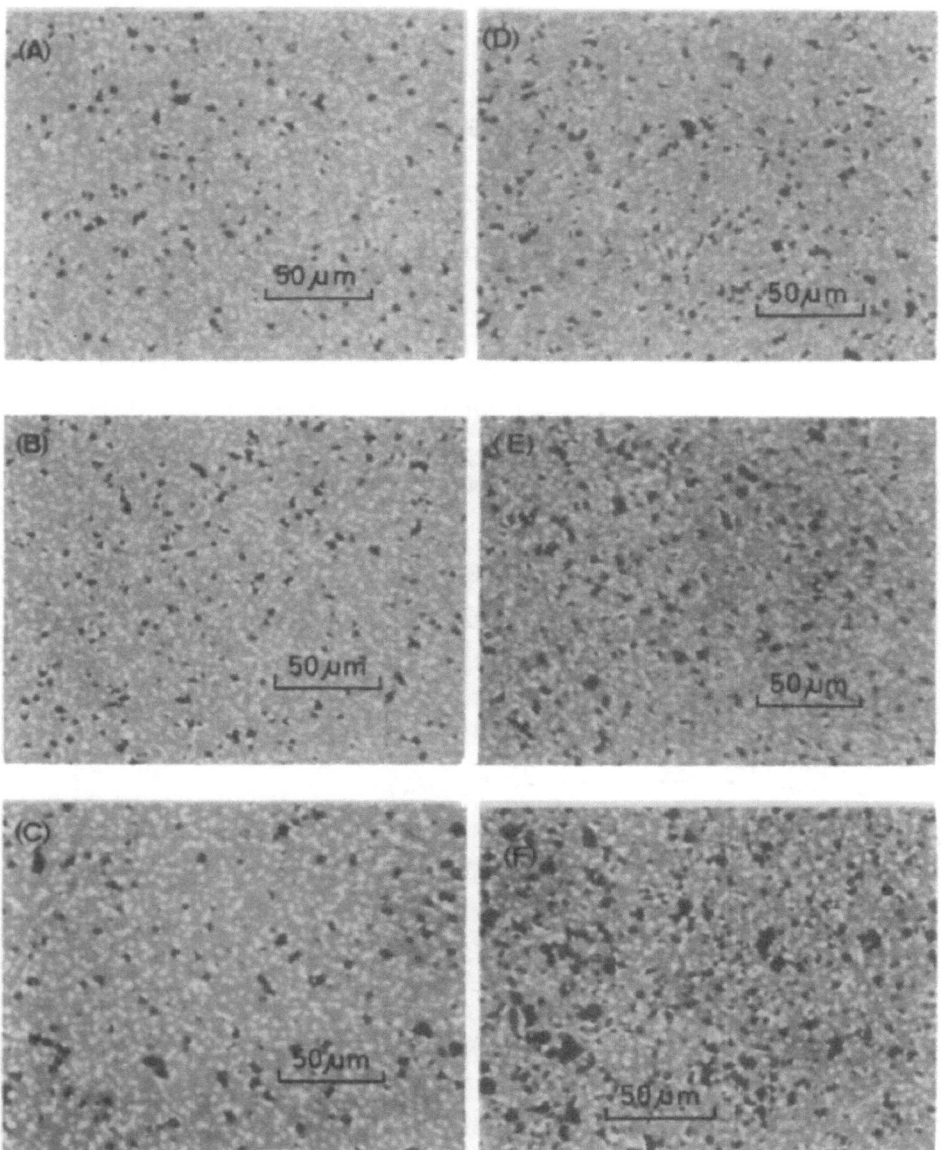

Fig. 2 - Optical microstructural of both aged samples.
(A) X=0.25 composition as sintered; (B) X=0.25
composition aged at 1400°C-200h; (C) X=0.25
composition aged at 1500°C-100h; (D) X=1
composition as sintered; (E) X=1 composition
aged at 1200°C--200h; (F) X=1 composition aged
at 1400°C-200h.

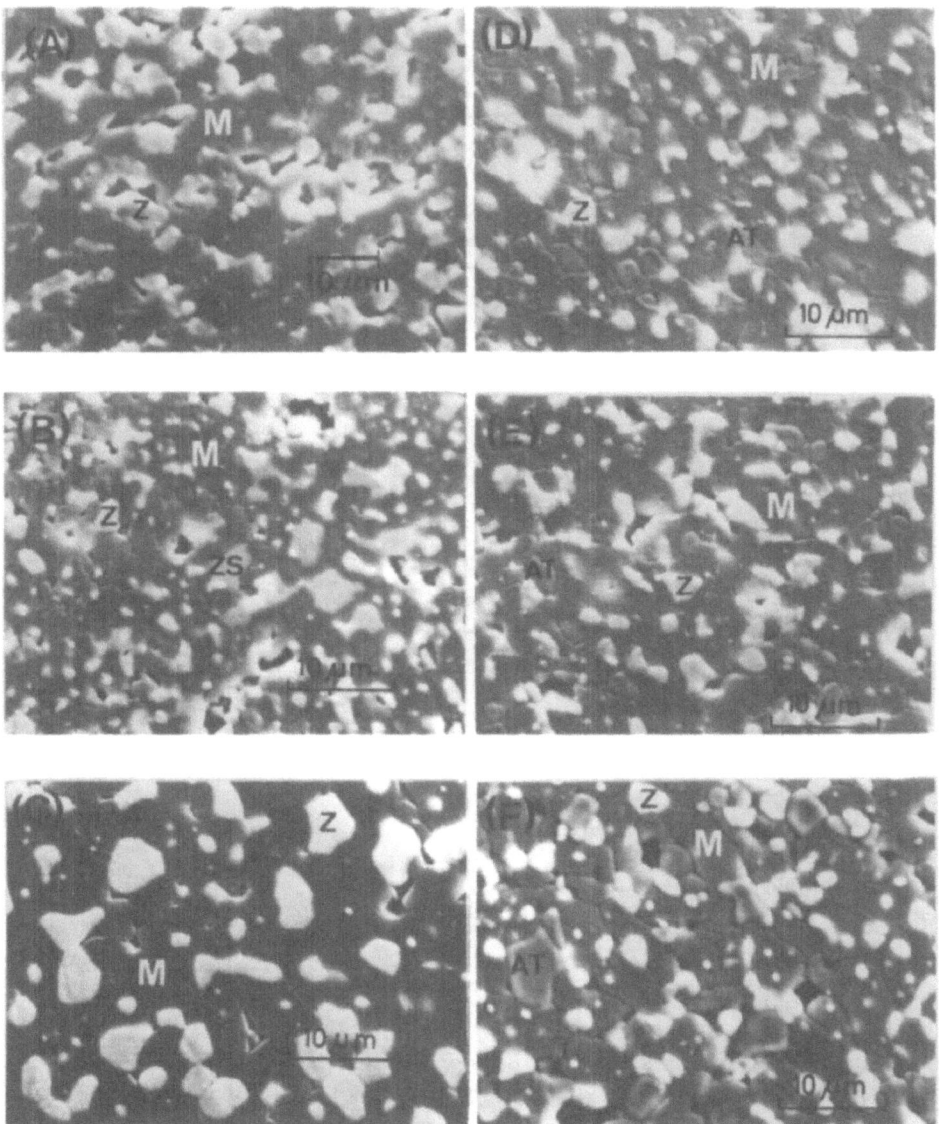

Fig. 3 - SEM micrographs of both aged samples.(A) X=0.25
composition as sintered; (B) X=.025 composition
aged 1200°C-100h; (C) X=0.25 composition aged
1500°C-100h; (D) X=1 composition as sintered;
(E) X=1 composition aged 1200°C--200h. (F) X=1
composition aged 1400°C-200h. ZS-zircon; Z-
zirconia; M-mullite; AT-aluminium titanate.

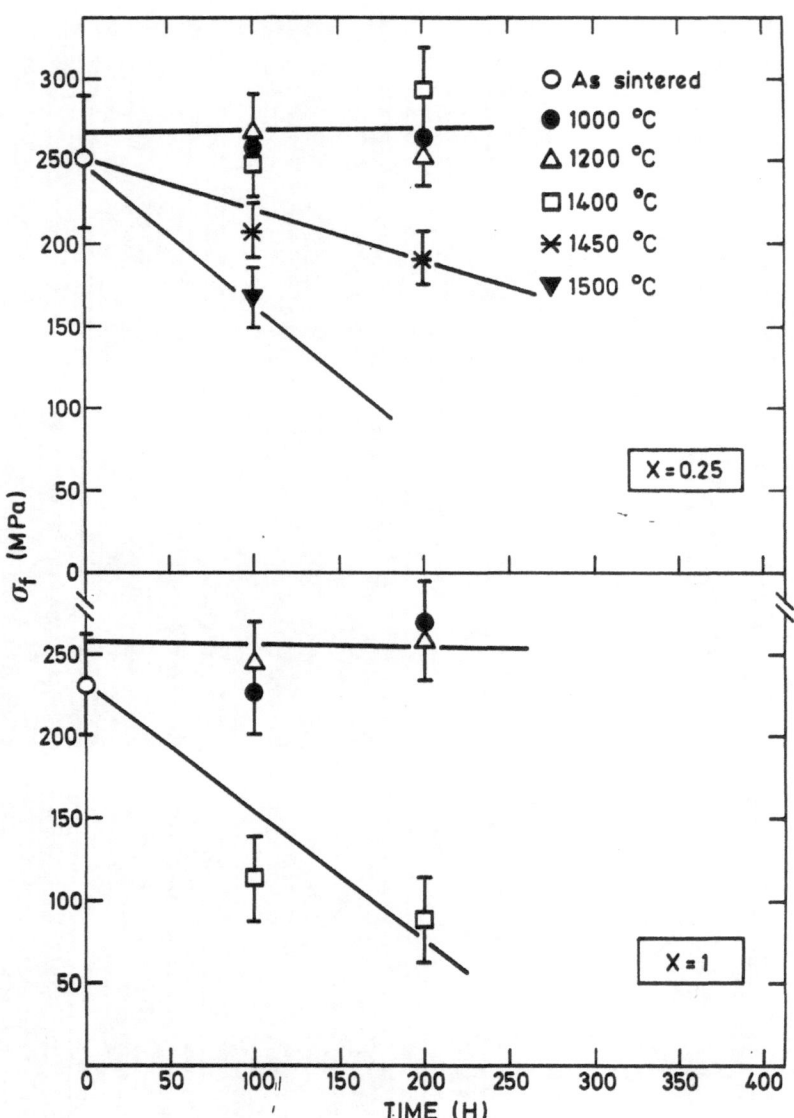

Fig. 4 - Bending strength vs time at different ageing treatments for both compositions.

Consequently at T ≤ 1450°C the compatibility tetrahedra could be $Z_{(ss)}$-ZS-Mullite$_{(ss)}$, as observed in Fig.5.

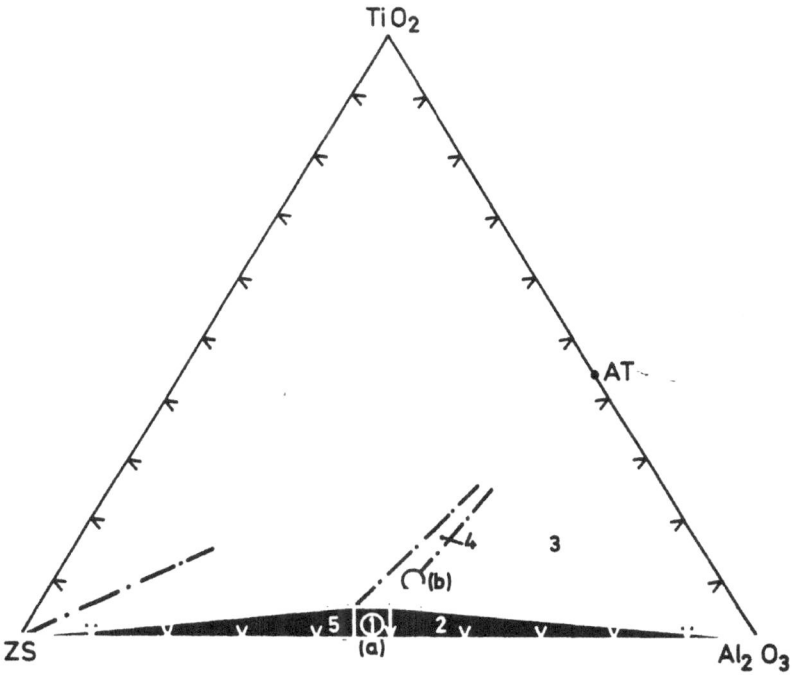

Fig. 5 – Compatibility tetrahedra sections with the inherent solid solutions. 1-$A_3S_{2(ss)}$-Z_{ss}; 2-Z_{ss} - -$A_3S_{2(ss)}$-A; 3-Z_{ss} -$A_3S_{2(ss)}$-A-AT_{ss} ; 4-Z_{ss} -$A_3S_{2(ss)}$ - -AT_{ss} ; 5-$A_3S_{2(ss)}$-Z_{ss}-ZS. (a) X=0.25; (b) X=1 comp.

The presence of zircon is very clear in the corresponding SEM micrographs shown in Fig.3(B).

With respect to the microstructural evolution of both samples during ageing, the following points can be stated:

  i) In both compositions the porosity appears to be constant for overall ageing treatments. No noticeable variation in the final density has been detected (Table I).

ii) A significant increase in the average pore size is
     detected in composition X=0.25 and X=1 after
     ageing at 1450°C and 1400°C respectively. This
     effect is more evident in composition X=1 (Fig.2)
     consequently, pores coalescence takes place at
     these temperatures.

iii) At ageing temperatures ≥ 1400°C zirconia and
     Aluminium Titanate (in X=1) grain growth takes
     place, as can be observed in Fig. 6 and Fig. 3 .

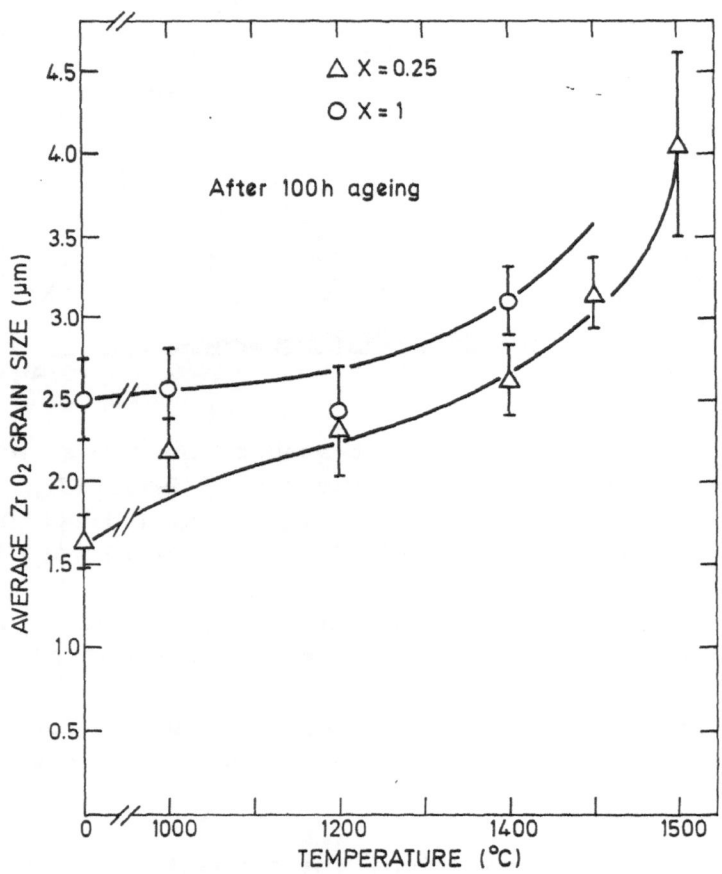

Fig. 6 - Average grain size of zirconia as a function of
         ageing temperature (held for 100h).

The strength degradation observed for compositions X=0.25 and X=1 at T ≥ 1450°C and T ≥ 1400°C respectively can be associated with pore coalescence and zirconia particle coarsening. These effects are more marked in sample X=1, were, in addition, Aluminium Titanate grain growth is also present to produce a negative effect in the σ_f value (Fig. 4).

On the other hand, the K_{IC} increases in the samples treated at temperatures ≥ 1400°C for both compositions (Table I). This is probably due to the contribution of microcracking-toughening as a consequence of the higher average grain size of zirconia particles.

This fact agrees with data previously reported by Srikrishna, et. al. (15).

TABLE I - FINAL DENSITY AND TOUGHNESS VALUES

PARAMETERS	Samples	As sintered	1200°C (200h)	1400°C (200h)	1450°C (200h)	1500°C (100h)
Final Density (g/cm3)	X=0.25	3.65 (±0.01)	3.64 (±0.01)	3.66 (±0.00)	3.66 (±0.01)	3.63 (±0.02)
	X= 1	3.58 (±0.02)	3.54 (±0.11)	3.62 (±0.01)	-	-
K_{IC} (MPa m½)	X=0.25	4.4 (±0.1)	4.5 (±0.2)	4.5 (±0.2)	4.7 (±0.3)	4.8 (±0.4)
	X= 1	4.0 (±0.2)	4.1 (±0.2)	4.4 (±0.2)	-	-

4 - CONCLUSIONS

To summarize, the following conclusions are drawn:

- No microstructural or mechanical degradation is observed for X=0.25 and X=1 compositions at temperatures below to 1450 and 1400°C respectively.

- Degradation is present in both composites at temperatures higher than 1450 and 1400°C and is related to pore coalescence and particle coarsening. Both effects are more evident in composition X=1.

- Because of the change in solid state compatibilities which takes place at T < 1450°C, the zircon is compatible with mullite$_{(ss)}$ and zirconia$_{(ss)}$. This change does not affect substantially the mechanical properties of the X=0.25 composition.

- An increase of K_{IC} values is present for both composites during ageing treatments at higher temperatures (> 1400°C).

AKNOWLEDGMENTS

The authors deeply appreciate discussions with their colleague P. Pena and would like to thank to Conceição Victor and A.M. Geraldes for their contribution to the preparation of sintered materials and microstructural observations and to Teresa Magalhães for X-ray diffraction patterns.

REFERENCES

(1) M.F. Melo, J.S. Moya, P. Pena and S. de Aza, J. Mater. sci. 20 (1985) 2711.

(2) M.F. Melo, J.S. Moya, Bol. Soc. Esp. Ceram. Vidr., 26 (1987) 3, 163-169.

(3) J. Rincon, J.S. Moya and M.F. Melo, Br. Ceram. Trans. J., 85 (1986) 201.

(4) Takaki Masaki, Int. J. High Technology Ceramics 2 (1986) 85-98.

(5) R.H.J. Hannink, K.A. Johnston, R.T. Pascoe and R.C. Garvie, in Advances in Ceramics 3: Science and Technology of Zirconia (1981) pp. 116-136.

(6) Masakazu Watanabe, Satoshi Lio and Isamu Fukuura, in Advances in Ceramics 12: Science and Technology of Zirconia II (1984) 391-8.

(7) Takaki Masaki, Yukio Murata, J. Mater. Sci. 22 (1987) 407-414.

(8) San-Yuan Chen, Hong-Yang Lu, J. Mater. Sci. 23 (1988) pp. 1195-1200.

(9) P. Pena and S. de Aza, Sci. of Ceramics, 12 (1983) pp. 201-208.

(10) Garvie, R.C. and Nicholson, P.S. J. Amer. Ceram., Soc. 55 (1978) 303.

(11) C. Richmon, Thesis Univ. Sheffield, 1971.

(12) P. Miranzo and S. Moya, Ceram. Internat., 10 (1984) 4, pp. 147-152.

(13) P. Pena, S. de Aza, J.S. Moya, Science of Ceramics 14 (1988) pp. 751-756.

(14) K. Okada, N. Otsuka, Science of Ceramic 14, Ed. D. Taylor, Inst. of Sci. of Ceramic, U.K. (1988) pp. 497-502.

(15) K. Srikrishna, G. Thomas and J.S. Moya, Zirconia 86 Third Int. Conf. on Science and Ceramics-Metal Systems. Plenum Press, N.Y. (1986) pp. 155-163.

PEROVSKITE AND FLUORITE-RELATED TERNARY PHASES IN THE SYSTEM CaO-ZrO$_2$-TiO$_2$

M.O. FIGUEIREDO[*] and A. CORREIA DOS SANTOS[**]

[*] Centro de Cristalografia e Mineralogia, IICT,
Al. Afonso Henriques, 1000 LISBOA

[**] Centro de Química-Física e Radioquímica, INIC, and
Departamento de Química, Universidade de Lisboa,
1200 LISBOA, PORTUGAL

ABSTRACT

Results of a phase equilibria investigation on the·system CaO-ZrO$_2$-TiO$_2$ are reported. Powder mixtures ˜ prepared with laboratory grade CaCO$_3$, ZrO$_2$, and TiO$_2$, were preheated at 900 and calcined at 1300C with intermediate regrindings. Structural characterization by room-temperature X-ray diffraction and electrical measurements on sintered powders up to 300C were performed over compositions along the perovskite series CaZrO$_3$ - CaTiO$_3$, the CaO - TiZrO$_4$ and ZrO$_2$ - CaTiO$_3$ joins and the pseudopyrochlore line, CaZr$_{3-x}$ Ti$_x$O$_7$. In the later, a fluorite-type solid solution with excess baddeleyite occurs close to the zirconia-rich side, while monoclinic zirconolite (ideally CaZrTi$_2$O$_7$) with free rutile appear in the titania-rich region. For intermediate compositions, a tetragonal solid solution akin to either the mineral calzirtite (Ca$_2$Zr$_5$Ti$_2$O$_{16}$) or pyrochlore is formed together with zirconolite and minor phases.

INTRODUCTION

Ternary oxides in the system CaO-ZrO$_2$-TiO$_2$ (C-Z-T) attracted interest as high-level radioactive waste immobilizers (1-3) and as potential solid electrolytes when anion deficient (4). Increased attention is now being focused on zirconia-based ceramics with additives like titania to improve toughness and wear resistance (5,6). Furthermore, the complex stability relations amongst natural calcium, zirconium, titanium oxides - the rare minerals calzirtite, zirconolite, polymignite and zirkelite (7) - also requires clarification, thus stressing the need for a reapprisal of the C-Z-T phase diagram (8).

Results of a phase equilibria investigation and structural study at 1300C along the pseudo-pyrochlore and the pseudo perovskite lines, $CaZr_{3-x}Ti_xO_7$ and $CaZr_{1-x}Ti_xO_3$ respectively, are reported. The $CaZr_{3-x}Ti_xO_7$ joins calcia $-$ $TiZrO_4$ and zirconia $-$ $CaTiO_3$ were also investigated.

CRYSTAL CHEMISTRY OF PEROVSKITES AND PYROCHLORES

Perovskites are double oxides with ideal cubic symmetry and one formula unit ABO_3 per unit cell ($a_o \sim 4A$). The atomic arrangement may be described by a more or less distorted cubic closest packing of oxygen anions and large **A** cations with smaller **B** cations in octahedral sites, or alternatively, by a framework of BO_6 octahedra sharing corners and leaving large interstices filled by **A** cations whose coordination numbers range from 9 to 12. The structural flexibility of such framework, achieved by octahedral tilting and distortion allows for a remarkable chemical compliance with respect to both cations ionic sizes and charges. Moreover, subsequent symmetry changes implying a bulk deformation of the structure and/or local distortions - like cation off-centering in BO_6 octahedra - disclose the appearance of interesting physical properties and a perceptible structural response to temperature and pressure changes.

Pyrochlores are also double oxides with ideal stoichiometry $A_2B_2O_7$ and cubic symmetry. There are eight formula units per cell, approximately doubling the lattice parameter relatively to perovskites. The atomic arrangement is also a framework of corner--shared BO_6 octahedra but with quite a distinct topology and more complex. In fact, the large interstices now contain an extra oxygen anion plus the **A** cations, which have a distorted cubic coordination (CN 7 or 8).

The pyrochlores consequently present a certain structural hindrance to distortion, being still chemically-adaptative through polytypism, a feature also recognized for perovskites.

There is however an important structural relationship between cubic pyrochlore and fluorite - which is also the basic structure type for both tetragonal and monoclinic zirconia. The atomic arrangement in pyrochlore may be indeed derived from a fluorite-type array with conveniently ordered anionic vacancies through a distortion process driven by the tendency of BO_6 octahedra towards geometrical regularity.

In the **C-Z-T** system there are two lines corresponding to the stoichiometries of perovskites and pyrochlores, with calcium playing the role of large **A** cations, titanium as small **B** cations and zirconium behaving as one or the other depending on the atomic proportions. To ascertain phase equilibria conditions, and stereochemical plus bonding effects, a preliminary study of those lines was started (9). In view of the differences from previously published results (8), a more complete study of the $CaO-ZrO_2-TiO_2$ system was undertaken.

EXPERIMENTAL

Polycrystalline compositions in the **C-Z-T** system (fig. 1) were synthetized by reaction sintering of laboratory grade $CaCO_3$, ZrO_2 and TiO_2, and also from previously synthetized $CaZrO_3$ and $CaTiO_3$ perovskites. Reactants were mixed and homogenized by hand-milling in an agathe mortar, pre-heated in alumina crucibles in air to ca. $900^{\circ}C$ to remove CO_2 and H_2O, then sintered at $1200^{\circ}C$ for 24h and subsequently at $1300^{\circ}C$ for 12 and 60h with intermediate regrindings.

Phase identification and structural characterization were achieved by X-ray powder diffraction usind a Debye-Scherrer camera and Cuk_{α} radiation.

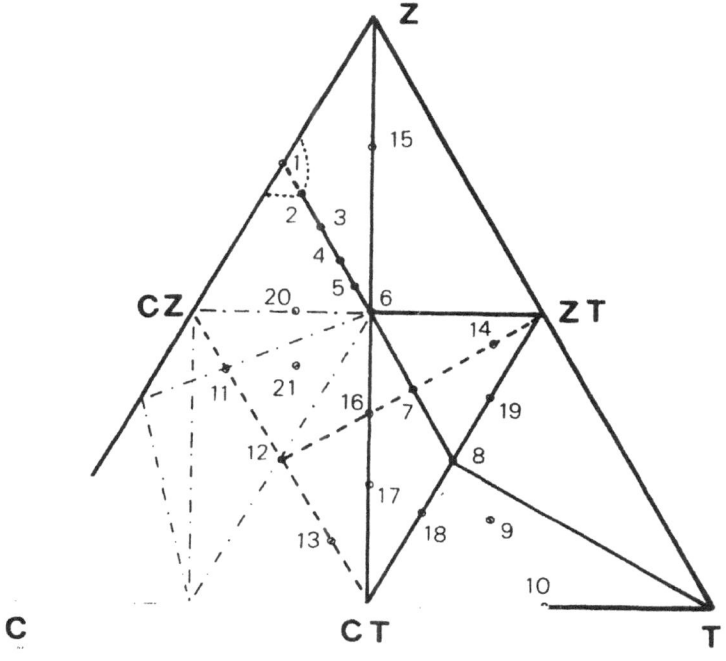

Fig. 1. $CaO-ZrO_2-TiO_2$ system at $1300^{\circ}C$.

Electrical conductivities were measured in air as a function of temperature using a two-point probe method. Pellets measuring 6 mm diameter, faces painted with silver paste, were mounted onto a sample holder which served to position them between Pt electrodes and Pt-10% Rh vs. Pt thermocouples within the furnace hot zone. Ohmic behaviour was confirmed.

RESULTS AND DISCUSSION

Along the pseudo-pyrochlore line, $CaZr_{3-x}Ti_xO_7$ (compositions 1 to 10, fig. 1), phases structurally akin to both fluorite and pyrochlore arrangements were found, with bulk symmetry decreasing along with zirconia content.

Mixture 1 does not fit the powder pattern of "phase ϕ_2" (10), being rather formed by uncombined monoclinic zirconia and a fluorite-type cubic solid solution (css). The same holds for composition 2, with less baddeleyite.

Mixtures 3 to 7 contain a tetragonal phase derived from the fluorite arrangement by the ordering of cations and anionic vacancies into a structure related to either the mineral calzirtite, $Ca_2Zr_5Ti_2O_{16}$ (11), space group $I4_1/acd$, or tetragonally distorted pyrochlore, sp. gr. $I4_1/amd$ (fig. 2), as this phase occurs almost pure for composition 6, a detailed structural study is in progress. Mixture 3 still contains traces of free ZrO_2 plus css, a phase that extends down to composition 4 where traces of $CaTiO_3$ are already apparent. Perovskite is still present in mixture 5 while in composition 7 titanium is totally incorporated in phases structurally related to pyrochlore: the tetragonal phase and monoclinic zirconolite, $CaZrTi_2O_7$ (12). Composition 8 is almost pure zirconolite and mixture 9 is a three-phase composition with free rutile and perovskite, the only phases present in mixture 10.

No solid solution was obtained between $CaTiO_3$ and $CaZrO_3$ along the pseudo-perovskite line $CaZr_{1-x}Ti_xO_3$. The high-Zr mixture 11 contains the above reported tetragonal phase and a Ca-rich non-identified phase. Intermediate composition 12 still holds the tetragonal phases now combined with a Ti-rich phase (Ca_3TiO_5 ?), while mixture 13 is a combination of these with zirconolite. No difference in phase constitution was found for compositions along the pseudo-perovskite line prepared with stoichiometric mixtures of oxides or of pure perovskites (CT and CZ).

A three-phase mixture of zirconolite, tetragonal phase and zirconium titanate, was obtained for

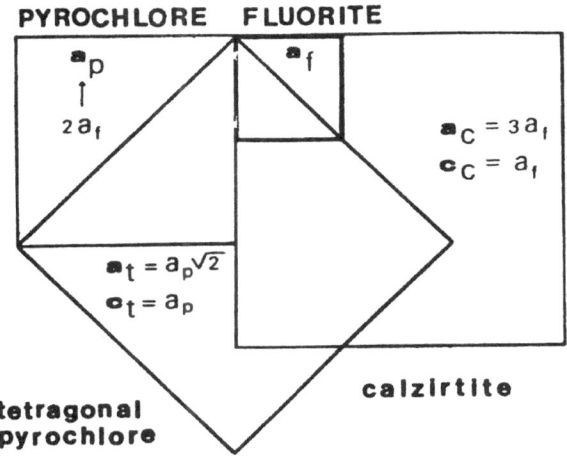

Fig. 2. Schematic relation between unit cells of
cubic and tetragonal phases.

composition 14, close to zirconium titanate along C-
-ZT line.

Synthetized compositions over the Z-CT join are
mixtures of either monoclinic zirconia and tetragonal
phase (c.15), or of this phase and perovskite (c.16
and 17). Intermediate compositions along the CT-ZT
join are also mixtures - perovskite plus zirconolite
(c.18), zirconolite plus zirconium titanate (c.19).

Extra Ca-rich and Zr-rich mixtures were also
synthetized to clarify the behaviour of $CaZrO_3$.
Composition 20 apparently gave rise to two cubic
fluorite-type solid solutions with different lattice
parameters, while composition 21 is mainly the
tetragonal phase with minor extra phases. These
results corroborate the complexity already recognized
for the $CaO-ZrO_2$ system (13).

Electrical conductivity data are shown in fig.3
which plots log ρ vs. 1/T for $CaTiO_3$, $CaZr_{0.2}Ti_{0.8}O_3$
and $CaZr_{0.5}Ti_{0.5}O_3$. The result is a linear variation,
the compositions behaving as semiconductors with
activation energies ca. 1.5 eV. The pseudo-pyrochlore
series showed higher values of resistivity at $335^{\circ}C$;
however, resistivity was found to decrease with
increasing Zr/Ti ratio possibly due to the presence
of the disordered cubic solid solution.

In spite of not being absolute resistivities -
because measurements were made over compact powders

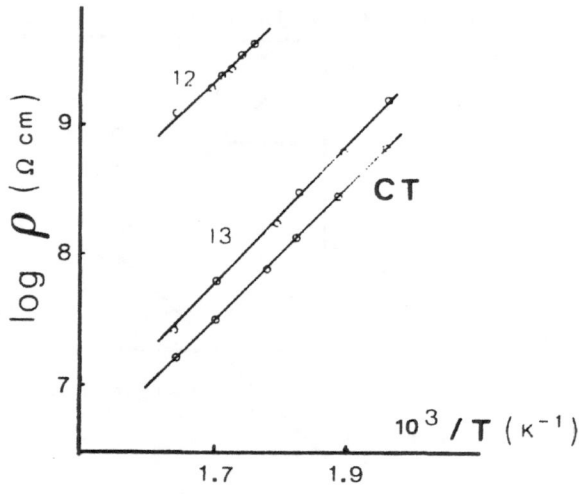

Fig. 3. Resistivity vs. 1/T plot
12: $CaZr_{0.5}Ti_{0.5}O_3$; 13: $CaZr_{0.2}Ti_{0.8}O_3$

with the inherent hindrances, - like porosity, grain boundary and contact resistance effects - the present results indicate that the study materials can be classified as electronic insulators. Some compositions are now being synthetized with dopants (Li and Na) adequate to the ionic transport in solid state. The influence of doping with rare earths is also under investigation.

The transfer of the actual phase equilibria results to the domain of mineral stability and paragenesis will be reported elsewhere.

REFERENCES

(1) A.E. Ringwood, S.E. Kesson, N.G. Ware, W.Hibberson and A. Major. Nature, **278**, 219 (1979).
(2) W.J. Buykx, D.J. Cassidy, C.E. Webb and J.L. Woolfrey. Ceram. Bull., **60**, 1284 (1981).
(3) R.A. Penneman and P.G. Eller. Radiochim. Acta, **32**, 81 (1983).
(4) J.M. Réau, J. Portier, A. Levasseur, G. Villeneuve and M. Pouchard. Mater.Res.Bull., **13**, 1415 (1978).
(5) C. Baudín and J.S. Moya. J. Amer. Ceram. Soc., **67**, C-134, (1984).
(6) J. Rincon, J.S. Moya and M.F. Melo. Brit. Ceram. Trans. **85**, 201 (1986).

(7) F. Mazzi and M. Munno.Amer.Miner., **68**, 262(1983).

(8) L.W. Coughanor, R.S. Roth, S. Marzullo and F.E. Sennet. J. Res. NBS, **54**, 191 (1955).

(9) M.O. Figueiredo and A. Correia dos Santos. Proc. Port. Mater. Soc. Meetg., **III**, 8pp. (1985).

(10) V.S. Stubican and S.P. Ray. J. Amer. Ceram. Soc., **60**, 534 (1977).

(11) H.J. Rossell. Acta Cryst., **B38**, 593 (1982).

(12) B.M. Gatehouse, I.E. Grey, R.J. Hill and H.L. Rossell. Acta Cryst., **B37**, 306 (1981).

(13) M. Hellman and V.S. Stubican. J. Amer. Ceram. Soc., **66**, 620 (1983).

STRUCTURE AND PROPERTIES OF ZIRCONIA-ALUMINA PLASMA SPRAYED COATINGS

M.C. FOUJANET, J.L. LUMET, J.L. DEREP, F. NARDOU*

CREA, ETCA, Arcueil, France
* Laboratoire de Céramiques Nouvelles, Université de Limoges, France

ABSTRACT

Dry mixed powders of zirconia partially stabilized with yttria (7-8 wt%) and alumina were plasma sprayed using argon-hydrogen atmospheric plasma jet. The microstructure of plasma sprayed zirconia-alumina composite coatings was analyzed by different methods. X Ray analysis and scanning electron microscopy provided the phase composition of the coatings. Microstructure study was complemented by transmission electron microscopy. Mechanical and thermal properties were investigated and correlated to the addition of alumina in usual YPSZ.

KEYWORDS

Plasma sprayed deposits, zirconia, alumina, coating characterization, hardness, Young's modulus, adhesive strength, annealing

INTRODUCTION

Ceramic deposited by air plasma spraying are widely used as Thermal Barrier Coatings (TBC's) for the protection of metal components in diesel engines and gas turbines. Utilities are extension of the service life, reduction or elimination of the cooling requirement, improvement of engine efficiency.

Thermal barrier coating system generally consists of a two layer coating system which is an insulating ceramic coating (0,6 mm) on to a MCrAlY (0,1-0,2 mm) bond coating [1]. The most commonly employed ceramic is Yttria Partially Stabilized Zirconia (YPSZ) due to its low thermal conductivity, high thermal expansion coefficient, thermodynamic stability, high cycling resistance. The bond coating prevents substrate oxidation, promotes ceramic adhesion and accommodates stresses induced by difference of thermal expansion coefficients in the system.

The aim of the study is to determine the influence of alumina addition in conventional YPSZ. Areas of investigation are microstructure, mechanical and thermal properties.

EXPERIMENTAL PROCEDURE

Metal substrate (50 mm diameter, 5 mm height) is Inconel 600. Bond coat is NiCrAlY. A plasma spray gun in air utilizing argon as primary gas and hydrogen as secondary gas is used to deposit ceramic and bond coatings.Characteristics of ceramic powders and plasma spraying conditions are mentionned in tables 1 and 2.

| Lot | PSZ | | | | Al_2O_3 | | |
	Manufac-turer	Mean diameter /µm	Y_2O_3/wt%	Particule form	Manufac-turer	Mean diameter /µm	Particule form
A	Magnesium Elektron	23	8	conglo-merated	SHMI	28	blocky broken
B	CEA/DMG	34	7	conglo-merated	CEA/DMG	34	-

Table 1 Characteristics of used powders

Lot	A wt% Al_2O_3	B
Plasma gas Ar/H₂ Flow rate (l/h)	2170/840----2460/480	
Electric power (kW)	42----30	25
Powder feed rate (g/min)	25	17
Spraying distance (mm)	120	70

Table 2 Plasma spraying parameters

RESULTS

Coating morphology

Microstructure is analyzed on polished samples using electron scanning microscopy. Coatings have a lamellar structure with high density of microcracks (Fig.1). It is easy to distinguish black lamellae of alumina, white lamellae of zirconia and black voids which are porosity.

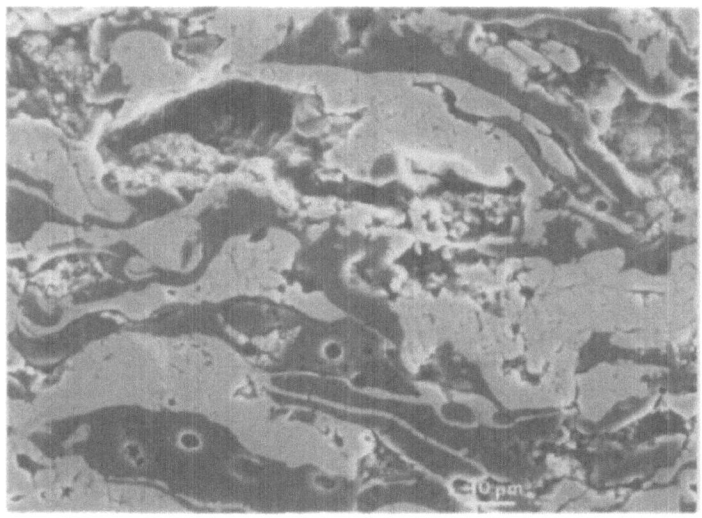

Fig.1. Cross section of the coating

Porosity measurements made on cross sections by image analyses indicate lower values for high power (6 % lot A) than for lower power (13 % lot B). Experimental studies [2] have shown that an increase of power leads to an increase of velocity with almost same surface temperature of particules as the dwelling time in the hot zone is almost the same. Higher particule velocities decrease porosity. Surface roughness of the coatings, similarly to porosity is higher for the B lot (Rt # 60 µm) than for the A lot (Rt # 43 µm).

Thus, coating morphology depends, to a large extent on the spraying parameters but also on the powder quality essentially governed by the grain size, the grain size distribution and the production process.

Pores are commonly elongated, perpendicular to the spraying axis with an elliptical shape which axis ratio varies from two to four.

EDS and image analyses of deposits confirm the decrease of alumina content during plasma spraying. Loss of alumina is issued from bad penetration of particules in plasma jet and evaporation.

Phases

Cristallographic study is carried out using Cu Kα radiation on the as sprayed side. (111), (400) planes [3] of zirconia and (113), (400) planes of alumina are chosen for qualitative and quantitative determination.

Major zirconia phase is the non transformable high yttria tetragonal phase [4]. Low monoclinic zirconia content is detected. It is lower for high power and sample cooling due to better melting and quenching of particules.

As Iwamoto's results [5], addition of alumina induces a decrease of monoclinic content due to its better heat conductivity.

The alumina most part is gamma phase with varying content of alpha according to quenching rate. During annealing, there is no significant evolution for zirconia but $\gamma \longrightarrow \delta*$ [6] and $\delta* \longrightarrow \alpha$ alumina transformations occur (Table 3). The first one is complete and the second one partial.

Lot	As sprayed					Annealed 1000°C 100 h					
	YPSZ			Alumina		YPSZ			Alumina		
	C	M	T'	α	γ	C	M	T'	α	δ*	γ
A	-	1-3	b	*	****	-	2-4	b	*	*	
B	-	1-2	b	*	*	-	1-2	b			

C for cubic, T' for tetragonal, M for monoclinic (mol%)
b for balance, * for present

Table 3 Phases distribution

Preliminary observations of the microstructure by transmission electron microscopy show the complexity of such coatings. Areas of fine equiaxed grains (Fig.2) are constituted of tetragonal zirconia as confirmed by X-ray diffraction with cubic zirconia. It is easy to distinguish tetragonal phase from cubic phase by (112) reflexion on diffraction patterns.

Tetragonal phase has a thin plate morphology (Fig.3). Intergranular residual porosity at triple point, along grain boundaries and facetted porosity appears in zirconia lamellae (Fig.4,5). Sometimes, twinned monoclinic zirconia is observed (Fig.6).

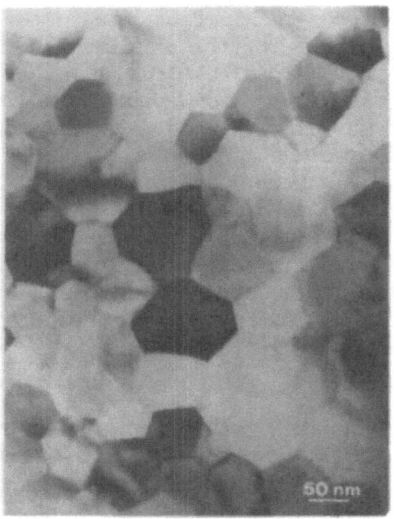

Fig.2. Equiaxed grains of tetragonal and cubic zirconia

93

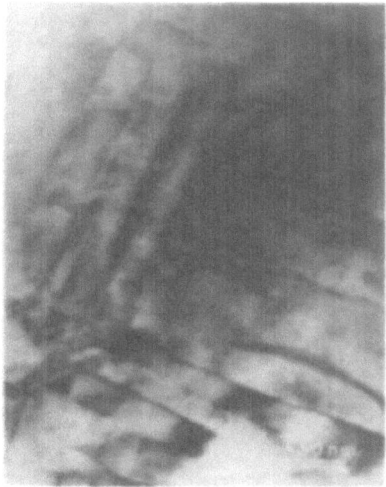

Fig.3. Thin plates of tetragonal zirconia

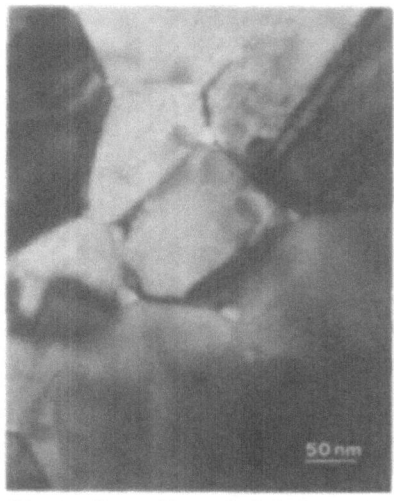

Fig.4. Porosity at triple point
in zirconia

Fig.5. Facetted porosity
in zirconia

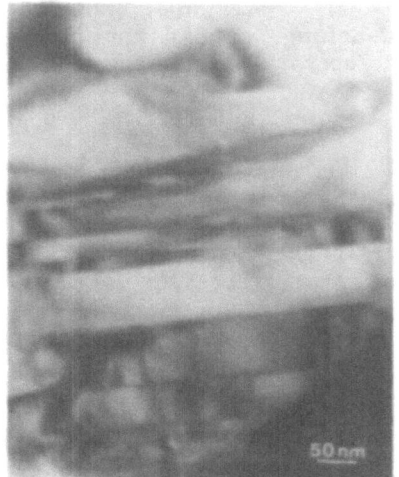

Fig.6. Monoclinic zirconia

Alpha alumina (Fig.7) has larger grains than gamma alumina. It contains entrapped porosity with dislocations. Gamma alumina contains similarly to zirconia facetted porosity. Main characteristic of gamma diffraction patterns is the occurence of diffuse elongated satellites corresponding to dark and white spots on bright field electron image (Fig.8). A domain structure (Fig.9) can be observed for a special crystal orientation.

94

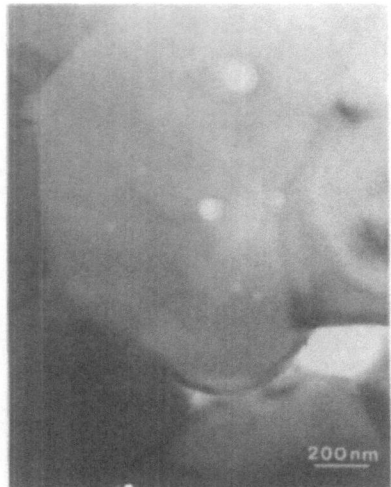

Fig.7. Alpha alumina

Fig.8. Gamma alumina

In regions of rapid quenching, appear amorphous and microcrystallized phases.

Mechanical and thermal properties

Hardness is correlated to coating strength and porosity which is directly influenced by spraying parameters (electric power, spraying distance, granulometry...)

For minimal porosity, hardness is higher than for more porous coatings (about 600 Hv_{900} lot A, 300 Hv_{900} lot B) (Fig.10).

There is no theorical law to explicit variation of hardness with porosity (P). Only an empirical law (Soroka and Serada) well fits experimental values:

$$H/H_0 \,\#\, \exp(-aP)$$

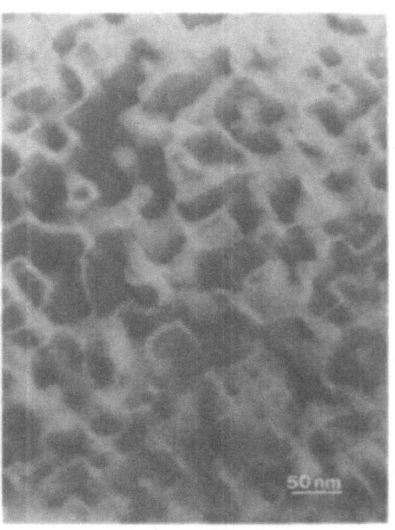

Fig.9. Domain structure in gamma alumina

The empirical constant, a, depending on the coating composition is about ten.

Young's modulus of the coatings is determined by a resonance method based on the propagation of ultrasonor waves in the material. Measurements give very low values (about 50 GPa for YPSZ 20 wt% Al2O3) but of the order of magnitude of such plasma coatings [7].

Those values are lower than expected for a dense material with the same content of porosity and an uniform distribution of closed pores. Such values can be explained by the high density of microcracks and the low real area of contact between lamellae.

Adherence of coating is quantified by a tensile adhesion test (Fig 11). Addition of alumina would improve adherence of the coating ($\sigma_R \# 20$ MPa) (Fig 12). Interpretation of test must be made carefully because failure mode and fracture stress depend on coating thickness.

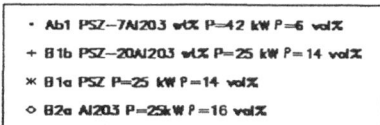

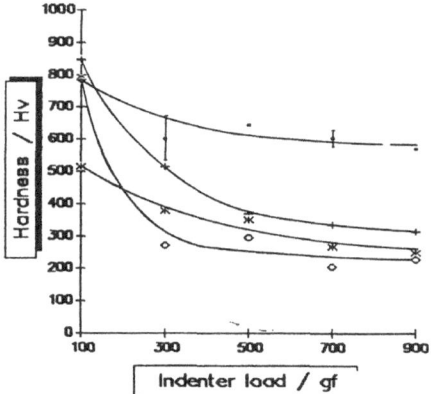

Fig.10. Evolution of hardness with indenter load for different YPSZ-alumina coatings

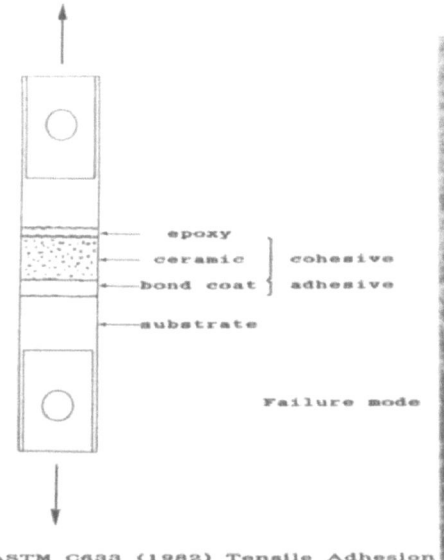

Fig.11. Failure mode [8]

Fig.12. Adhesive failure

Thermal test which consists of heating coatings with solar radiation has shown a lower heating for YPSZ–alumina coating than for usual YPSZ coating (Fig.13)

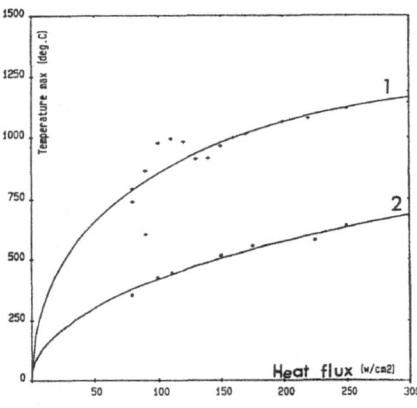

Fig.13. Heating of YPSZ (1)
YPSZ–alumina (2) coatings

CONCLUSION

From results obtained in this study, main following conclusions can be established:

* Coating microstructure depends on spraying parameters, essentially electric power and spraying distance which largely influence porosity and thus hardness.

* Coating composition depends on powder characteristics. A suitable choice of alumina is necessary to avoid loss in alumina content during spraying.

* Distribution of phase content is influenced by power input and sample cooling. Major phases are non–transformable tetragonal zirconia and metastable gamma alumina. Monoclinic content decreases with power increase and optimization of cooling.

* Contrary to zirconia, alumina phase composition is very sensitive to spraying conditions.

* Addition of alumina in usual PSZ is source of decrease in monoclinic zirconia content.

* Low values of Young's modulus for the as sprayed state are obtained (50 GPa). Those results demonstrate that mechanical properties (hardness, stren h, Young's modulus) are governed by the typical microstructure whicn consists of lamellar grains, pores, low real surface of contact between lamellae and high density of microcracks.

The observed properties of plasma sprayed zirconia–alumina composites are similar to those of usual YPSZ. These relatively bad coatings properties against the one expected are explained by the high microstructure heterogeneity and the very low layer density (i.e. porosity and microcracks network). The convenient futur studies orientations are to improve the phases distribution and to optimize the porosity. Some possibilities are :

− the use of a powder in which each particule is an alumina YPSZ mixing. Some tests have been realized for a such powder (i.e. 80 wt % Al_2O_3 20 wt% ZrO_2) and the results are promising.

− the post treatments of the as sprayed coatings which can be, with the accurency to substract properties integrity, a classical annealing or a surface treatment by means of CO_2 laser beam (using the photon–phonon transformation) or of YAG pulsed laser (shock wave sintering).

REFERENCES

1- P.A. SIEMERS, R.L. MEHAN, "Mechanical and Physical Properties of Plasma Sprayed Stabilized Zirconia", Cer. Eng. Sci. Proc., 3 (9-10) (1983) 828-840

2- A.VARDELLE, M. VARDELLE, P. FAUCHAIS, Plasma Chem. Plasma Process, 2 (3) (1982) 255-291

3- R.A. MILLER, J.L. SMIALEK, R.G. GARLICK, "Phase Stability in Plasma Sprayed Partially Stabilized Zirconia Yttria", Adv. Cer., 3 (1981) 241-253

4- N.R. SHANKAR, G.C. BERNDT, H. HERMAN, "Phase Analysis of Plasma Sprayed Zirconia Yttria Coatings", Cer. Eng. Sci. Proc., 3 (9-10) (1983) 784-791

5- N. IWAMOTO, N. UMESAKI, M. KAMAI, "Characterization of Plasma Sprayed Alumina Zirconia Coatings", Intl. Symp. on Plasma Chemistry, Eindhoven, (1985) 1143-1148

6- D. FARGEOT, "Etude des Phases Métastables de l'alumine projetée au chalumeau à plasma", Thèse, Limoges (1987)

7- P. BOCH, P. FAUCHAIS, D.LOMBARD, B. ROGEAUX, M.VARDELLE, "Plasma Sprayed Zirconia Coatings", Adv Cer, 12 (1984) 488-502

8- C.C. BERNDT, "Determination of Material Properties of Ceramic Coatings", Advances in Thermal Spraying, Pergamon Press, (1986) 149-158

ZIRCONIA BASED ELECTROCHEMICAL CELLS TO STUDY THE REACTION KINETICS BETWEEN MATERIALS AND OXYGEN AT HIGH TEMPERATURE

D. Gozzi, G. Carnevale, P.L. Cignini, L. Petrucci

Dipartimento di Chimica, Universita' di Roma "La Sapienza"
P.le Aldo Moro 5, 00185 Roma, Italy

and

M. Tomellini

Istituto di Chimica Universita' della Basilicata
Via N. Sauro 85, 85100 Potenza, Italy

ABSTRACT

Preliminary results obtained by a solid state electrochemical technique to study the high temperature interaction between materials and oxygen are described. This , *in situ* technique, was applied to tantalum and copper foils between 800 and 1000 C at oxygen partial pressures greater than 1.10^{-14} Pa in order to test its potentialities in the study of the gas/solid reaction kinetics. The technique seems promising for the study of the oxidation kinetics of metal under time changing oxygen pressure.

INTRODUCTION

In the study of high temperature corrosion of metal a particular attention to the measurement of the oxidation kinetics was devoted[1]. The kinetics of the metal-oxygen interaction is dependent from the oxygen activity and temperature, as well as from the sample surface. The solid-gas reaction is generally studied by thermogravimetric techniques, where the kinetic data are obtained through the measurement of the sample weight change. However in this experimental set-up it is no easy to set finely the gas phase composition and flux. Conversely, using the evolution on time of consumed oxygen gas at sample surface, the oxidation kinetic could be also investigated. This is possible by making use of a particular technique based on the coupling of two solid state electrochemical cells, with yttria stabilized zirconia as electrolyte. In this paper the new electrochemical technique is presented with some results concerning the high temperature oxidation of tantalum and copper, at low oxygen activity.

EXPERIMENTAL

The technique is based on the coupling of an oxygen generator and oxygen sensor both as electrochemical cells, with YSZ as electrolyte. A schematic view of the electrochemical cells is reported if Fig.1. The YSZ tubes are assembled in such a way as to create, between the "oxygen generator" and the "oxygen sensor", a microchamber in which the sample under study is contained. Through holes in the microchamber a flux of high purity argon can circulate for washing purposes. The cell assembly was placed in a horizontal high vacuum alumina furnace. In the microchamber a continuous oxygen flux is produced by the oxygen generator by applying a constant current density.

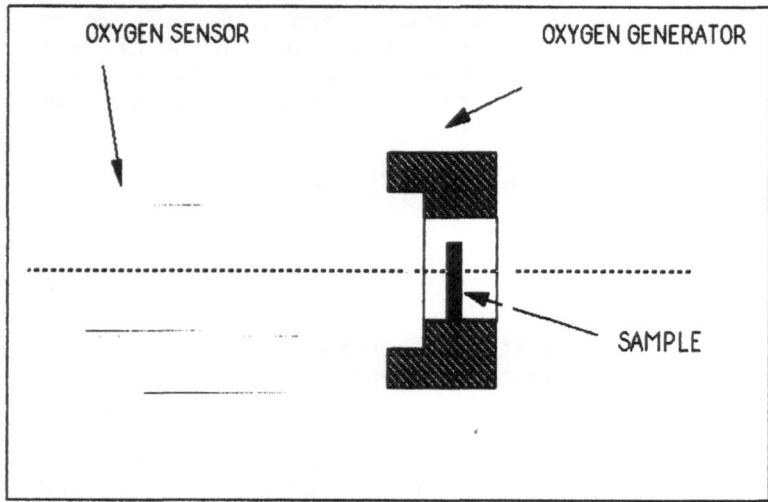

Fig.1 View of the electrochemical cell. The oxygen generator, oxygen sensor, microchamber and sample are shown.

A digital differential electrometer was used to measure the oxygen sensor emf and a set of instruments interconnected was used to drive and monitor the oxygen pump polarizations. Samples were cutted in the appropriate dimensions and cleaned in an ultrasonic bath with a non aqueous detergent.

RESULTS AND DISCUSSION

As previously reported[2], kinetic data of oxidation can be obtained, by using the above mentioned technique, through the measurement of the sensor emf when the pump is polarized. In fact when a current density, $i(t)$, is applied to the pump, the oxygen flux produced, Jp, is given according to:

$$J_p = \frac{t_0}{4F} i(t) \qquad O_2 \frac{moles}{m^2 s} \qquad (1)$$

where t_0, F and 4 are, respectively, the oxygen vacancy transport number in YSZ, the Faraday's constant and the number of electron of the charge transfer. By assuming $t_0 = 1$, the average value of the oxygen pressure in the microchamber, $P(t)$, can be obtained by the sensor emf, $E(t)$, as follows:

$$P(t) = P_r e^{-\frac{4FE(t)}{RT}} \qquad (2)$$

In the volume where the reaction between metal and oxygen occurs, the oxygen flux balance is:

$$\frac{dn_0}{dt} = |J_p| A_p - |J_s| A_s - |J_g| A_g \qquad (3)$$

where J and A are fluxes and areas, while p, s, and g stand for "pump", "sample" and "getter". The J_g term is referred to any interaction between oxygen and surrounding materials. By considering two different experiments carried out with and without sample, the oxidation rate dn_s/dt, is given by the equation[2]:

$$\frac{dn_s}{dt} = B\left[\dot{E}_s(t)\,e^{-\sigma E_s(t)} - \dot{E}_b(t)e^{-\sigma E_b(t)}\right] \qquad (4)$$

where the variables with the b subscripts are referred to the experiment without sample. In experiment performed without sample the quantity $dP(t)/dt$ should be constant on time if Jp = constant as well as J_g . In fact in this condition $J_s = 0$, so that dn_o/dt in Eqn.3 is also constant. Finally because of dn_o/dt is proportional to $\dot{P}$, it follows that: $\dot{P}$ = constant. In the Fig.2 the dP/dt function, evaluated through the measurement of the sensor emf, as recorded in a "blank" experiment, is reported. As immediately appears, good agreement between experimental data and the theoretical predictions previously given, is obtained. Therefore, in our experimental conditions, the YSZ ionic transport number is constant in the explored oxygen pressure range, as well as J_g.

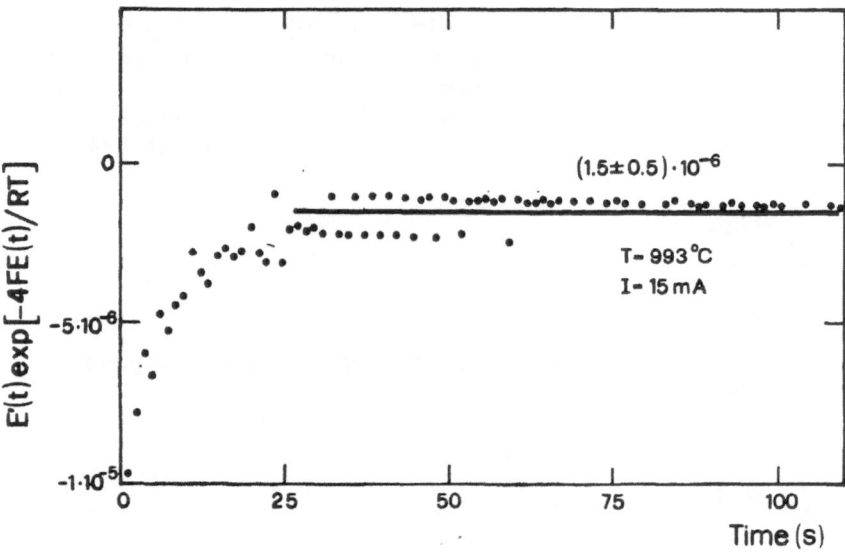

Fig.2 Experimental trend of the dP/dt function obtained through the measurement of the sensor emf.

In Fig.3 the sensor emf vs time curves, recorded during the oxidation of tantalum and copper at various temperatures, are shown. The oxygen pressure values reached in the chamber, when the current is switched off, are dependent to the reactivity of the sample. Referring to Fig.3, the starting pressure read in the tantalum experiment is lower than that of copper oxidation, in agreement with the related thermodynamic data from which: $P_{Cu/Cu_2O} = 4.58 \cdot 10^{-4}$ Pa at T=1113 K and $P_{Ta_2O_5} = 2.2 \ 10^{-26}$ Pa at T=1073 K[3]

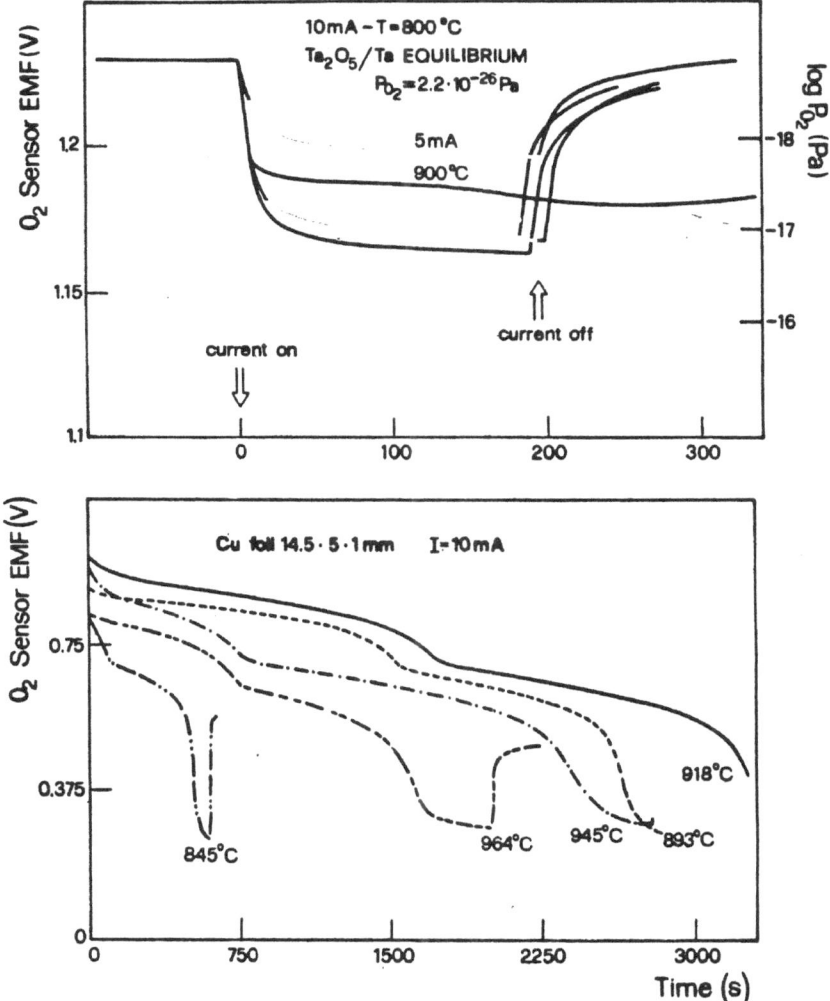

Fig.3 Sensor emf vs time measured during the oxidation of tantalum and copper at different temperatures.

The emf values recorded during the copper oxidation, have been interpreted on the basis of eqn.4 in order to obtain kinetic data. The calculated oxidation flux has been reported in Fig.4 together with the experimental pressure value. By using the literature[4,5] data, the reaction rate have been also calculated on the framework of the Wagner's theory[6], and compared to the experimental flux in Fig.4. As reported in Fig.4, the oxygen flux experimentally consumed by the sample is lower than the other fluxes.

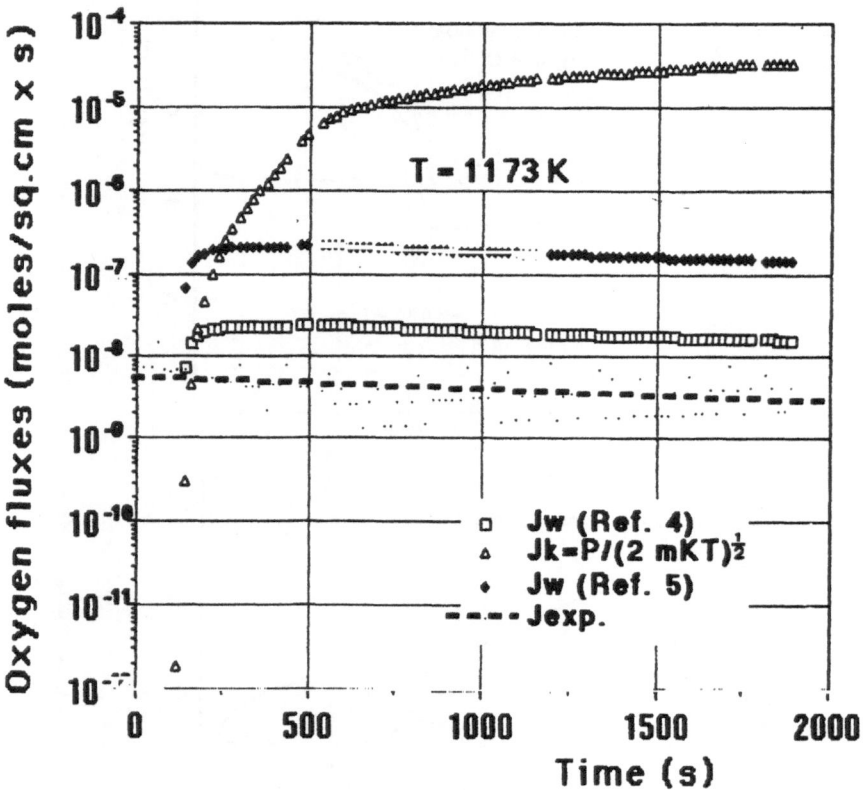

Fig.4 Comparison between literature and experimental oxidation rate for copper. J_k is the oxygen flux at the surface, given by Knudsen's equation.

If the Wagner mechanism is still considered, this implies that the oxygen activity at the sample surface is lower than the oxygen activity in the gas phase. Otherwise, the rate limiting step could be the adsorption of oxygen at the surface, with respect to the transport in the solid phase. Experimental evidences in agreement with the last interpretation are reported in another work[7]. In conclusion, the new technique here presented allows the study of metal oxidation at low oxygen pressure, through the determination of the reaction rate. In addition, the experimental configuration, in which the oxygen is electrochemically produced at constant rate, seems to allow oxidation with a rate limiting step different from the expected one by Wagner's theory

ACKNOWLEDGEMENTS

This work was carried out with the partial financial support of the Progetto Finalizzato Energetica II of the National Research Council under contract 87.02218.59

REFERENCES

1. Per Kofstad, High Temperature Oxidation of Metals, J. Wiley & Sons, New York, 1966: High Temperature Corrosion, Elsevier, (1988)

2. D. Gozzi, P.L. Cignini, G. Carnevale, L. Petrucci and M. Tomellini, **High Temperature & High Pressures, 20** (1988) in press

3. **JANAF** Thermochemical Tables Suppl. J. Phys. Chem. Ref. Data 1975 , 4 156;99

4. J.P. Baur, D.W.Bridges and W.M. Fassel Jr. **J.Electrochem. Soc., 103,** (1956) 273

5. W.J. Tomlinson and J. Yates, **J.Phys.Chem.Solids, 38,** (1977) 1205

6. C. Wagner, **Z. Physik. Chem.,B21,** (1933) 25

7. D. Gozzi, M. Tomellini, G. Carnevale, P.L. Cignini and L. Petrucci, **J. Electrochem. Soc.** in press

TETRAGONAL ZIRCONIA POLYCRYSTALS IN THE MgO-Y₂O₃-ZrO₂ SYSTEM (MgY-TZP).

K.Haberko, W.Pyda, M.Kuraś, M.Bućko

Institute of Materials Science, Academy of Mining and Metallurgy, Cracow, Al.Mickiewicza 30, Poland

The MgO-Y₂O₃-ZrO₂ system was investigated from the point of view of preparation of fully tetragonal or nearly tetragonal zirconia polycrystals. The composition range studied covered Y₂O₃ concentrations from 2 to 5 mol.% and MgO concentrations from 0 to 8.0 mole %. Twelve mixtures were prepared by the coprecipitation technique with subsequent hydrothermal treatment of the gels.

Phase composition measurements of the samples sintered at 1350 and 1400°C revealed in all cases tetragonal ZrO₂ s.s. as a dominant phase. Some samples were fully tetragonal. Fracture toughness, K_{Ic}, ranged from 3 to 15 MPam$^{1/2}$.

It was found that the MgO additives increase critical grain sizes compared to the samples with Y₂O₃ alone.

1. INTRODUCTION

Tetragonal zirconia polycrystals (TZP) belong to the group of ceramic bodies of the highest strengths and fracture toughness at least at room temperature. Their properties are related to the martensitic tetragonal to monoclinic transformation of zirconia solid solutions at the tip of the propagating crack [1-3].

The TZP bodies were obtained in the zirconia systems with the following alloying components: Y₂O₃ [4-7], CeO₂ [8-11] and recently CaO [12,13]. Phase diagram studies of the system MgO-Y₂O₃-ZrO₂ [14,15] as well as some other observations within this system [16] indicate it as a potential source of the TZP bodies. The purpose of the present investigation was to study this possibility.

2. EXPERIMANTAL PROCEDURE

In the present work the technique based on hydrothermal treatment of the coprecipitated Mg, Y and Zr
hydroxides was applied. The technique is known to give
fine crystallite powders composed of soft agglomerates
[17]. Compacts of such powders show uniform pore size
distribution and hence, result at low sintering temperatures in dense and fine grained bodies, both features
being important from the point of view of the TZP-s.
    The powder preparation technique described in
detail elsewhere [17] comprises the following steps:(i)
dissolution of $ZrOCl_2$, $MgCl_2$ and YCl_3 in distilled
water, (ii) pouring the solution into vigorously stirred NH_4OH water solution (final pH=9.0), (iii) sepa-
rating the coprecipitated hydroxides from the mother
liquor by filtering and (iv) drying at 120°C. The
coprecipitated gels were hydrothermally treated at
250°C for 4hr in distilled water or NaOH water solution
(1.7M). After hydrothermal cristallization the powders
were repitedly washed with water to remove NH_4Cl or
NaOH, dried, ground and uniaxially compacted under
196MPa into discs of 14 mm diameter and 3 mm thickness.
Sintering was performed in air in an electrically
heated furnance with SiC heating elements at 1350 and
1400°C with 2 hr soaking time at the peak temperature
and 6°C/min rate of temperature increase. The studied
range of chemical compositions is shown in table 1.

Table 1. Range of chemical compositions

Specimen	Composition (mole %)		
	ZrO_2	Y_2O_3	MgO
1	98.0	2.0	—
2	96.5	3.5	—
3	95.0	5.0	—
4	96.5	2.0	1.5
5	95.0	2.0	3.0
6	92.5	2.0	5.5
7	90.0	2.0	8.0
8	95.0	3.5	1.5
9	92.5	3.5	4.0
10	90.0	3.5	6.5
11	92.5	5.0	2.5
12	90.0	5.0	5.0

The Porter and Heuer [18] formula was applied to find the volume fraction of the monoclinic phase, M, both in powders as well as in sintered samples:

$$M = \frac{1.603 \cdot \overline{I}(111)_M}{1.603 \cdot \overline{I}(111)_M + I(111)_{T,C}} \qquad (1)$$

where I denotes integral X-ray line intensities of indicated reflections and subscripts M, T and C stand for monoclinic, tetragonal and cubic phases, respectively. Relative volume fractions of the cubic and tetragonal phases in the sintered samples were obtained from the Miller et al. [19] relation:

$$\frac{C}{T} = A \frac{I(400)_C}{I(400)_T + I(004)_T} \qquad (2)$$

The term A=0.94 was used the same as in Ref.12. Measurements based on equations (1) and (2) allowed to calculate the volume fractions of all three phases.

Fracture toughness was determined on as-received surfaces by Vickers indentation under 176 N load. Since in all cases the ratio l/a < 2.5 (where l is the fracture length and a is the half width of the indent) the Palmqvist crack model and relevent Niihara's [20] equation could be applied.

Transmition electron microscopy was used to characterize the powder morphology. In one case the replica taken from the as-received surface allowed to calculate the "true" grain size distribution by the Saltykov method [21].

3. RESULTS AND DISCUSSION

3.1. Powder characteristics

The data presented in Fig.1. and table 2 demonstrate the strong influence of the crystallization environment on the phase composition of the resultant powders. The process performed in NaOH (1.7M) water solution results in powders of much higher monoclinic phase fractions in contrast to that performed in water. The influence of MgO concentration in the coprecipitated gel on the monoclinic phase content is negligible, and that of Y_2O_3 could be noticed only with the powders crystallized in NaOH solution.

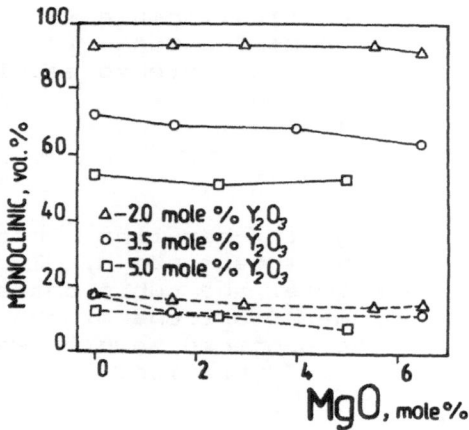

Fig.1. Monoclinic phase content in powders vs. MgO concentration. Solid lines correspond to the samles crystallized in NaOH solution, dotted lines to those crystallized in H₂O.

Specimen	Crystallization environment	
	NaOH	H₂O
1	93.5	17.8
2	72.1	18.0
3	53.9	12.1
4	94.2	16.3
5	93.1	14.8
6	93.2	13.1
7	91.4	15.1
8	69.5	11.0
9	68.7	12.0
10	63.4	12.4
11	50.9	11.8
12	52.7	8.8

Table 2. Monoclinic phase content (vol.%) in powders

Transmission electron microscopy (Fig.2) reveals deep differences between the powders processed in both crystallization environments. The powders crystallized in NaOH show two populations of particles: very small isometric particles and much larger elongated ones (Fig.2a). As substantiated by the sharper X-ray line profiles (Fig.3) the larger particles have monoclinic symmetry and the smaller ones are tetragonal or cubic. The powders crystallized in water (Fig.2b) are composed of small (~15nm) isometric crystallites. Morphology of the phases present in these powders is undiscernable. The similar observations on the influence of the crystallization environment were made with pure zirconia [22] and some zirconia systems [23].

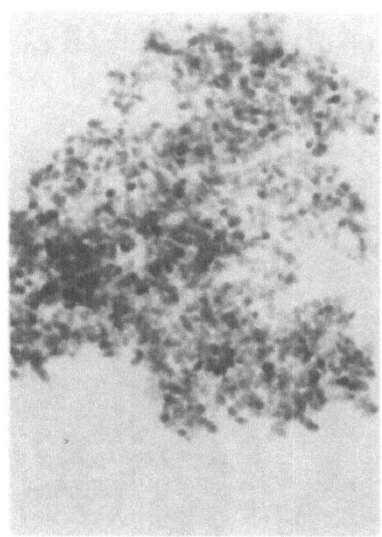

a)   b)

Fig.2. Transmission electron micrographs of the powder
4, a) crystallized in NaOH and b) crystallized in H₂O.
Note change of the scale.

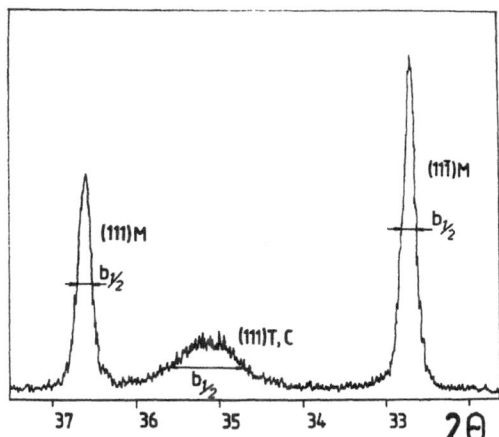

Fig.3. X-ray pattern of the powder 9 crystallized in
NaOH water solution. Half-breadth intensity of the
monoclinic phase profile is smaller indicating its
larger crystallite size.

3.2 Characteristics of sintered samples

Phase compositions of the sintered samples are shown in Figs 4-6. No other phases were observed except of the monoclinic, tetragonal and cubic zirconia solid solutions. Generaly, the MgO additives lead to the increased tetragonal phase contents and the decreased contents of the monoclinic one. This phenomenon is observed at each Y_2O_3 level both in the case of powders crystallized in NaOH and in H_2O. The powders crystallized in NaOH result in sitered samples of higher cubic phase content than those crystallized in water. In relation to the phase diagram of the system [15], we can state that the powders crystallized in NaOH, composed of relatively large, elongated monoclinic particles and very small, isometric tetragonal and/or cubic ones, reach during sintering the state closer to

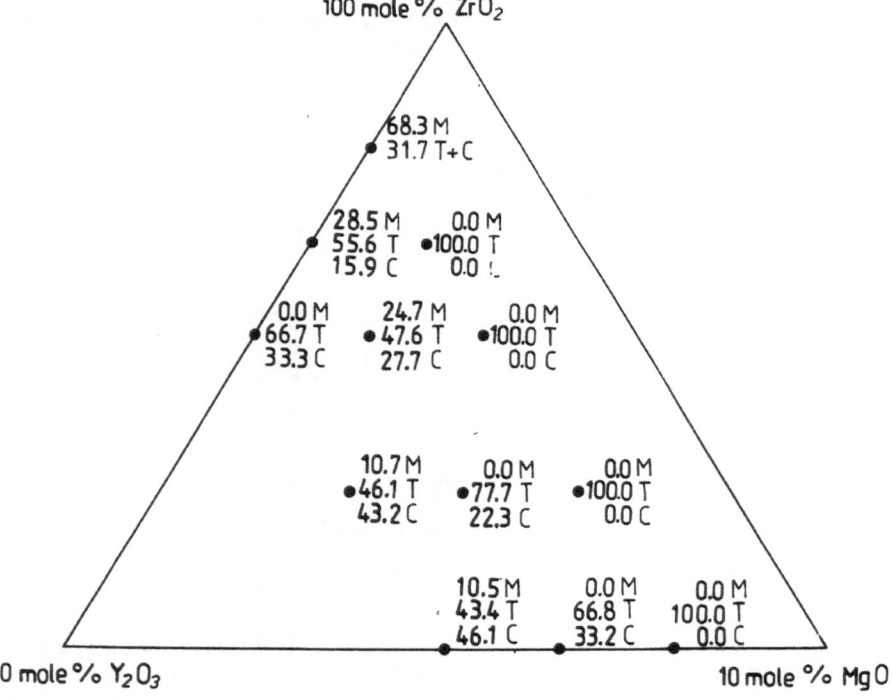

Fig.4. Phase composition of the samples sintered at 1350°C. The starting powders crystallized in NaOH water solution. M, T and C stand for monoclinic, tetragonal and cubic phases, respectively, expressed in vol.%.

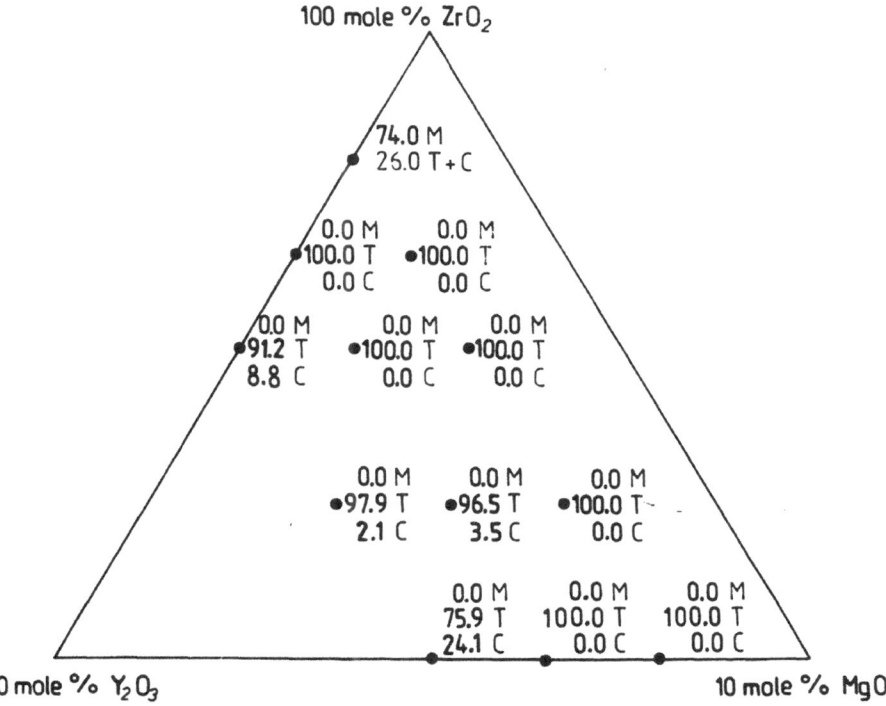

Fig.5. Phase composition of the samples sintered at 1350°C. The starting powder crystallized in water. Descriptions on the graph as in Fig.4.

the phase equilibrium. It requires redistribution of the alloying components that are initially uniformly spread over the coprecipitated gel [24]. Most probably this process occures already during crystallization under hydrothermal conditions in the presence of NaOH. This is not the case as for crystallization in water, or at least not to such an extent.

Results in Figs 4-6 indicate that with compositions lacking of cubic phase or laen of it, the MgO additives allow to retain tetragonal phase, otherwise not stable at room temperature. It suggests that MgO increases the critical grain size, that is the size of the tetragonal grains below which they do not transform to the monoclinic phase during cooling.

The critical grain size was assessed on the basis of the grain size distribution curve in the sample 4 sintered at 1400°C (Fig.7). The method described

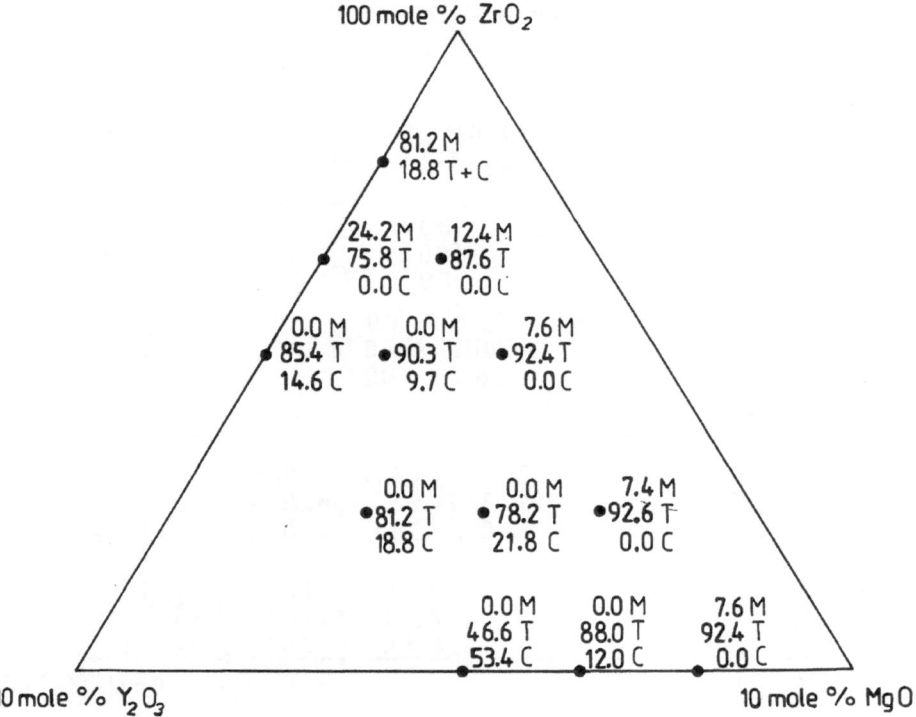

100 mole % ZrO$_2$

81.2 M
18.8 T+C

24.2 M   12.4 M
75.8 T   87.6 T
0.0 C   0.0 C

0.0 M   0.0 M   7.6 M
85.4 T   90.3 T   92.4 T
14.6 C   9.7 C   0.0 C

0.0 M   0.0 M   7.4 M
81.2 T   78.2 T   92.6 T
18.8 C   21.8 C   0.0 C

0.0 M   0.0 M   7.6 M
46.6 T   88.0 T   92.4 T
53.4 C   12.0 C   0.0 C

10 mole % Y$_2$O$_3$ 10 mole % MgO

Fig.6. Phase composition of the samples sintered at 1400°C. The starting powder crystallized in water. Descriptions on the graph as in Fig.4.

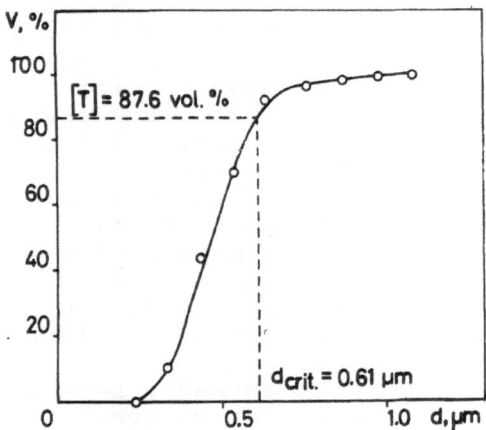

V, %

100

[T] = 87.6 vol. %

80

60

40

20

d$_{crit.}$ = 0.61 μm

0   0.5   1.0   d, μm

Fig.7. Grain size distribution of the sample 4 sintered at 1400°C.

earlier [12] is based on the plausible assumption that the tetragonal grains are the smallest grains in the system. The measurement indicates the critical grain size value of $d_{crit} = 0.6$ μm which should be compared to 0.2 μm found by Lange [25] for the alloy of Y_2O_3 content of 2 mole %. The studied composition 4 contains 2 mole % Y_2O_3 and 1.5 mole % MgO.

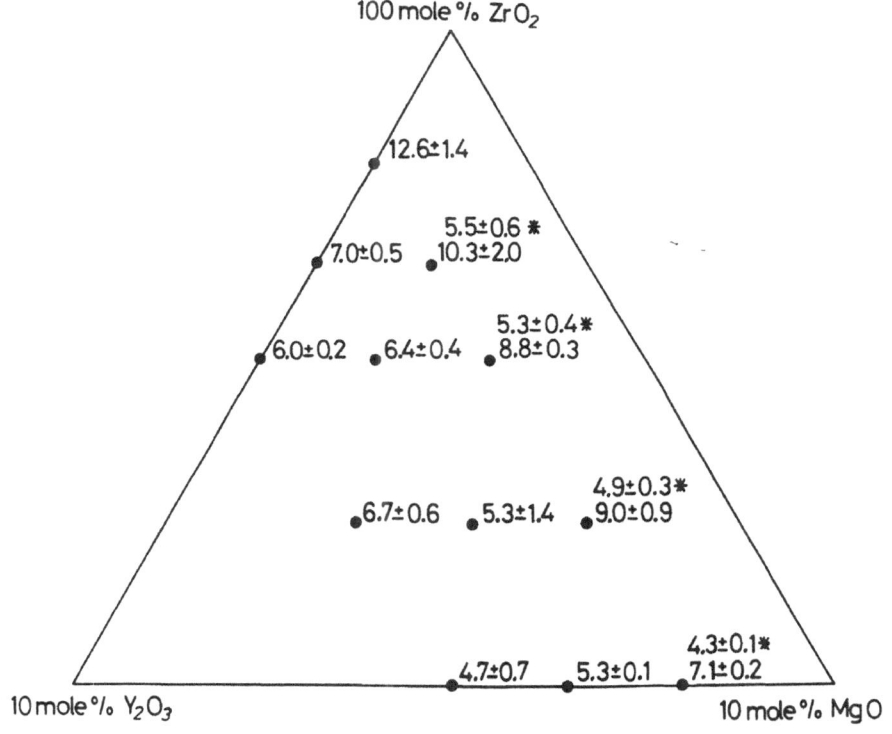

Fig.8. Fracture toughness, K_{Ic} (MPam$^{1/2}$), vs. chemical composition. Samples sintered at 1350°C. * — samples made of powders crystallized in NaOH, the remaining samples made of powders crystallized in water.

Figs 8 and 9 illustrate effect of the chemical composition of the sintered samples on their fracture toughness. The highest values are comparable to those given in the literature for the TZP bodies and are observed in case of the bodies of the lowest Y_2O_3 content (2 mole %) and the MgO concentration ranging from 1.5 to 8.0 mole % sintered at 1400°C. These bodies show some fractions of the monoclinic phase most probably present mainly in the surface layer. As known the

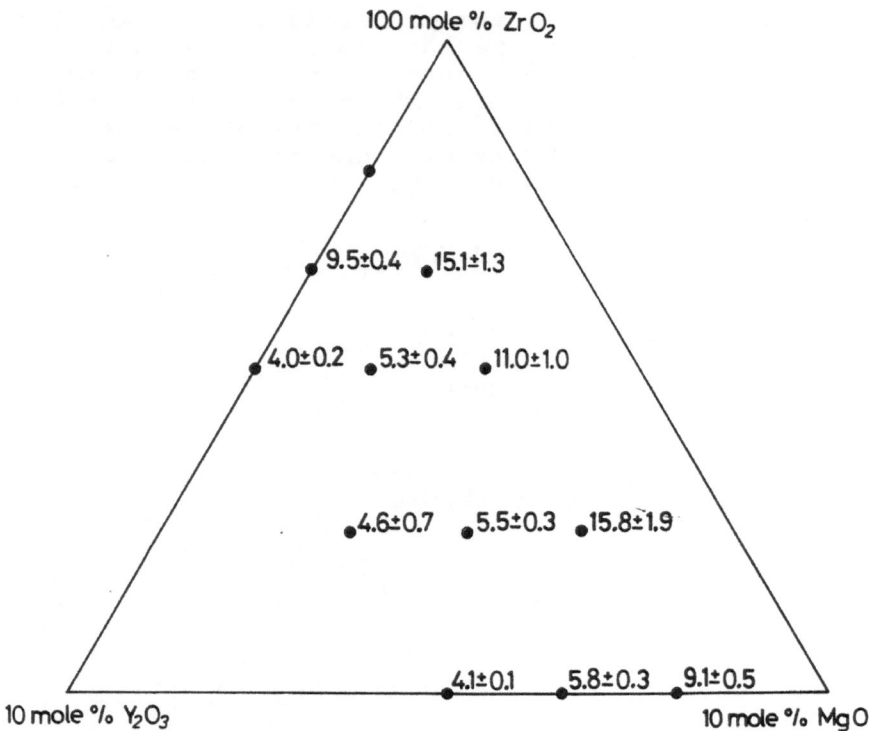

Fig.9. Fracture toughness, K_{Ic} (MPam$^{1/2}$), vs. chemical composition. Samples sintered at 1400°C. Samples made of powders crystallized in water.

Table 3. Sintered samples density

| Specimen | Apparent density (g/cm³) | | |
| | Crystallized in NaOH Sintered at 1350°C | Crystallized in H₂O Sintered at | |
		1350°C	1400°C
4	5.32	5.49	5.60
5	5.65	5.84	5.89
6	5.37	5.77	5.83
7	5.43	5.67	5.75

* confidence interval of apparent density on confidence level of 0.95 varies between 0.10 - 0.15 g/cm³

tetragonal to monoclinic phase transformation results in compressive stresses in the surface layer which suppress the crack propagation [26,27] and increase fracture toughness determined by indentation.

The lower K_{Ic} values obtained in case of the bodies made of the powders crystallized in NaOH can be explained by their lower density as substantiated by the data of table 3. The lower sinterability of these powders should be attributed to their much greater particle sizes and elongated particle shapes.

4. SUMMARY

It was demonstrated that fully tetragonal zirconia polycrystals of K_{Ic} up to 15 MPam$^{1/2}$ could be obtained in the MgO-Y$_2$O$_3$-ZrO$_2$ system by applying hydrothermally crystallized micropowders.

Better results were found in case of the powders crystallized in water. They reach higher densyties becase of their fine (~15 nm) and isometric particles. The powders crystallized in NaOH water solution show much bigger, elongated particles of the monoclinic phase together with fine tetragonal and/or cubic crystallites.

Introducing of 1.5 mole % MgO into the body of Y$_2$O$_3$ concentration of 2 mole % changes the critical grain size from 0.2 to 0.6 μm.

References

1. D.L.Porter, A.G.Evans, A.H.Heuer, Acta Met., **27** (1979)1649.
2. A.G.Evans, A.H.Heuer, J.Am.Ceram.Soc., **64**(1980)241.
3. A.G.Evans, N.Burlingame, M.Drory, W.M.Kriven, Acta Met., **29**(1981)447.
4. T.K.Gupta, Sci.Sintering, 10(1978)205.
5. T.K.Gupta, J.H.Bechtold, R.C.Kuznicki, L.H.Cadoff, B.R.Rossing, J.Mater.Sci., **12**(1977)2421.
6. T.K.Gupta, F.F.Lange, J.H.Bechtold, J.Mater.Sci., **13**(1978)1464.
7. K.Haberko, R.Pampuch, Ceramics.Int., **9**(1983)8.
8. K.Tsukuma, M.Shimada, J.Mater.Sci., 20(1985)1178.
9. K.Tsukuma, Am.Ceram.Soc.Bull., **65**(1986)1386.
10. S.Meriani, Proc. of the 12th Int. Technical Colloquium on Ceramics Processing, Rimini, sept.1987.
11. T.W.Coyle, W.S.Coblenz, B.A.Bender, J.Am.Ceram.Soc.

71(1988)C-88.

12. W.Pyda, K.Haberko, Ceramics Int., **13**(1987)113.
13. R.Pampuch, W.Pyda, K.Haberko, Ceram. Int. **14**(1988) 245.
14. J.S.Toropov, S.J.Pliner, D.S.Rutman, G.A.Taksis, A.F.Maurin, Ognieupory (in Russian), N° 11, (1979) 49-52.
15. J.R.Hellmann, V.S.Subican, J.Am.Ceram.Soc. **66**(1983) 265.
16. R.R.Lee, A.H.Heuer, J.Am.Ceram.Soc. **70**(1987)208.
17. K.Haberko, W.Pyda, Advances in Ceramics, vol.12, N.Claussen, M.Rühle, A.H.Heuer (eds), The American Ceramic Society, OH, 1984, 391-98.
18. D.L.Porter, A.H.Heuer, J.Am.Ceram.Soc., **62**(1979)298
19. R.A.Miller, J.Smialek, R.G.Garlick, Advances in Ceramics, vol.3, The American Ceramic Society, Columbus, Ohio, 1981, p.241.
20. K.Niihara, J.Mater.Sci.Let., **2**(1983)221.
21. S.A.Saltykov, Stereometric Metallography, 3rd edition, Moscow, 1970.
22. H.Nishizawa, N.Yamasaki, K.Matsuoka, H.Mitsushio, J.Am.Ceram.Soc., **65**(1982)343.
23. K.Haberko, M.Bucko, M.Haberko, M.Jaśkowski, W.Pyda, Proc. XXXVIII Berg- und Hüttenmanischer Tag, Freiberg, June 1987.
24. K.Haberko, Rev. int. Htes Temp. et Refract., **14** (1977) 217.
25. F.F.Lange, J.Mater.Sci., **17**(1982)240.
26. U.Dworak, H.Olapinski, G.Thamerus, Proc. Int. Conf. Science of Ceramics, K.J. de Vries (Ed.), vol.9, p.543.
27. K.Haberko, R.Pampuch, Ceram. Int., **9**(1983)8.

ZIRCONIA POWDER FOR TZP-CERAMICS
Ti-Y-TZP

H. Hofmann[*], B. Michel[*], L.J. Gauckler[**]

[*] Swiss Aluminium Ltd., Neuhausen,Switzerland

[**] Swiss Federal Institute of Technology, Zürich, Switzerland

1. INTRODUCTION

Tetragonal Zirconia Polycrystal-ceramics (TZP) are known for their excellent properties. The coprecipitation of these powders produces very fine particles that give, after the sintering process, an excellent uniform structure with a grain size smaller than 300 nm and very interesting mechanical properties.

Unfortunately, the 3 mol-% Y-TZP shows a strong degradation when exposed to a humid atmosphere at 200 to 500°C. This phenomenon is typical for a Y-TZP but can be retarded with the new Ti-Y-TZP powder described in the following report.

2. EXPERIMENTAL PROCEDURE

The synthesis of the Ti-Y-TZP powder with a sub-micron crystal-size, controlled particle size distribution and homogeneous chemical composition is essential for optimal properties of the sintered material. This powder is obtained by coprecipitation of the different elements. The process is shown below:

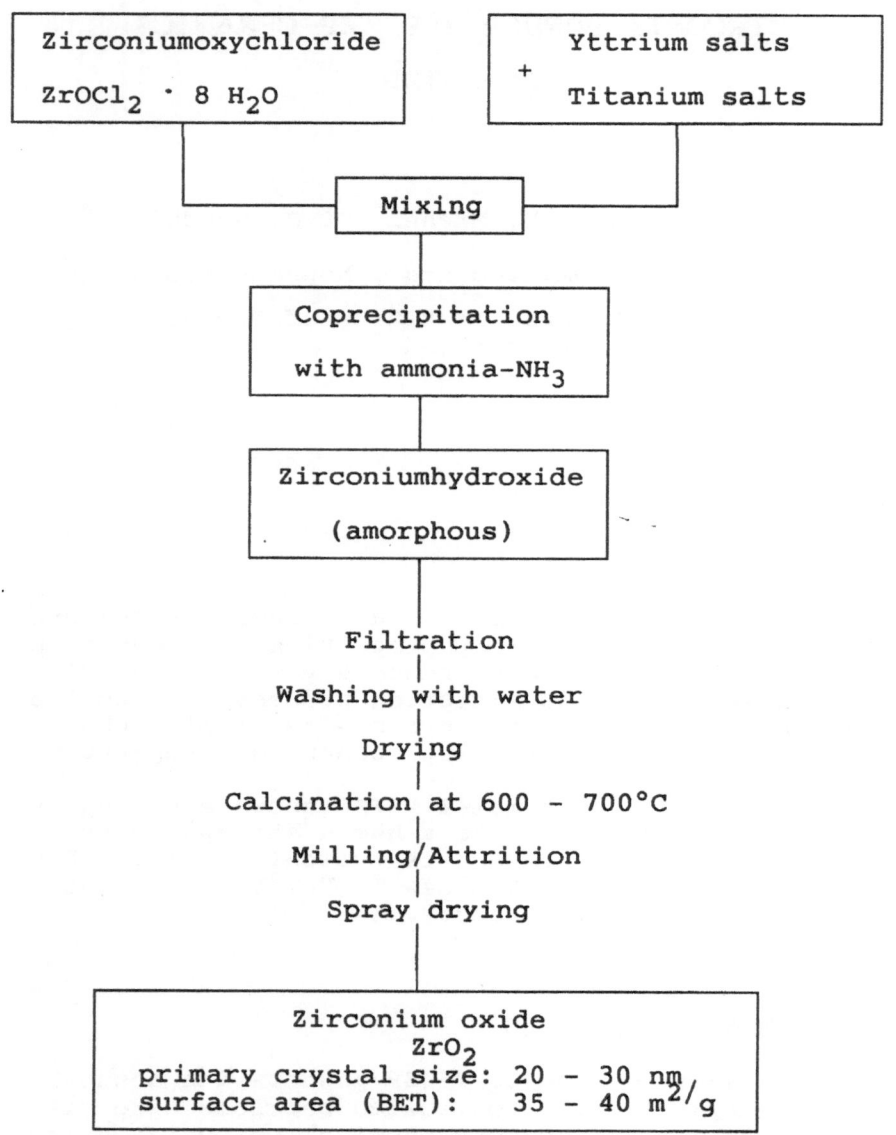

The purity of the Ti-Y-TZP powder depends on the quality of the raw materials. The different steps of the preparation had to be strictly controlled during the entire process to insure the lowest level of impurities.

3. CHARACTERIZATION OF THE Ti-Y-TZP-POWDER

3.1 Particle morphology

Scanning electron microscopy (SEM) and transmission electron microscopy (TEM) were used to estimate the particle size and morphology. Specific surface area was measured by the N_2 gas adsorbtion (BET) method.

The primary crystal size is about 20 - 30 nm (Fig. 1) and the specific surface area 35 - 40 m^2/g.
The size is uniform and can be influenced by the pH-value during the precipitation of the hydroxide.

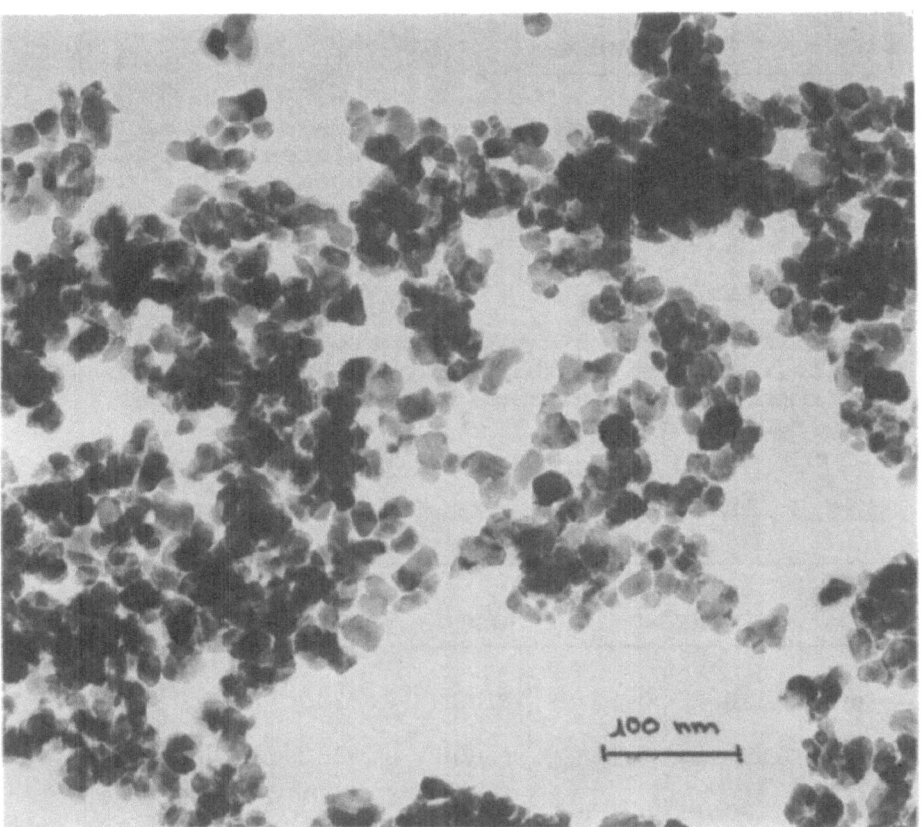

Figure 1 Ti-Y-TZP powder.

3.2 Chemical Analyses

The rigorous process guarantees a high and uniform quality of the Ti-Y-TZP powder (Table 1).

Table 1: chemical composition

TiO_2	8.00	wt %
Y_2O_3	2.50	wt %
HfO_2	1.300	wt %
SiO_2	0.070	wt %
Al_2O_3	< 0.005	wt %
MgO	< 0.002	wt %
Fe_2O_3	< 0.010	wt %
CaO	0.010	wt %
Cl	< 0.100	wt %

The major impurities are near or below the detection limits and derive from the zirconiumoxychloride. We could not detect any contamination from the preparation process.

3.3 X-Ray Measurements

X-ray investigations for the phase analysis and lattice parameter determinations (Table 2) were performed using a Siemens diffractometer with Cu Kα radiaton.

Table 2: lattice parameters

	a (nm)	c (nm)	c/a	V (nm^3)	ß (°)	Lit
t-ZrO_2-pure	0.5082	0.5189	1.021	0.1340	90	[1]
Y-TZP	0.5116	0.5157	1.008	0.1350	90	[1]
Ti-Y-TZP	0.5066	0.5195	1.025	0.1333	90	

The solution of the Ti^{4+} cation in the ZrO_2 unit cell produces a lengthening of the primitive Bravais-lattice. This is exactly the opposite effect that Y^{3+} cation produces. Therefore the c/a-ratio also increases greatly.

The main phase present in the powder is the tetragonal. Sometimes a small amount of monoclinic phase is detectable.

4. SINTERING PROCESS

Different sintering tests were done in the dilatometer and with an electrical furnace. For both experiments, the heating rates were between 3 and 30 K/min, the sintering temperatures were between 1200 and 1500°C and the sintering times were between 5 minutes and 5 hours.

Results of the sintering experiments with the dilatometer are shown in Fig. 2. The beginning of shrinkage is at 950°C and a dense sample is reached at 1300°C. Higher temperatures provoke the formation of pores. Sintering times > 3 h at 1300°C decrease the stability of Ti-Y-TZP ceramic due to grain growth and therefore a t → m tranformation was observed during cooling.

The optimal structure (grain size < 300 nm) as in Fig. 3 is obtained with a strict control of the sintering process.

In this new composition, the TiO_2 stabilizes the tetragonal phase and the Y_2O_3 limits a possible exaggerated grain growth.

124

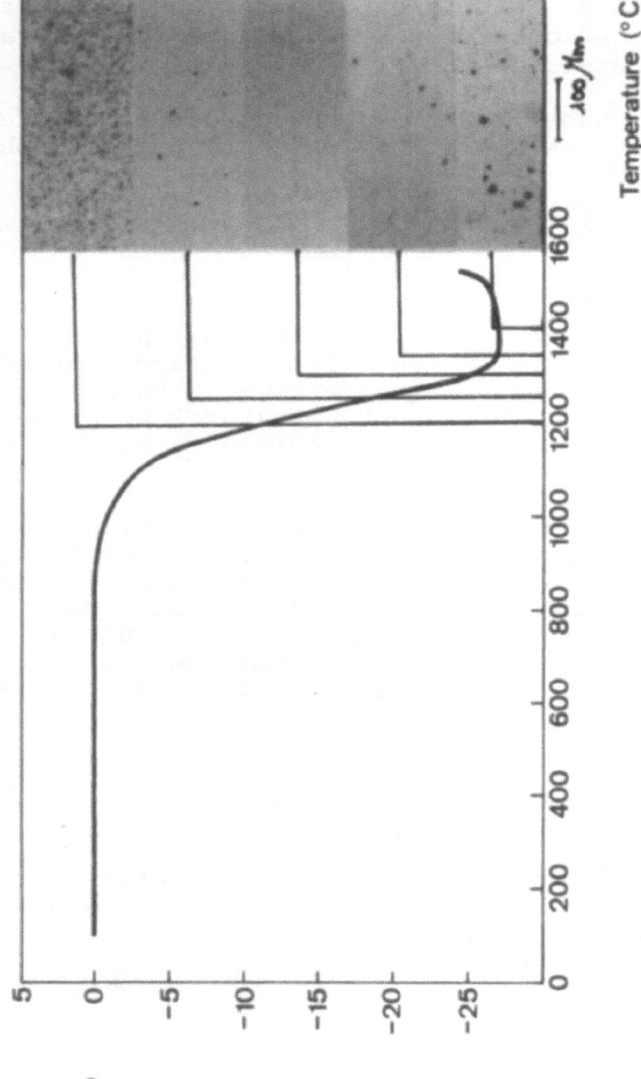

Figure 2 Sintering experiment with dilatometer
(10 K/min. – 5 minutes).

Figure 3 Sintered sample.
              (1300°C - 1 hour)

5. <u>MECHANICAL PROPERTIES</u>

A series of sintering tests were done to investigate
the influence of heating rates and sintering temperatures on the density, hardness and $K_{I}c$
(Figs. 4, 5, 6).

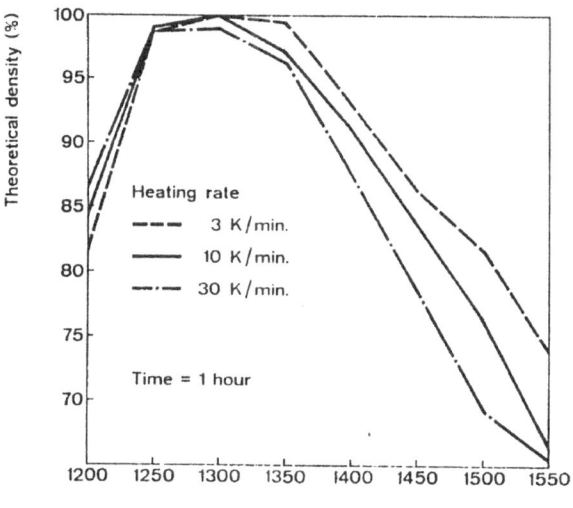

Figure 4 Influence of heating rates and sintering
              temperatures on the density.

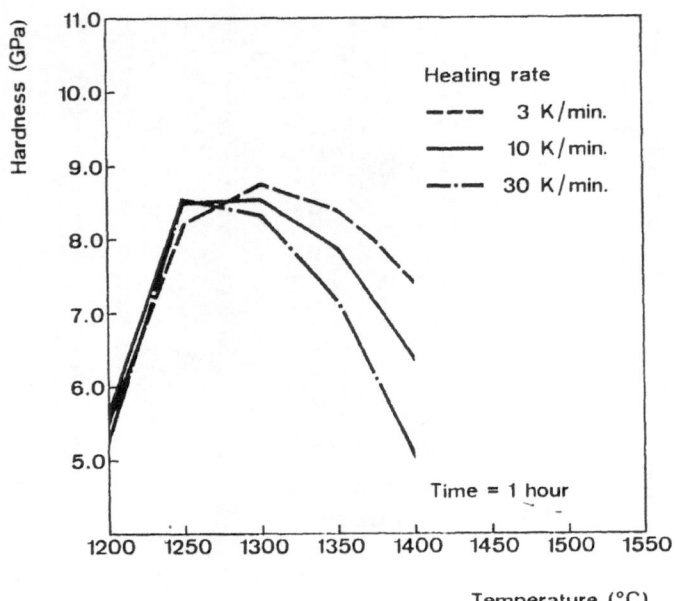

Figure 5 Influence of heating rates and sintering temperatures on the hardness.

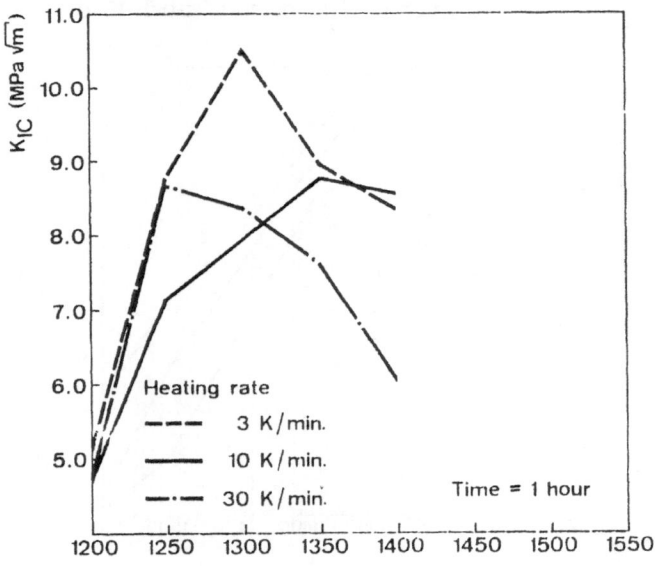

Fig. 6 Influence of heating rates and sintering temperatures on the K_{IC}.

The complete densification (Fig. 4) is obtained, as already mentioned, between 1250 and 1350°C with a heating rate of 3 - 10 K/min. The full theoretical density of this Ti-Y-TZP ceramic is 5.7 g/cc.

For the hardness, the heating rate does not play an important role. However, in contrast, the K_{IC} value is very sensible to the heating rate. This is easy to understand because the tetragonal/monoclinic phase ratio strongly influences the value of the toughness [2].

6. STABILITY IN HUMID ATMOSPHERE

The degradation of Y-TZP ceramic in the temperature range of 150° to 400°C in air and above 100°C in water is a problem for broad applications. The degradation is caused by the transformation of the tetragonal to the monoclinal phase which provokes micro- and macro-cracking, initially at the surface and then through the entire piece [3-4]. A strict control of the grain size can decrease this instability but the choice of the stabilizer is decisive.

With the use of TiO_2 in place of Y_2O_3, the tetragonal phase remains longer stable because the tendency for TiO_2 to form an hydoxid is less than that of the Y_2O_3.

A comparison test of a 3 mol-% Y-TZP with the Ti-Y-TZP under extreme conditions (temperature: 250°C, pressure: 38 bar, medium: water) shows that the life time was increased by a factor of 10 for the TZP with Ti (Fig. 7).

7. CONCLUSIONS

The Ti-Y-TZP powder is already fully dense, with high mechanical properties, at a sintering temperature of 1300°C. The new Ti-stabilizer and the small grain size guarantee a better stability in humid atmosphere.

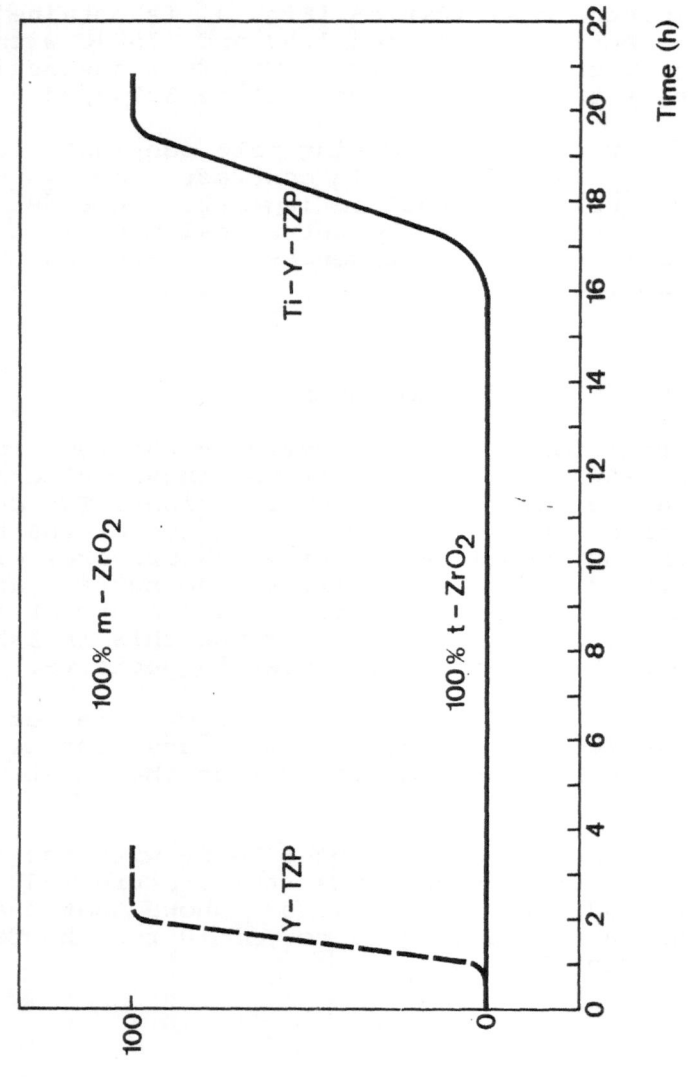

<u>Figure 7</u>   Stability in humid atmosphere of a 3 mol-% Y-TZP ceramic compared to a Ti-Y-TZP.

REFERENCES

[1] R.-R. Lee and A.H. Heuer, "Morphology of Tetragonal ZrO_2 in a Ternary (Mg, Y)-PSZ", J.Am.Ceram.Soc., 70 [4] 208-13 (1987).

[2] T. Masaki, "Mechanical Properties of Toughened ZrO_2-Y_2O_3 Ceramics", J.Am. Ceram.Soc., 69 [8] 638-40 (1986).

[3] T. Sato and M. Shimada, "Transformation of Yttria-Doped Tetragonal ZrO_2 Polycrystals by Annealing in Water", J.Am.Ceram. Soc., 68 [6] 356-59 (1985).

[4] M. Yoshimura, T. Noma, K. Kawabata and S.Somiya "Role of H_2O on the degradation process of α-Y-TZP", J.Mater.Sci.Let., 6 465-467 (1987).

ELECRON MICROSCOPY INVESTIGATION OF THE CRYSTALLIZATION PROCESSES OF AMORPHOUS ZrO$_2$ FILMS PREPARED IN SUPERHIGH VACUUM.

Kutelia E.R., Shalamberidze O.P., Golodze N.A., Maisuradze N.I., Dzigrashvili T.A., Tsivtsivadze D.M. and Khakhanashvili K.G.
Georgian V.I. Lenin Polytechnical Institute, Tbilisi, USSR.

Zirconium dioxide ZrO$_2$ deserves the attention of industrial engineers for its overall electrophysical properties. In particular, a significant attention has lately been paid to the problem of ZrO$_2$ thin films' application in solid-state microelectronics (1). ZrO$_2$ thin-film layers prepared by sputtering on cold substrates typically turn out to be amorphous and subsequent heat treatments led to their partial or complete crystallization (2,3). Accordingly, the purpose of this work has been to reveal the mechanism of atomic arrangement of completely amorphous ZrO$_2$ films prepared in superhigh vacuum by the laser evaporation method, and to investigate the characteristic properties of their initial crystallization stages when heated by the electron beam.

Films with thickness of about 1000 $\overset{\circ}{A}$ were prepared by sputtering on fresh NaCl breaches at room temperature in $5 \cdot 10^{-9}$ torr vacuum, during target evaporation from polycrystal ZrO$_2$ heated by neodymium-doped glass laser pulses ($\lambda = 1.06$ μ). After being separated from the substrate, these films were investigated with the OPTON EM- 10 C/CR transmission electron microscope at 100 kW accelerating voltage.

Heating of local film areas right in the electron microscope was carried out by the method of condenser diagram shift from electron beam axis. The temperature of the heated area was regulated by electron beam focusing, which enabled to observe and record the crystallization process dynamics with a step-by-step photography.

To investigate a short-range order in the initial amorphous ZrO$_2$ films the method of integral analysis was used. With the aid of Fourier transform of coherent electron scattering intensity it enabled to calculate the radial distribution function (RDF) of atomic density, which for multicomponent systems takes the following form:

$$\sum 4\pi R^2 \rho \, (R) = \sum\sum K_i K_j 4\pi R^2 \rho_{ij}(R) = 4\pi R^2 \rho_0$$

$$(\sum K_j)^2 + \frac{2R}{\pi} \sum K_i^2 \int_0^{s_m} S_i(S) \sin(SR)\, dS$$

where K_i and K_j are scattering powers of Zr and O atoms, $K_O = 1$, $K_{Zr}=3.34$, ρ_{ij} is a radial junction of j atoms distribution around i type atoms, ρ_O is a medium atomic density.

The interference function i (s) (Fig.1a) was found according to the result of multiple photometric measurement of electron diffraction patterns (shown in the corner of Fig. 1a) in different directions. Photometric measurement was carried out on the Leits MPV-3 microphotometer in automatic regime accurate within 0.05%. Two maximums are shown on the curve i(s) (Fig. 1a) at $S_1 = 2.47$ $\overset{\circ}{A}{}^{-1}$ and $S_2 = 3.98$ $\overset{\circ}{A}{}^{-1}$, corresponding to interplane spacings $d_1 = 2.54$ $\overset{\circ}{A}$ and $d_2 = 1.56$ $\overset{\circ}{A}$, which are very close to $d_{200} = 2.535$ $\overset{\circ}{A}$ and $d_{113} = 1.55$ $\overset{\circ}{A}$ parameters of a ZrO_2 tetragonal lattice.

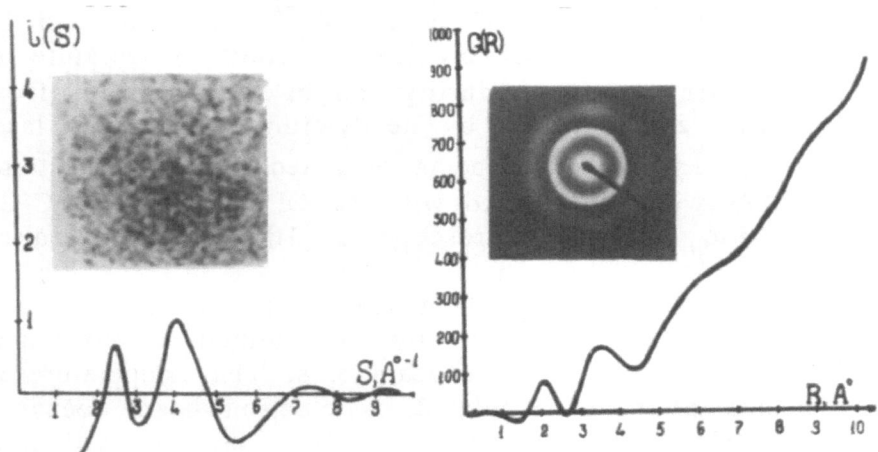

Fig.1: a) Interference function i (s) and b) corresponding RFD curve for amorphous ZrO_2 (High resolution electron microphotography and corresponding microdiffraction are also shown here)

The curve RDF, calculated according to the formula (1), is shown in the Fig. 1b. The coordination maximums are arranged at $R_1 = 2.15$ Å and $R_2 = 3.55$ Å. The corresponding areas under the maximums equal $Q_1 = 40.65$ and $Q_2 = 170$. By taking into account the atomic radii for Zr and O to be correspondingly equal to $R_{Zr} = 1.6$ Å and $R_O = 0.6$ Å, we get that 2.15 Å spacing coincides with Zr=O spacing. By assuming atoms of the other kind to be the only nearest neighbours, we have the number of oxygen atoms around Zirconium atoms to equal:

$$N_{Zr,o} = \frac{Q_1}{2K_{Zr} \cdot K_o \cdot 1} = 6.08$$

and the number of Zirconium atoms around oxygen atoms to equal:

$$N_{o,Zr} = \frac{Q_1}{2K_{Zr} \cdot K_o \cdot 2} = 3.04$$

The spacing $R_2 = 3.55$ Å seems to correspond to the nearest spacings Zr=Zr. The coordination number in this case $N_{Zr,Zr} = 7.56$. In the ZrO_2 tetragonal lattice the shortest spacing Zr-O equals 2.21 Å, and Zr-Zr = 3.58 Å, which agrees with the interatomic spacings obtained from the RDF. Therefore, this analysis shows that the ZrO_2 amorphous film obtained by the above method has a short-range structure, corresponding to an atomic coordination in the ZrO_2 tetragonal lattice. Dimensions of ordered areas in the initial films do not exceed 30 Å. It means that ZrO_2 polymorphic modifications in an amorphous state have different atomic structures, and, consequently, phase transitions are possible between them.

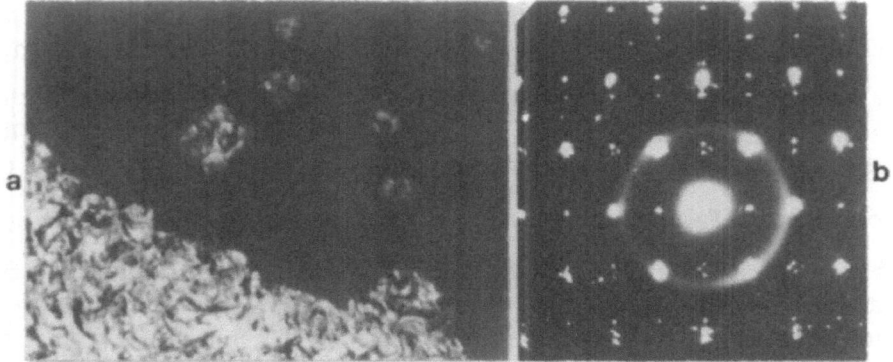

Fig. 2. Electron microphotogrphy of partially cristallized amorphous film ZrO_2 (a) and nucleating center microdiffraction in amorphous matrix (b), 4the position of a selecting diaphram is shown by an arrow) ˍ ˍ

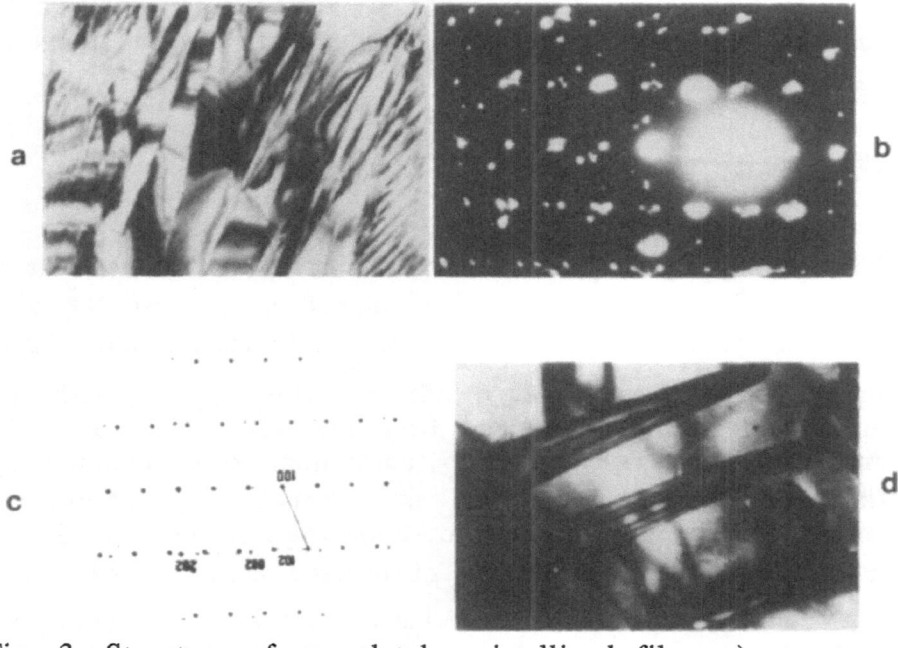

Fig. 3. Structure of completely cristallized film: a) structure of monoclinic and tetragonal phase intermitting lamelae; b) microdiffractionof the given area, $\{010\}_m \parallel \{010\}_t$ zone axis; c) electron microphotography of wedge shaped microtwins; d) microdiffraction of an area with twins, $\{010\}_m \parallel \{010\}_t$ zone axis.

The pecularity of the amorphous ZrO_2 film structure most likely predetermines the initiation of the crystallization process. It originates via many crystallization centers (Fig.2a), where both monoclinic and tetragonal ZrO_2 are being simultaneously formed (Fig. 2b). By heating the film crystallization proceeds to completion with the formation of intermitting lamellae and packs of monoclinic and tetragonal modified crystals with the characteristic morphology of martensite transformation products (Fig. 3a,b). The electron diffraction pattern analysis (Fig.2b and 3b) showed two different modifications existing simultaneously. Both transformation and modification twins are very often observed in crystallized products (Fig. 3c,d). Consequently, the coordinated crystal growth by selecting two corresponding orientation alternatives and twinning relationships leads to a maximal compensation of elastic stresses. As a result, a crystallized cracks free thin film of Zirconium dioxide is obtained.

REFERENCES
1. Ben-Dor L., Elshein A., Halabi S., Pinsky I; and Shappir I., L.Electron. Mater., 13, n.2, 263-272, (1984).
2. Kellong R., Nestechiometry, Moscow, Mir Publishers, (1977).
3. Farabaugh E.W., Sanders D.M., Wilke M.E., Hurwitz S.A. and Haller W.K., "Preparation of thin amorphous films by beam evaporation from multiple sources", VS.Per.commer.Nat.Bur.Stand.Spec.Publ., (1983), N638, laser Induced Damage in optical Mat. Proc. Symp. Soylder, Colo, Nov. 17-19 1981, 451-458.
4. Movchan B.A., Tavadze Ph. N., Okrosashvili M.N., Shalamberidze O.P. and Kutelia E.R., PSEM Journal, P.9, 68-74, (1978).
5. Shalamberidze O.P., Kutelia E.R., Tavadze Ph.N., Golodze N.A. and Khakhanashvili K.H., Reports of USSR Academy of Science, V. 278, 152-154, (1984).

HIGH TEMPERATURE MECHANICAL BEHAVIOUR OF MULLITE-ZIRCONIA COMPOSITES OBTAINED BY REACTION SINTERING

A. LERICHE, P. DESCAMPS, F. CAMBIER
C.R.I.B.C. - Belgian Ceramic Research Center
4, Avenue Gouverneur Cornez, B-7000 Mons - BELGIUM

Abstract

Mullite-zirconia composites were prepared by reaction sintering with the TiO_2 and $TiO_2 + Y_2O_3$ as additives with the aim to reduce the intergranular amorphous phase volume usually present in reaction sintered materials.
The mechanical and microstructural properties of the materials are presented and discussed in comparison with previous results obtained for materials containing MgO as additive. The mullite-zirconia materials prepared with TiO_2 presents a finer microstructure characterized by equiaxed mullite and intergranular zirconia grains with small amount of low viscosity intergranular glassy phase. The reduction of glassy phase volume and fluidity compared to MgO-composites induces the existence, from 600° C up to 1200° C maximum temperature testing, of a large plastic zone in front of the crack absorbing the crack propagation energy.

1. INTRODUCTION

It has been largely shown that Reaction Sintering of alumina and zircon is an easy and cheap route to obtain a good dispersion of zirconia grains in mullite matrix (1-10). The reaction sintered mullite zirconia composites can be prepared by different methods :
- the two-step method (1) which separates the densification and the reaction step by special firing cycle including successive plateaux at 1400 and 1600° C
- the sintering of ultra-rapidly quenched zircon-alumina mixtures (11) which occurs at about 1600° C.
- the sintering in presence of a few percent of oxides (3, 7-9, 12) which form with silica transitory liquid phase favouring the simultaneous occurrence of densification and reaction at low temperature. The composites prepared by this last method always contain an unnegligeable amount of residual glassy phase mainly located at grain boundaries. So, the good mechanical properties of these materials (flexural strength above 300 MPa , critical stress intensity factor between 3 and 4 MPa√m) dramatically decrease above 800° C.
The aim of this work was to reduce as low as possible the amorphous phase volume by using titania and titania-yttria as additive and by carrying out annealing

treatment of sintered materials and to test their mechanical behaviour at high temperature.
Three compositions were chosen in the system $ZrO_2-SiO_2-Al_2O_3-TiO_2$ from previous studies (12) because of their high tetragonal zirconia content. These compositions are located on a straight line between mullite (A_3S_2) and aluminium titanate (AT) of the phase diagram $Al_2O_3-SiO_2-TiO_2$ with the same molar content of ZrO_2 and of SiO_2 (9). These compositions correspond to x values of 0.7, 0.85 and 1.0 following the equation [1]

$$2 \; ZrSiO_4 + 3 \; Al_2O_3 + x \; (TiO_2 + Al_2O_3) \rightarrow 2 \; ZrO_2 + 3Al_2O_3 \; 2SiO_2 + x \; Al_2TiO_5 \quad [1]$$

We added adequate amount of yttria to stabilize tetragonal zirconia (3 $^m/_o$ of zirconia) to the intermediate composition (x = 0.85) and increasing amount of yttria and alumina (y = 0.037, 0.110 , 0.183) to form yttrium aluminate ($Y_3Al_5O_{12}$) at the grain boundaries following the equation [2]

$$2 \; ZrSiO_4 + (3 + x + \tfrac{5}{3} y) \; Al_2O_3 + x \; TiO_2 + (y + 0.06) \; Y_2O_3 \longrightarrow$$

$$2 \; ZrO_2.0.06 \; Y_2O_3 + 3 \; Al_2O_3.2 \; SiO_2 + x \; Al_2TiO_5 +$$

$$\tfrac{y}{3} (5 \; Al_2O_3.3 \; Y_2O_3) \quad [2]$$

The studied compositions expressed in weight percent are listed in table I.

TABLE I : Studied compositions in weight percent

Composition		Al_2O_3	$ZrSiO_4$	TiO_2	Y_2O_3
X = 0.7		47.2	45.8	7	
X = 0.85		47.5	44.3	8.2	
X = 1.0		47.7	42.8	9.4	
X = 0.85	y=0.037	46.9	42.7	7.9	2.5
	y=0.110	46.8	41.3	7.6	4.3
	y=0.183	46.4	40.2	7.4	6

2. EXPERIMENTAL PROCEDURES

- The alumina*, zircon**, and additives *** powders were
mixed in alcohol by stirring during 15 minutes. The suspension
containing 50 percent of dry matter was attrited during
3 hours with yttria stabilized zirconia balls of 0.5 mm diameter.
After drying, the powder mixtures were shaped as cubes of
3 x 3 x 3 cm³ by isostatic pressing under 170 MPa during
3 minutes.
For the composition x = 0.85 without yttria, sieving of a part
of the dry attrited powder was carried out in order to
eliminate agglomerates formed during the processing.
After a prefiring of 2 hours at 1050° C, the cubes were
cut as platelets (3 x 3 x 0.5 cm³) which were thermally
treated (soaking temperature : 1450 to 1600° C, time :
15 to 240 minutes). The compositions containing titania
and yttria were sintered in two steps : - plateau of
2 hours at 1260° C which allows a densification of the
components without reaction.
                                        - plateau of
30 to 240 minutes at different temperatures (1350 to 1500° C)
for reaction achievement.
This special firing cycle was determined from dilatometric
curve which shows that in presence of yttria the reaction
occurs from 1350° C (50° C lower than for the composition
without Y_2O_3).
- The sintered materials were characterized by classical
methods. The open porosity and bulk density were measured
by Archimede's method. The crystalline phase contents
were determined by X Ray Diffraction by using the quantitative
method of external standard (13). The relative contents
in different allotropic zirconia phases (monoclinic M,
tetragonal T and cubic C) were measured on bulk samples
and calculated from modified Porter and Heuer relation
(14) for T/M and C/M and from Schmid relation (15) for
T/C.

---

* Al_2O_3 A 16SG Alcoa, Particle size = 0.4 μm
** $ZrSiO_4$ Opazir Quimicos Minerales S.A., after milling
          of 24 hours : particle size = 0.26 μm
*** TiO_2 R-SM2-Tioxide. Particle size = 0.7 μm
     Y_2O_3 BDH Chemicals Ltd. Particle size = 3 μm

The samples were cut in parallelipipeds of 5 x 3 x 30 mm³
for mechanical characterization.
The Young modulus (E) was determined at room temperature by
a sonic method and at high temperature from the load displacement
curve, the flexural strength (σ_F) by a 3-point bending test
with a span of 15 mm and a crosshead displacement of 0.1 mm/min
(ZWICK 1474 testing machine) and the critical stress intensity
factor (K_{IC}) with the same 3 point bending equipment following
the single etched notched beam technique with a notch width
of ~ 190 µm.

3. RESULTS AND DISCUSSION

3.1. Materials with titania

The sintering kinetic results have shown a temperature of
1500° C is necessary for firing of the compositions with titania
to obtain a complete dissociation of zircon. Moreover, the
temperature and firing duration increase induces a dramatic
reduction of relative tetragonal zirconia content because of
grain coarsening. According to these data, TiO_2 materials were
sintered at 1500° C during 15 minutes. Their physical,
mineralogical and crystallographical properties are presented
in the table II.

TABLE II : Physical, mineralogical and crystallographical
properties of mullite-zirconia composites prepared
with TiO_2 (open porosity ~0)

COMPOSITION	ρ BULK	WEIGHT %					$T_{R\,BULK\,SAMPLE}$ %
		$ZrSiO_4$	Al_2O_3	MULLITE	ZrO_2*	Al_2TiO_5	
X = 0.7	3.7	2	13	53	32	-	24
x = 0.85	3.6	-	8	55	30	7	17
x = 0.85 SIEVED	3.7	-	8	56	28	8	53
x = 1	3.7	-	12	47	26	14	32

* monoclinic + tetragonal zirconia

The TiO_2 content increase induces an increase of aluminium
titanate content in the sintered bodies. Moreover, the
composition (x = 1), located at the limit of secondary
crystallisation field of mullite and alumina (9), presents
lower content of mullite conjugated to enhancement of alumina
content.

Concerning the sieved material (x = 0.85), it presents a higher relative tetragonal zirconia content (53 % instead of 17 % for unsieved powders) whereas the size of zirconia grains is larger.
A typical example of microstructure obtained after polishing and thermal etching is presented in figure 1.

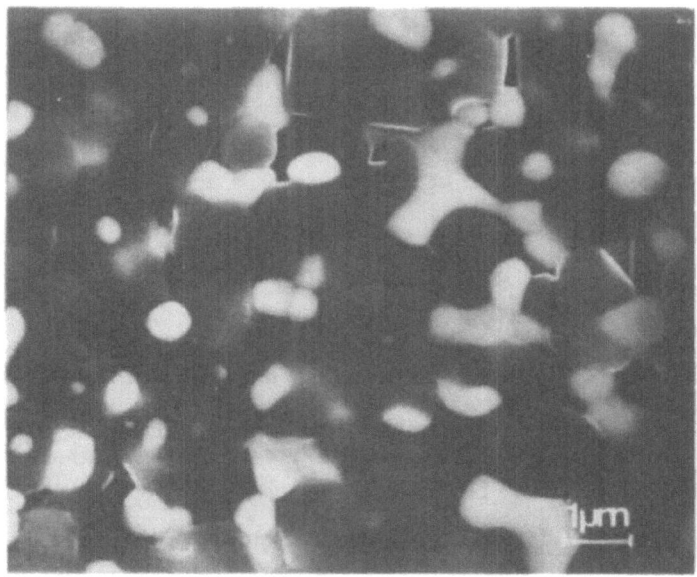

FIGURE 1 : Typical Microstructure of reaction sintered mullite zirconia composites in presence of titania (composition x = 0.85).

The microstructure is characterized by a dense mullite matrix constituted of fine equiaxed grains (1-3 μm) and an intergranular dispersion of submicronical zirconia grains.

The mechanical properties at room temperature are summarized in table III.

TABLE III : Room temperature mechanical properties of reaction
sintered mullite-zirconia composites containing
TiO_2

	$E_{(GPA)}$	N	$\sigma_{F(MPA)}$	N	K_{IC} $(MPAM^{\frac{1}{2}})$	N	A_C (M)
x = 0.7	206 ± 6	20	284 ± 59	7	3.0 ± 0.1	2	70
x = 0.85	201 ± 5	20	238 ± 37	7	3.1 ± 0.5	4	110
x = 0.85 (SIEVED)	210 ± 4	20	287 ± 47	8	3.2 ± 0.3	8	80
x = 1.0	203 ± 5	20	289 ± 72	8	3.9 ± 0.9	4	110

N = trial number.

From these data, it can be deduced that :
- the different materials present quite similar values for
elastic modulus and critical stress intensity factor
- the most rich titania material (x = 1) presents a wide
dispersion of flexural strength (21 %) and of critical stress
intensity factor values (23 %), although this last parameter
has a higher mean·value compared with materials with lower
TiO_2 contents.
The sieved material (x = 0.85) presents improved flexural
strength compared to the unsieved one and therefore smaller
critical defect size (a_c) calculated from Irwin relation (16)
and confirmed by scanning electron microscopy observation.
Because of the high value of K_{IC} obtained for the material
(x=1), it can be noted that the reduction of critical defect
size (110 μm) by sieving could lead to a significant
improvement of the flexural strength.
The figures 2 to 5 present the mechanical behaviour at
high temperature of TiO_2 materials. The σ_F curves show a
parallel evolution to this obtained for mullite-zirconia
with MgO as additive (18). Indeed, σ_F decreases slowly and
continuously up to 700° C (600° C in the case of MgO) then
increases up to 800° C and falls down above 800° C. However,
the stress intensity factor evolution with temperature is
quite different, the slow decrease up to 800° C being followed,
in the case of TiO_2, by a sharp increase up to 1200° C, the
maximum test temperature. In the case of MgO addition, the
K_{IC} curve presented the same evolution as σ_F curve.

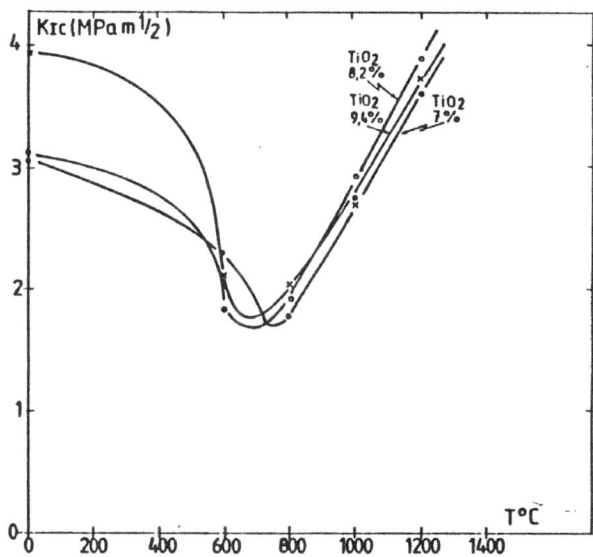

FIGURE 2 : K_{IC} values versus temperature for materials
with TiO_2

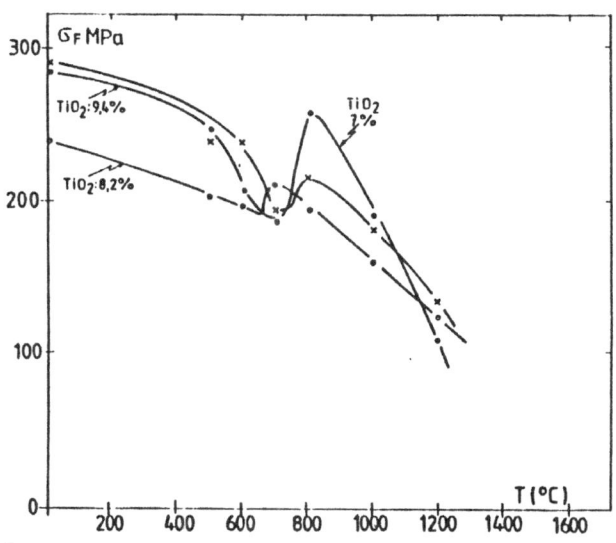

FIGURE 3 : σ_F values versus temperature for materials
with TiO_2

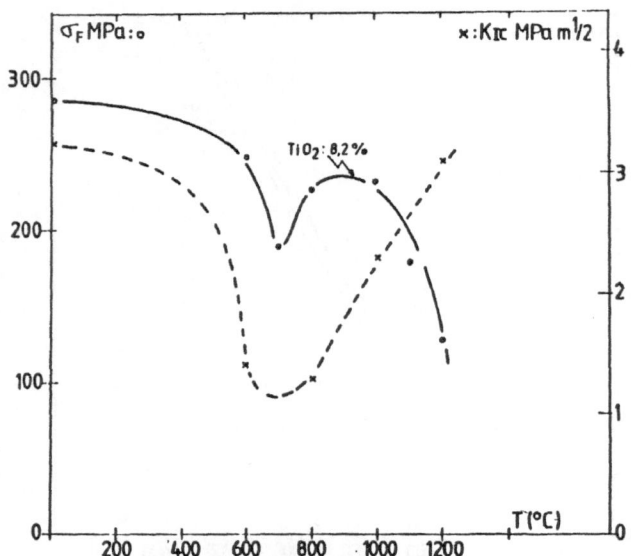

FIGURE 4 : K_{IC} and σ_F values versus temperature for sieved material with TiO_2 (x = 0.85)

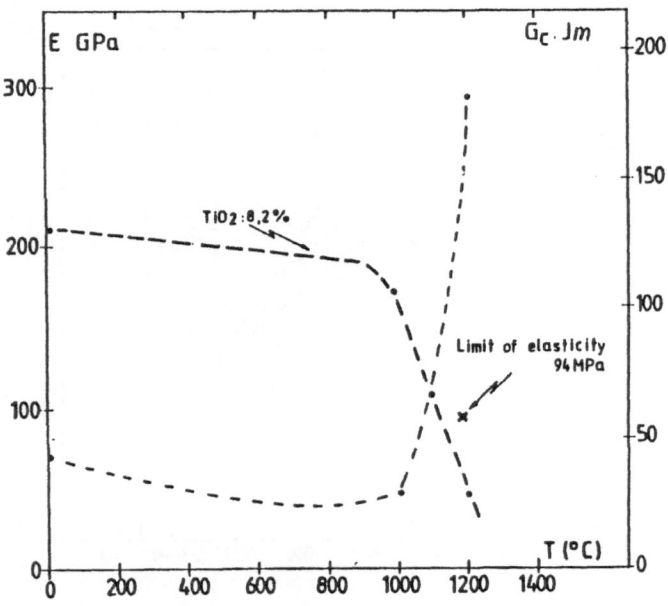

FIGURE 5 : Young modulus values (E) and toughness values (G_c) versus temperature for sieved material with TiO_2 (x = 0.85).

The phenomenon of sharp increase of K_{IC} has already been observed for TiC particle reinforced alumina composites (17). It has been assumed that the formation of a plastic zone in front of crack absorbs the energy necessary for crack extension.

Such an explanation can also be proposed for the mullite-zirconia composites with TiO_2.

Orange et al (18) have suggested that the presence of large amount of glassy phase in MgO reaction sintered materials induces the formation of plastic zone ahead of the crack tip and a stress relaxation in the grain boundary phase. From 800° C (above the glass transition temperature), the glassy phase becoming fluider, grain boundary sliding occurs and becomes more important than plastic relaxation which provokes matrix decohesion and mechanical properties decrease.

In the case of TiO_2 composites, by comparison with MgO ones, the amount of glassy phase is dramatically reduced and its viscosity can be reasonably expected to be higher according to the higher valence of titanium ions (19). Therefore, it can be suggested that energy dissipation due to relaxation phenomena in the plastic zone remains effective at temperature much higher than 800° C.

The evolution of elastic modulus (E) and of toughness (G_c) with temperature, but also the low value of the limit of elasticity (94 MPa) at 1200° C, determined from load displacement curves, are also significative of the presence of a large plastic zone at the crack tip (figure 5) and then, are in accordance with the above explanation.

The fracture faces analysis has revealed that for all temperatures, the rupture is transgranular through mullite grains and intergranular for zirconia grains. However, in the case of sieved materials, it has been observed that the fracture faces at 1200° C present two distinctive zones (figure 7 **a**); one (light) which corresponds to an inter-granular fracture mode (figure 7 **C** the crack goes around zirconia and mullite grains) and the other (dark) which corresponds to a transgranular fracture mode (figure 7 **b**) presence of clivage steps). It seems therefore that at 1200° C for this material the crack slowly propagates (subcritical crack growth) until the rupture becomes catastrophic.

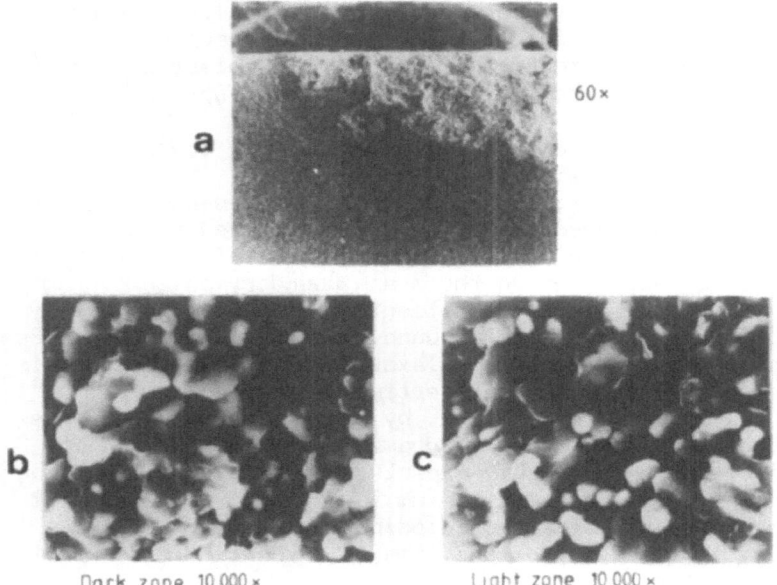

FIGURE 7 : Rupture faces of sieved material (x=0.85)
at 1200° C.

Rincon, Moya and Melo (20) have previously studied microstructure of mullite-zirconia with TiO_2 content (x = 1.0).
They observed solubility of TiO_2 both in mullite and in zirconia
and formation of solid-solutions between zirconia, mullite
and titania. They also observed presence of aluminium
titanate and especially of glassy phase containing important
amount of titania mainly located at triple-point.

The fracture face analysis which clearly show that fracture
is intergranular at the level of zirconia grains, even at low
temperature and the evolution of σ_F and K_{IC} curves with
temperature lead to prove that the presence of glassy phase of
high viscosity solid solution effects is the main fracture
controlling the mechanical behaviour at high temperature.

3.2. Materials with titania - yttria

The kinetic results have shown that reacted composites presenting a high stabilized zirconia content are obtained after 1 hour soaking at 1400° C. The physical, mineralogical and crystallographical characteristics for the three tested compositions are listed in table IV.

TABLE IV : Physical, mineralogical and crystallographical properties of mullite-zirconia composites prepared with TiO_2-Y_2O_3 (open porosity ~ 0).

COMPOSITION x =0.85 SIEVED	ρ_{BULK}	WEIGHT %					BULK SAMPLE		GROUND SAMPLE	
		$ZrSiO_4$	Al_2O_3	MULLITE	ZrO_2*	Y_2SiO_5	T_R(%)	C_R(%)	T_R(%)	C_R(%)
Y = 0.037	3.8	1	6	56	33	4	77	10	55	10
Y = 0.110	3.8	3	18	38	36	5	54	35	22	50
Y = 0.183	3.9	1	18	38	36	7	44	46	25	50

* monoclinic + tetragonal + cubic zirconia.

It is noted that the compositions containing 3 and 5 % of yttria are probably located in alumina secondary crystallisation field as previously seen for the addition of titania ; an another side, no aluminium titanate has been detected. The addition of yttria induces the formation of yttrium silicate instead of expected yttrium aluminate and a very high level of zirconia stabilization as tetragonal and cubic phase (about 90 % for bulk samples and about 70 % for ground samples).

The mechanical properties listed in table V are not improved even after annealing treatment of 20 hours at 1300° C. The X ray diffraction analysis of annealed samples shows decrease of residual glassy phase and crystallisation of yttrium silicate.

TABLE V : Room temperature mechanical properties of reaction
sintered mullite-zirconia composites containing
$TiO_2-Y_2O_3$

COMPOSITION	E (GPa)	N	σ_F (MPa)	N	K_{IC} ($MPam^{\frac{1}{2}}$)	N	A_C (μm)
x= 0.85 y=0	210 ± 4	20	287 ± 47	8	3.2 ± 0.3	8	80
x= 0.85 y=0.037	209 ± 3	28	277 ± 28	9	3.1 ± 0.2	10	78
x= 0.85 y=0.110	213 ± 3	26	233 ± 34	10	2.4 ± 0.4	10	66
x= 0.85 y=0.183	225 ± 3	12	265 ± 36	12	3.0 ± 0.2	10	80
x= 0.85 y=0.183 AFTER ANNEALING	-	-	250 ± 32	8	-	-	-

N = number of trials

In fact, the addition of yttria and alumina in the aim
to form garnet lead to the formation of yttrium silicate
phases and residual glassy phase containing silica and
titania. Therefore, we did not mesure any improvement
of mechanical properties at room temperature and we did not
evaluate the mechanical behaviour of these materials at
high temperature.

4. CONCLUSION

There are major microstructure differences between mullite-
zirconia materials when prepared by reaction sintering
with TiO_2 or MgO as additives.
Materials with TiO_2 are characterized by fine equiaxed
mullite grains, intergranular zirconia and reduced quantity
of amorphous phase whereas materials with MgO present
elongated cross linked mullite grains, inter and
intragranular zirconia and a large volume of intergranular
amorphous phase. In spite of these differences of microstructure, both composites present almost the same
mechanical properties at room temperature ($\sigma_F \sim$ 300 MPa, K_{IC}
higher than 3 MPa $m^{\frac{1}{2}}$). However, their mechanical behaviour
at high temperature is quite different. Indeed, only the
flexural strength presents a similar evolution whereas

the critical stress intensity factor of TiO_2 materials does not vary as the flexural strength but increases up to 1200° C, maximum test temperature. Such a behaviour could be explained by the existence, in materials containing TiO_2, of a significant plasticity zone in front of crack. This phenomenon can be explained by the low volume of amorphous phase presenting a high viscosity which allows to absorb the required energy for the crack propagation. On the contrary, in the case of previously studied materials with MgO content, a higher volume of amorphous phase presenting a lower viscosity induces a matrix decohesion and consequently the K_{IC} falls down above 800° C. Concerning the addition of yttria as tetragonal phase stabilizer (3 mole %) and of increasing amount of yttria and alumina in order to reinforce grain boundaries by formation of crystallized yttrium phases, it can be concluded that it does not induce any improvement of mechanical properties in spite of the increase of stabilized tetragonal and cubic zirconia content. Indeed, yttrium silicate phase (Y_2SiO_5) is formed instead of the expected yttrium aluminate $(Y_3Al_5O_{12})$.

Acknowledgement

The authors thank Amal Chafai (ISIC - Mons) who prepared the materials.

5. REFERENCES

(1) N. CLAUSSEN, J. JAHN, "Mechanical properties of sintered in situ reacted mullite-zirconia composites". J. Am. Ceram. Soc. 63, p 228 (1980).

(2) J.S. WALLACE, N. CLAUSSEN, "Development of phases in situ-reacted mullite-zirconia composites". Surfaces and interfaces in ceramic and ceramic-metal systems. Edited by Joseph Pask and Anthony Evans. Plenum Publishing Corporation, 1981.

(3) P. PENA, J.S. MOYA, S. DE AZA, E. CARDINAL, F. CAMBIER, C. LEBLUD, M.R. ANSEAU, "Effect of magnesia additions on the reaction sintering of zircon/alumina mixtures to produce zirconia toughened mullite". J. Mater. Sci. Letters 2,pp 772-774 (1983).

(4) A. LERICHE, L. DAPRA, M.R. ANSEAU, C. LEBLUD, F. CAMBIER
"Microstructure and mechanical properties of reaction
sintered zirconia toughened mullite ceramics". in
Mechanically, Chemically and Thermally induced failure
in engineering materials. Ed. by J.D. Bolton and
S. Hampshire Proc. of the 2nd Irish Durability and
Fracture Conference. NIHE-Limerick 28-29 March 1984.

(5) A. LERICHE, F. CAMBIER, R.J. BROOK, "Study of Some
Factors Influencing the Microstructural Development
of Mullite Zirconia Composites obtained by Reaction
Sintering". Special ceramics 8 British Ceramic
Proceedings nº 37. ed. by S.P. Hewlett and D. Taylor
Publ. by Institute of Ceramics Shelton, Stoke-on-Trent
U.K. (1986).

(6) F. CAMBIER, C. BAUDIN DE LA LASTRA, P. PILATE,
A. LERICHE, "Formation of Microstructural Defects in
Mullite-Zirconia and Mullite-Alumina-Zirconia
Composites obtained by Reaction Sintering of Mixed
Powders", Br. Ceram. Trans. J. 83, 196-200, 1984.

(7) P. PENA, P. MIRANZO, J.S. MOYA, S. DE AZA
"Multicomponent Toughened Ceramic Materials Obtained
by Reaction Sintering, Part. I : System Zirconia-
Alumina-Silica-Calcia" J. Mater.Sci. 20, pp 2011-22,
1985.

(8) P. MIRANZO, P. PENA, J.S. MOYA, S. DE AZA
"Multicomponent Toughened Ceramic Materials Obtained
by Reaction Sintering". Part II : System Zirconia-
alumina-silica-magnesia" J. Mater. Sci. 20
pp 2702-10,1985.

(9) M.F. MELO, J.S. MOYA, P. PENA. S. DE AZA
"Multicomponent Toughened Ceramic Materials Obtained by
Reaction Sintering". Part III : System Zirconia-alumina-
silica-titania". J. Mater. Sci. 20, pp 2711-18,1985

(10) P. BOCH, J.P. GIRY
"Preparation and properties of reaction-sintered mullite-
zirconia Ceramics" - Mater. Sci. Eng. 71, p 39-48 (1985)

(11) P. PILATE, P. DESCAMPS, F. CAMBIER
" Fabrication of high temperature resistant ceramic using
ultra rapidly quenched powders" - 7th Simcer. International
Symposium on Ceramics - Bologna (Italie) 14-17 décembre 1988.

(12) A. LERICHE, M. VIVEY, F. CAMBIER "Preparation of
     Mullite-Zirconia Ceramics by Reaction Sintering"
     Study of their Thermomechanical Properties" contrat
     n° SUT-116-B. Programme de Recherche de CEE "Materials
     Sector-Substitution and Materials Technologies and
     Ceramics 1982-85."

(13) M. DELETTER, A. LERICHE and F. CAMBIER "A Linear
     Model for Both Qualitative and Quantitative X-Ray
     Analysis" submitted to publication in Powder Diffraction

(14) A. LERICHE - "Influence des Paramètres d'Elaboration
     de Composites Mullite-Zircone sur leur Microstructure".
     Ph.D.. Thesis - Université de l'Etat à Mons -
     Belgium - 1986.

(15) H.K. SCHMID "Quantitative Analysis of Polymorphic
     Mixes of Zirconia by X-Ray Diffraction." J. Am. Ceram.
     Soc. 70   [5] 367-76 (1987).

(16) A. JAYATILAKA "Fracture of Engineering brittle
     Materials". Ed. Appl. Sci. Pub. 1979

(17) W. Grellner. H. HUBNER, B. ILSCHNER and F.W. KLEINLEIN
     "On High Temperature Strength of a two-phase Al_2O_3-
     base material". Science of Ceramics 10, Ed. H. Hausner
     Publ. Deutsche Keramische Gesellschaft p 513-519 (1980)

(18) G. ORANGE, G. FANTOZZI, F. CAMBIER, C. LEBLUD,
     M.R. ANSEAU, A. LERICHE "High temperature Mechanical
     Properties of Reaction Sintered Mullite-Zirconia and
     Mullite-Alumina-Zirconia Composites" J. Mater. Sci.
     20, pp 2533-40 (1985).

(19) G. URBAIN "Viscosité et structure de silicoalumineux
     liquides" Rev. Int. Htes Temp. et Réfract. pp 133-145,
     1974

(20) J.M. RINCON, J.S. MOYA, M.F. DE MELO "Microstructural
     Study of Toughened ZrO_2/Mullite Ceramic Composites
     obtained by Reaction Sintering with TiO_2 Additions"
     Br. Ceram. Trans. J. 85, pp 201-206, 1986.

Crack resistance curve of ZrO2/Al2O3
with long natural sharp cracks

T.Liu, G.Grathwohl, Institut fuer Werkstoffkunde-2
Universitaet Karlsruhe, F.R.Germany

Abstract

Crack resistance curve of two ZrO2/Al2O3 with very fine
grain size was investigated in 3 point bending. Long
natural sharp crack by bridge technique was used as
precrack. Crack extension was measured with electric
potential method. The specimens were tested in two conditions i.e. with and without the transformation zone
around the crack tip after the bridge precracking.The
results were analysed with both stress and energy con-
cept.A falling R-curve and a maximum of the R-curve instead of an increasing R-curve were measured in both of
the materials. Experimental results indicate also a
marked difference between stress derived and energy derived fracture toughness. The reults were discussed
with the stress induced phase transformation and the
subcritical crack growth.

1 Introduction

Many ceramic materials show the so called R-curve behavior: an increasing crack resistance is measured when
the crack is propogating through the material. Several
mechanisms are discussed in order to explain this effect, for example, wake effect /1/, crack branching /2/
and transformation toughening /3,4/ for Mg,Ca-PSZ. It
is one of the intention of this work to measure the R-
curve of fine grained Y2O3-stabilized ZrO2/Al2O3 materials. Because of its small grain size a new technique to
introduce natural sharp cracks had to be applied. Since
these ceramics suffer considerably from stress corrosion /5,6/, it is important to take the process of subcritical crack growth into account.

2 Experimental

The natural sharp cracks were produced by the bridge
technique /7/. Crack extension was measured with the
electric potential method. The insulating ZrO2/Al2O3
bending specimen with the bridge precrack was coated
on one side with a conductive TiC-layer by physical
vapour deposition. The crack extension could then be
determined in the 3-point bending test by measuring
the change of potential as shown in Fig.1. The accu-

PVD – TiC-layer

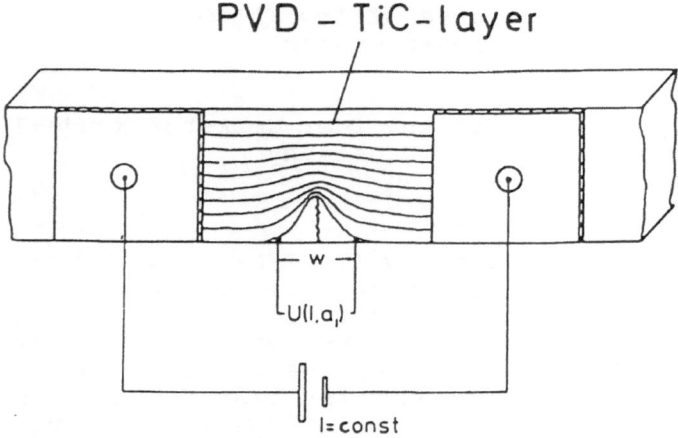

Fig.1 Electric potential method for
crack length measurement

racy was about ± 20 μm.Several specimens were annealed in vacuum after precracking, so that the transformation zone around the crack tip disappeared.The test
procedure and the analysis are shown in Fig.2. The results were analysed with both stress and energy con-
cept. Several properties of the investigated materials
are shown in table 1.

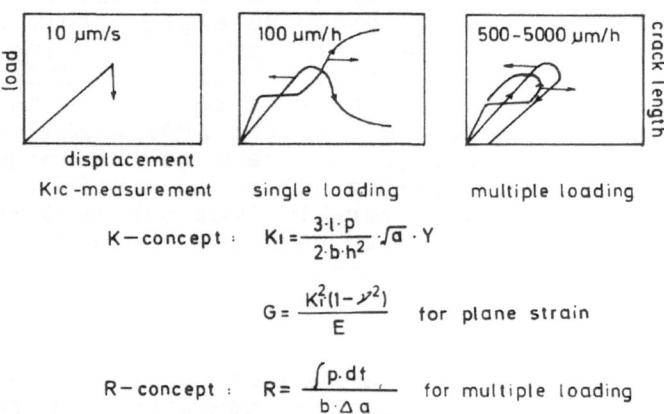

Fig.2 Test procedure and analysis

materials	Y-PSZ	TZ-3Y20A
phase at sin-tered surface	80% t + 20% c	20% Al2O3 + 80% TZP
phase at frac-ture surface	32% m + 48% t + 20% c	20% Al2O3 + 10% m + 70% t
grain size (μm)	t: 0.3; c: 3	Al2O3: 0.4 ZrO2: 0.4
density (g/cm^3)	5.81	5.45
E-modulus (GPa)	223	251
poisson's ratio	0.3	0.3
strength (MPa)	640	780
MI-K_{IC}* (MPa$\sqrt{m}$)	5.7	5.2
K_{IC}** (MPa$\sqrt{m}$)	4.9	4.6

    *   : microindentation method

    ** : 3 point bending with loading rate of 10 μm/s
           (precracking by bridge method)

Tab.1 Characteristics of the investigated materials

3 Results and discussion

Schematically Fig.3 shows the load-load point displacement curves of both materials in different states. In
both conditions an offset of the unloading line from
the origin has been observed, which is caused by the
inelastic strain due to the further stress induced
phase transformation during the crack extension. Specimens with a transformation zone reveal further a discrete point where the slope of the loading-unloading
curves changes. It is supposed that this point corresponds with the complete opening of the bridge precrack
which is only possible after the compensation of the
residual stresses within the transformation zone.If
this transformation zone disappears during a high temperature annealing process, the loading and unloading
line becomes again linear. Fig.4 shows two stress derived crack resistance curves of the Y-PSZ material.
Three characteristics are significant: First: slower
loading rate leads to lower K1-values. Second: the
crack resistance curve goes through a maximum in the

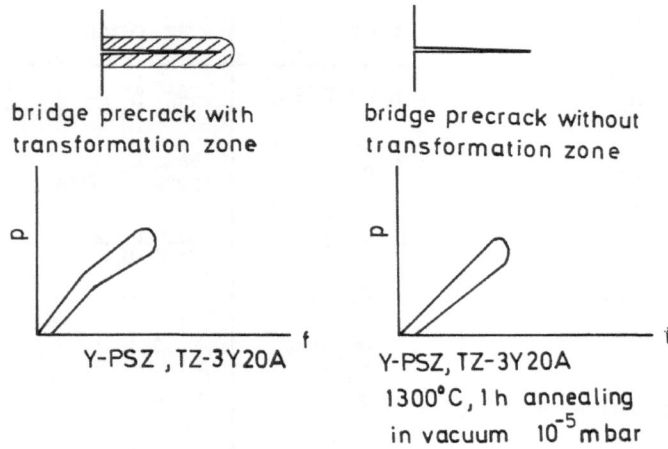

bridge precrack with   bridge precrack without
transformation zone   transformation zone

Y-PSZ , TZ-3Y20A   Y-PSZ, TZ-3Y20A
                              1300°C, 1 h annealing
                              in vacuum 10^{-5} m bar

Fig.3 Types of load-load point displacement curves

single loading test. Third: maximum K1-values can also be observed in the individual loading/unloading cycles in the multiple loading tests, again with the
tendency to a falling R-curve at longer cracks.These
observations can be related to the subcritical crack
growth. The stable crack growth in air is always a-
ccompanied by the subcritical crack growth. The slower the loading rate, the longer the time during which
the crack can grow under subcritical condition. This
is why lower K1-values are measured at slower loading

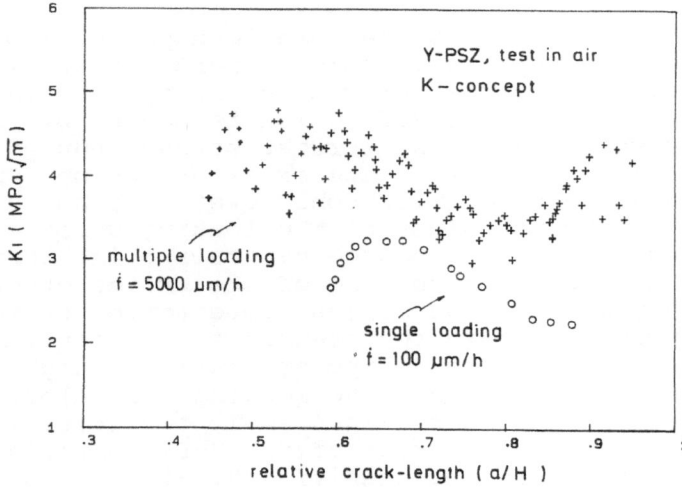

Fig.4 Stress derived crack resistance curves of Y-PSZ

rates. The same reason is found to be responsible for
the type of these curves. The initial fracture toughness starts at a lower level, because the crack grows
already at relatively small applied stress intensity
factors due to subcritical crack growth. The crack
length and velocity are then increasing which leads to
higher measured K1-values. Finally, at longer cracks
the crack velocity decreases again and smaller K1-va-
lues are measured. In the multiple loading test the
overall crack resistance is rather similar in its profile to the single loading test. The maximum in each
loading/unloading cycle is also due to the subcritical
crack growth. From this reason tests were conducted in
vacuum in order to exclude this subcritical crack growth due to environmental attack. The results are shown
in Fig.5. The K1-values measured in vacuum are nearly

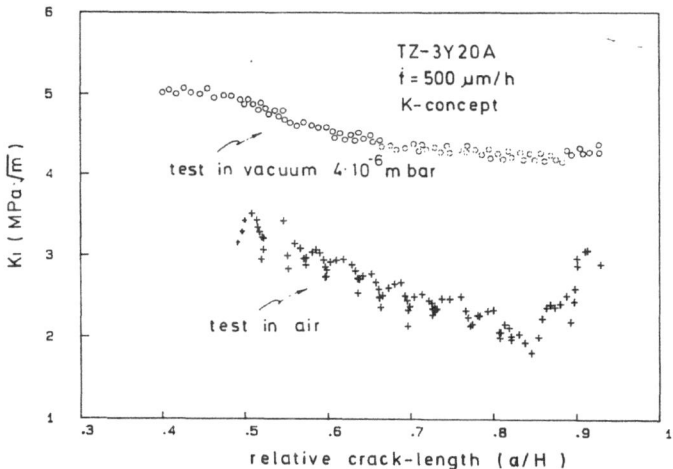

Fig.5 Stress derived crack resistance
curves of TZ-3Y20A

twice the values in air. Also the individual maxima
disappear in the vacuum curve and the K1-values are
less decreased at longer cracks. The stress derived
crack resistance is compared with the energy derived
crack resistance as shown in Fig.6. Higher crack resistance values are obtained according to the R-concept
which is caused by the irreversible processes during
the fracture process due to the stress induced phase
transformation. While the K-concept represents the fracture toughness under the assumption of linear elastic
theory, the energy derived crack resistance is the sum
of this linear elastic energy with additional energy

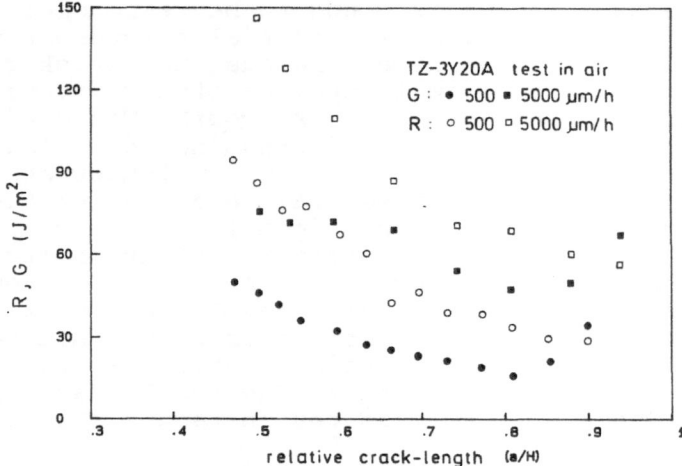

Fig.6 Stress derived and energy derived crack resis-
      tance curves of TZ-3Y20A with transformation
      zone around the bridge precrack

terms of the irreversible processes. If the specimen
was annealed in vacuum after precracking, where the
transformation zone around the crack tip disappears,
we get similar tendency of the R-curve (Fig.7) to the
specimen with the transformation zone (Fig.6), except
that the difference between the two concepts is much

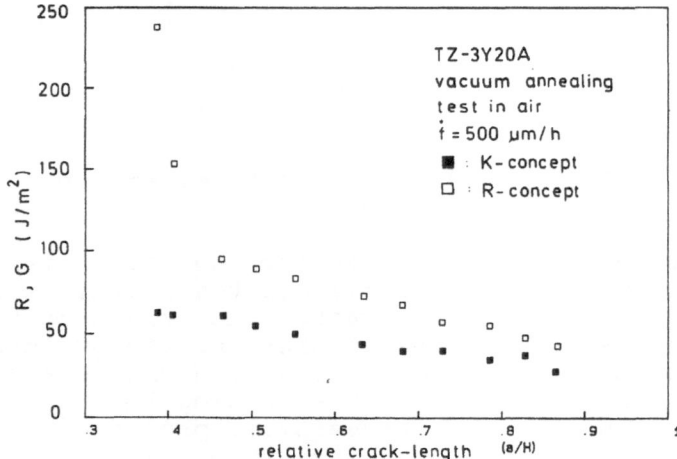

Fig.7 Stress derived and energy derived crack resis-
      tance curves of TZ-3Y20A without transformation
      zone around the bridge precrack

larger, particularly at the initiation of the crack extension.It is believed that this larger difference is again caused by the irreversible processes during the fracture process due to the stress induced phase transformation, which is more significant at the initiation of the crack extension with the formation of the transformation zone.

4 Conclusion

ZrO2(Y)/Al2O3 materials with very small grain size do not exhibit any increasing R-curve behavior. The R-curve measurement is strongly influenced by the subcritical crack growth. This is shown clearly from the dependence of the crack resistance on the loading rate and atmosphere. The measured falling R-curve as well as the maximum of the R-curve is also supposed to be due to the subcritical crack growth. Experimental results indicate a large difference between stress derived and energy derived crack resistance, which is caused by the irreversible processes during the fracture process due to the stress induced phase transformation in combination with the inelastic strain.

References

/1/ R.Steinbrech, R.Knehans, and W.Schaarwaechter "Increase of crack resistance during slow crack growth in Al2O3 bend specimens" J. Mater. Sci. 18 (1983) 265-270

/2/ M.Sakai, J-I.Yoshimura, Y.Goto, and M.Inagaki "R-Curve Behavior of a Polycrystalline Graphite: Microcracking and Grain Bridging in the Wake Region" J. Am. Ceram.Soc. 71 (8) (1988) 609-16

/3/ D.B.Marshall, A.G.Evans,and M.Drory "Transformation Toughening in Ceramics" Fracture Mechanics of Ceramics, Vol.6 (1983) 289-307

/4/ L.R.F.Rose, and M.V.Swain "Two R Curves for Partially Stabilized Zirconia" J. Am. Ceram. Soc. 69 (3) (1986) 203-207

/5/ L.Li, and R.F.Pabst "Subcritical crack growth in partially stabilized zirconia (PSZ)" J. Mater. Sci. 15 (1980) 2861-2866

/6/ J.E.Ritter, and J.N.Humenik "Static and dynamic fatigue of polycrystalline alumina" J. Mater. Sci. 14 (1979) 626-632

/7/ R.Warren, and B.Johannesson "Creation of stable cracks in hardmetals using bridge indentation" Powder Metallurgy Vol.27 No.1 (1984) 25-29

ALUMINA - ZIRCONIA CERAMICS:
PREPARATION AND PROPERTIES

E. Lucchini and S. Maschio
Istituto di Chimica Applicata ed Industriale
Università di Trieste - ITALY

INTRODUCTION

The flo-deflocculation method of ceramics processing which takes advantage of the electrostatic repulsion forces between the oxides particles in an aqueous environment is very attractive, becoming an effective route leading to the preparation of alumina zirconia composites containing a relative large amount of micronized zirconia inclusions.

There is a critical value of the zirconia particles size below which it is possible to keep them (t) tetragonal in the alumina matrix at room temperature.[2] The toughening mechanism related to the presence of this metastable phase, susceptible of a stress induced transformation (t-m), is more effective than the one based on the microcraking introduced by the spontaneous t-m transformation.[3]

In an aqueous suspension by changing the pH it is possible to positively or negatively charge the particles surfaces, defloculate the powders and break down the aggregates. By further pH adjustment, the so called "zero point" of charges can be eventually met and the floculation of the powders becomes feasible. By this mean the previously dispersed homogeneous particles distributio may become frozen down for further processing. In this manner well mixed powders without large aggregates can be obtained without tedious and contaminant mixing.

In this paper the influence of some parameters of the flo-defloculation process on the properties of alumina zirconia composites is investigated.

EXPERIMENTALS

In a previous paper[1] we have reported that the best results could be obtained starting from suspensions containing 2.5 Vol. % of solid phase, therefore this precentage was mantained through all the experiments performed in this work.

Weighed quantities of alumina (ALCOA A16) and two commercial zirconia powders, hereafter named A and B were dispersed in a "single-mix" by vigorous stirring for ten minutes in a water-hidrochloric acid solution at pH=2.

After some rest time the pH was controlled, eventually restored and the suspension stirred again for ten minutes. The powders were flocculated by adjusting the pH to 8 by NH_4OH additions.

The supernatant was then poured off, the remnant solids were centrifuged as a "cake", washed with generous water in order to remove the chloride ions.

The quantity of zirconia in the composites (15 %vol.) was fixed taking into account the fact that higher amounts of non-doped ZrO_2 lead to materials with a low percentage of tetragonal phase [2].

After drying the powders were pressed in bars (pressure = 110 MPa, cross section 25 mm^2) and sintered at 1580°C for one hour. The percentages of untransformed tetragonal phase were derived from the X-ray intensity measurements according to the Garvie Nicholson equantion [4]. Polished and thermally etched samples were examined by SEM in order to control the microstructure .

The bend strength was determined by the four point method on "as fired" samples.

The K_{Ic} was also determined by the four point method on polished and indented specimens (Vickers penetrator) using the equation proposed by Chantikul et al.[5].

The densities of the sintered bodies were obtained from water displacement measurements.

RESULTS AND DISCUSSION

In Figg. 1 and 2 the percentages of tetragonal zirconia, determined on the sintered bodies, as a function of the suspensions rest time before deflocculation, are reported. As one can see the percentage of tetragonal ZrO_2 initially increases until a "maximum" is reached after two hours, successively the content of the unstable phase decreases until a steady state is reached after 10 hours.

The percentage of the tetragonal phase in the composites is an evidence of the electrostatic repulsive forces between the particles (or at least between the ZrO_2 particles) in the suspensions.

In fact in order to obtain at room temperature ZrO_2 inclusions in the tetragonal form, the large aggregates formed by the Van der Waals attractive forces must be destroyed, therefore the electrostatic repulsive energy between the particles in the suspensions have changed with time as indicated by the different contents of metastable ZrO_2 in the composites. In order to explain this behavior the phenomena related to the interactions between particles' surfaces and solutions must be considered in a deeper detail.

Parks and de Bruyn [6] have suggested that the charging of the surface of an oxide in water occurs through the formation of hydroxo complexes, according to the following equilibrium.

In our case at pH=2 surfaces of both alumina and zirconia particles are positively charged.

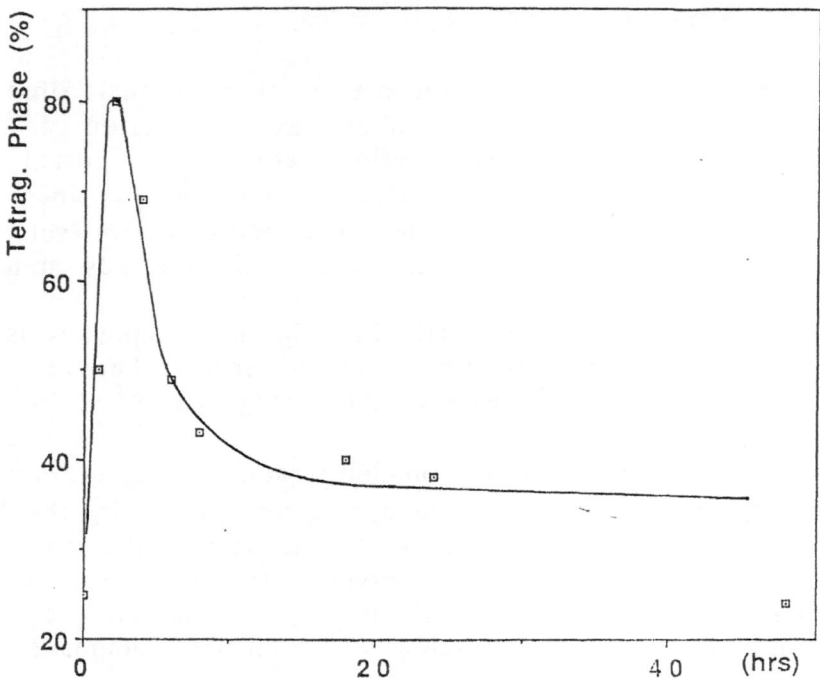

Fig. 1: Tetragonal fraction as a function of the rest time for samples prepared by ALCOA A16 and A-ZrO$_2$.

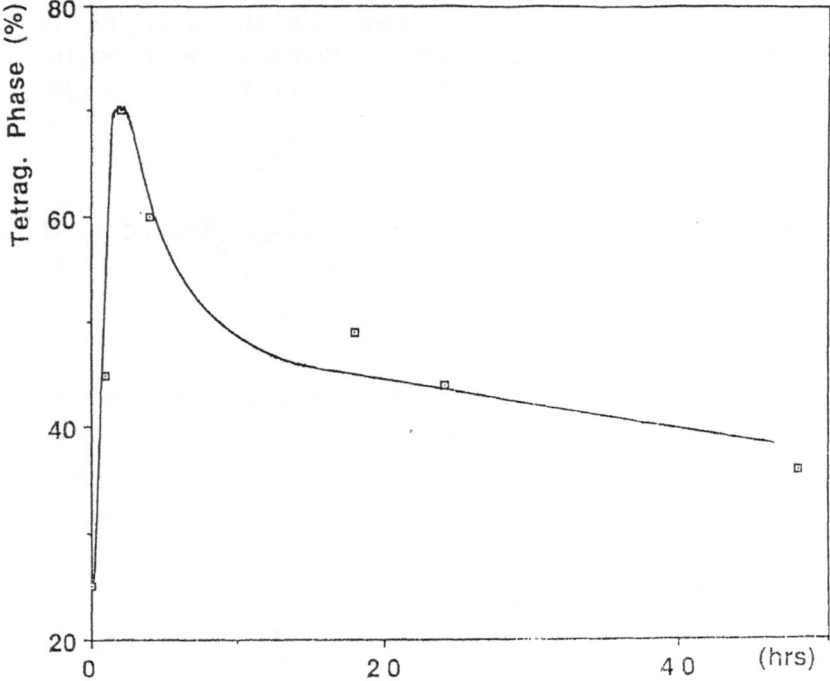

Fig. 2: Tetragonal fraction as a function of the rest time for samples prepared by ALCOA A16 and B-ZrO$_2$.

The whole system is electrically neutral, therefore the surface charges must be exactly balanced by the opposites ones in the liquid phase.

Because of the coulombic forces the counterions (in our case essentially Cl^-) will tend to concentrate in the vicinity of the solid surface. A resulting double layer will be formed by the charged solid surface's layer and by a layer of counterions in the adjacent solution.

The concentrations of counterions will be higher near to the particles surfaces (Stern layer) and lower at the back diffuse zone (Gouy layer).

When two particles with the double layers of the same sign approach each other the two layers begin to interfere to give rise to repulsion forces between particles.

If the particles are spherical forms the repulsion energy is given by the Verwey and Overbeek equation [7]:

$$V_R = \frac{\varepsilon \; a^2 \Psi_0^2}{H_0 + 2a} \beta \quad exp(-KH_0)$$

where ε is the dielectric constant of the medium, a is the particles radius, H_0 the distance between the centers of the two spheres, Ψ_0 is the surface potential, β is a factor which allows for the loss of spherical simmetry and K is the Debye-Hüeckel parameter given by the expression:

$$K^2 = \left[\frac{4\Pi \; e^2}{\varepsilon \; kT} \right] \sum_i n_i z_i^2$$

where k is the Boltzmann constant, n_i the number of ions per cm^3, Z the valency of counter ions and e the electronic charge.

If the interactions are relatively strong Ψ_0 has to be replaced by the Stern potential Ψ_D. In practical terms, the Stern potential is usually replaced by the electrokinetic of zeta (ζ) potential that can be evaluated, at different pH values, by measurements of the electrophoretic mobility of the disperse particles.

The total potential energy is the sum of repulsive and attractive terms:

$$V_T = V_R + V_A$$

From the Verwey and Overbeek equation it is clear that V_R depends on Ψ_0 (or Ψ_D or zeta potential) which in turn depends on σ_0 (charge density on the particle surfaces). The V_R depends also on the particles radius and decreases exponentially with the ions concentration.

The initial increase of the tetragonal zirconia with the rest time can be attributed to a relatively slow formation of an hydroxil layer on the ZrO_2 particles and therefore to a slow increase of Ψ_0 (or Ψ_D or zeta potential). Therefore the term V_R the total potential energy balance is increased and the powders aggregates are broken down allowing the obtainment of a high quantity of zirconia inclusion below the critical value and therefore a high quantity of ZrO_2 in the tetragonal form.

After a maximum, reached after two hours, the increase of the rest time causes a decrease of the tetragonal zirconia. In the previous[1] work we have attributed this trend to a "crowding" effect of the counterions. This hypothesis can be explained also taking into account the Verwey and Overbeek equation. During the rest time the counterions migrate in the diffuse layer and therefore the Debye-Hueckel parameter K is increased. As a consequence the term V_R is decreased and the tendency of the ZrO_2 particles to the reaggregation is raised.

From figg. 1 and 2 it is also possible to argue that the percentage of tetragonal zirconia in the composites prepared with B powders is lower than that obtained with A powders. This behavior is not unaspected because there is a large difference in BET surfaces between ZrO_2 powders: A= 28 m^2/g and B= 50-60 m^2/g The densities and the mechanical properties of some representation samples are reported in table I.

Table I

A Powders

% Tetr. ZrO_2	Density(%)	K_{Ic} (MPa m)	σ (MPa)
80	97	5.5	340
23	96	4.5	250

B Powders

70	98	4.6	260
60	98	4.5	245
44	95.5	3.5	140

The densities decrease with the increase of the content of the monoclinic phase. This trend can be attributed to the presence of a large quantity of microcracks due to the tetragonal-monoclinictransformation and /or to a non homogeneous distribution of ZrO_2 in the microstructure. In fact the presence of well dispersed zirconia inclusions inhibits the grain growth and promote the sintering processes.

The mechanical properties of the composites prepared with ZrO_2 B were generally inferior respect those prepared with ZrO_2 A containing approximatively the same percentage of the tetragonal phase.

The examination of the microstructure have allowed to explain this behavior. In fact, the zirconia inclusions, in the samples prepared with ZrO_2 B, are formed by large particles

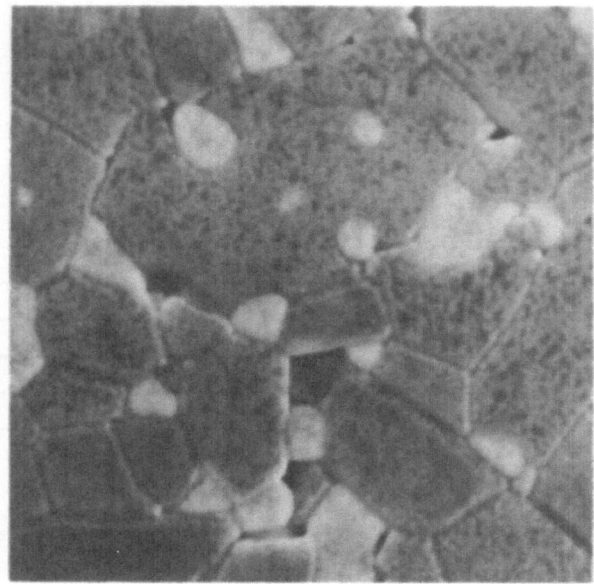

Fig. 3: SEM micrography of a polished and thermally etched sample, prepared by Alcoa A16 and B-ZrO$_2$ powders.

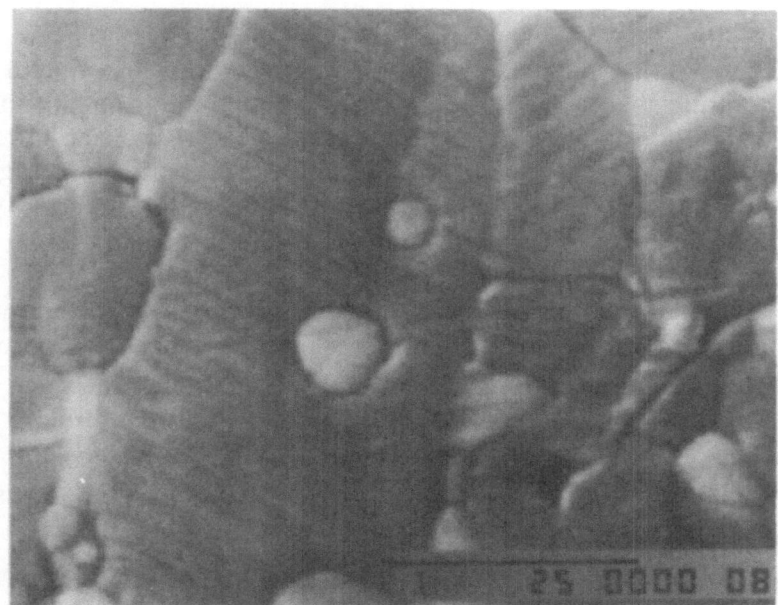

Fig. 4: SEM micrography of a polished and thermally etched sample prepared via double-mix procedure.

(monoclinic) and by very small particles (tetragonal) many of them found not BETWEEN the alumina grains but IN the alumina grains (see fig. 3). In this latter case the tetragonal inclusions are very stable and do not transform when interact with a propagating crack, therefore they do not participate to the toughening mechanism.

The toughness of the resulting composites is due essentially to the presence of microcracks around the transformed monoclinic inclusions.

The hypotheses that the poor mechanical properties of the composites prepared with B powders are due to the very small zirconia inclusions was confirmed by the following experiences.

The composites preparation method was modified: according to a double-mix procedure, the ZrO_2-A and the alumina powders were dispersed separately in two different beakers with the above mentioned procedures. After a rest time of 24 hours the supernatants were pured out, mixed in suitable ratios and flocculated. In this manner the large aggregates and particles present in both powders were removed. The characteristics of the composite prepared with these powders are reported in tab.II.

Table II

% Tetr. ZrO_2	Density %	K_{Ic} (MPa m)	σ (MPa)
100	96.5	4.5	300

By comparing tables I and II, it is possible to see, in the latter, the zirconia is completely in the metastable form but its mechanical properties are lower than those obtained for the composite containing only 80% of the tetragonal phase, prepared with the same powders, but according to the other single-mix procedure. The analysis of the microstructure (Fig.4) confirmed that a high quantity of zirconia inclusions are in intragranular positions, therefore in a very stable condition.

CONCLUSION

The flo-deflocculation method allows the obtainment of alumina-zirconia composites with good mechanical characteristics without long and contaminant mixings.

The rest time before the deflocculation is an important step in the powders treatment and strongly influences the microstructure and mechanical properties of the composites.

The use of very fine powders does not increase the amount of tetragonal zirconia and the mechanical properties because, at the same time large aggregates and very small particles are present in the flocculated powders are present.

The removal of large aggregates by sedimentation is also not effective because the very small zirconia tetragonal inclusions are very stable and do not transform when they interact with propagating cracks.

REFERENCES

1) Lucchini E. and Maschio S. , 2nd Conf. Eng. Mat., Bol., June 88, in Press.
2) Lange F.F., J.Mat.Science, Vol 17, p.247, (1982)
3) Faber K. T., Advances in Ceramics Vol. 12, P. 293, (1983)
4)Garvie R.C. and Nicholson P.S., J of Am. Ceram. Soc., Vol5, p. 303, (1972)
5) Chantikul P.,Anstis G.R. and Lawn B.R.and Marshall D.B., J. of Am. Ceram. Soc., Vol. 64, 539-43, (1981)
6) Parks G.A. and De Bruyn P.L., J Phys. Chem. ,Vol. 66, p. 967,(1962)
7) Verweg E.S.W. and Overbeek J.Th.G., in Theory of the Stability of Lyophobic Colloids, Elsevier Publ, Co, Amsterdam, (1948)

SINTERING AIDS FOR Ce-TZP

S. Maschio*, E. Bischoff°, O. Sbaizero*, S. Meriani*

*Istituto di Chimica Applicata e Industriale
Università di Trieste - ITALY

°Max Planck Institut fur Metallforschung
Institut fur Werkstoffwissenschaften
Stuttgart - F.R. GERMANY

ABSTRACT

Ceria-stabilized zirconia seems to become as interesting as the more studied yttria-zirconia system. Its critical grain size is larger and it does not undergo low temperature moisture degradation; but still its grain size control remains very important because the mechanical properties are strongly influenced by the size of the tetragonal grains.
In this work additives for the pressureless sintering of Ce-TZP, have been studied. The temperature at which the shrinkage begun did not decrease using MnO_2 or CuO, but at the same temperature the sintering rate was higher; MnO_2 and CuO were also very effective in improving both the tetragonal phase content retained at room temperature and the mechanical properties (hardness, bending strength and toughness)

1. INTRODUCTION

The ceria-zirconia system is being developed as an alternative to the more investigated yttria-zirconia because of its better performance in a moist environment[1], its lower price, the wider range of solubility in ZrO_2 and the very good mechanical properties[2]. Furthermore it is interesting for the presence of two tetragonal phases[3] which might give an additional improvement to the mechanical properties through crack deflection mechanism.
Unfortunately the use of CeO_2 showed a limited sinterability and even the use of expensive methods like hot- or hot-isostatic pressing to fully densify the poor sinterable material were not suitable because Ce is reduced from Ce^{+4} to Ce^{+3} during the process and zirconia undergoes the tetragonal (t) to monoclinic (m) transformation.
The aim of this investigation is to find out the effectiveness of additives (CuO, MnO_2) for the sintering of Ce-TZP according to a pressureless process, keeping in mind that in order to get outstanding mechanical properties the material must contain almost 100% of tetragonal phase and

the density must be as close as possible to the theoretical density to avoid the presence of defects which may provide easy routes for crack initiation.

2. METHODS

0.05 M solution of $CeCl_3 \cdot nH_2O$ was used as starting material. Cerium carbonate was precipitated by adding urea and then boiling the solution for 3 hrs. The white salt was washed with H_2O, dried and calcined at 750 °C.

The resulting oxide was granulated through 75 μm screen and mixed with ZrO_2 supplied by Toyo Soda in order to give a 12 mol% CeO_2-TZP.

The powders were mixed and homogenized by the flo-deflocculation route[4] using HCl for acidification and NH_3 solution for flocculation at pH 8. After flocculation the clear supernatant was removed and the wet powder was washed several times using distilled water, then dried.

In order to study the influence of additives on the sinterability, the dried Ce-TZP powder was divided in three sets:
- As-fabricated powder, labelled Ce-TZP
- As-fabricated powder + CuO (0.3 mol%), labelled Ce-TZP+Cu
- As-fabricated powder + MnO_2 (0.3 mol%), labelled Ce-TZP+Mn

Powders were ball milled 30 hrs with binder and lubricant, dried, granulated through 75 μm screen, uniaxially pressed at 200 MPa to form pellets 10 mm in diameter.

The green compacts, about 45% of the theoretical density, were sintered in air at 1450 and 1530 °C, respectively, for various periods of time.

Dilatometric tests were performed with a Netzsch dilatometer in air and oxygen, with a heating rate of 10°K/min.

The following measurements were made on the sintered bodies:

Archimede's method to measure densities,

X-ray diffraction for phases identification and the Garvie and Nicholson method[5] to assess the tetragonal to monoclinic ratio,

SEM pictures and the intercept method to estimate the average grain size,

SEM and TEM analysis to examine the microstructure as well as the additives' influence.

Modulus of rupture measurements were made using a 4-point bend jig with 30 mm outer span, 10 mm inner span and a cross-head speed of 0.1 mm/min.

Hardness was measured by Vicker's indentation with loads ranging from 90 to 200 N

The fracture toughness was measured by indentation strength in bending (ISB) technique using a Vicker's indenter, applying 200 N load, then breaking the specimen by the 4-points bending test above described.

3. RESULTS AND DISCUSSION

Fig. 1 shows the sintering plots for the different powders. They behaved in the same way as regards the thermal expansion of the pressed samples.

Moreover the onset shrinkage temperature, 1120 °C, remained unchanged by the addition of either Cu or Mn but the powders behaved differently after the initial shrinkage stage; the Ce-TZP displayed a two step densification (at 1220 °C) whereas powders with either CuO or MnO_2 densified in a single step. SEM analysis on Ce-TZP samples fired at 1220 °C revealed that the particles were already sintered leaving large intergranular voids which were difficult to eliminate hindering further densification. Continuing the heating it is possible to increase the shrinkage but at the same time anomalous grain-growth takes place.

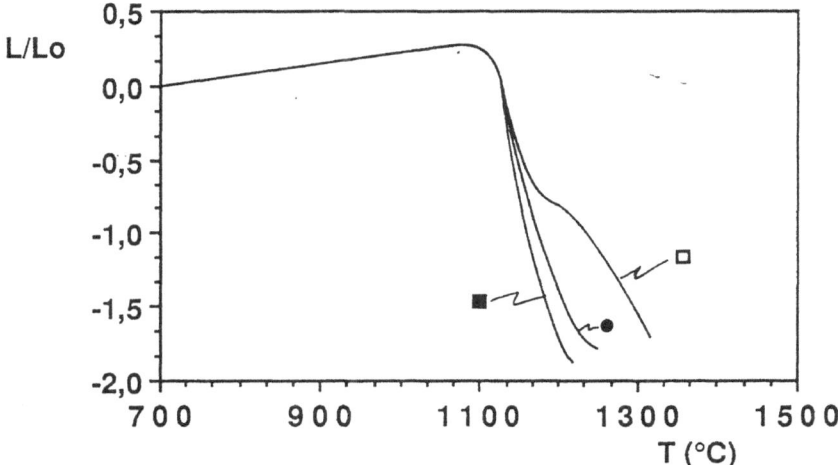

Fig. 1 Dilatometric curves for the different starting powders, (□) Ce-TZP, (■)Ce-TZP+Cu, (●) Ce-TZP+Mn.

Densification curves for the pressureless sintering in air at 1450, 1530°C are reported in Fig. 2, (separate samples were used to obtain the individual data points); the Ce-TZP compact required very high temperature and long time to reach good density while either CuO or MnO_2 additions enhanced densification as well as the amount of tetragonal phase retained at room temperature. CuO had the strongest effect whereas MnO_2 was slightly less effective but both remarkably changed the amount of tetragonal phase i.e. after 2 hours at 1450 °C a fully tetragonal body could be obtained.

From the Fig. 2 it is possible to see that for samples sintered either at 1450 or 1530°C the maximum in the density was reached after 4 hours at 1450 and 2 hours at 1530°C, afterwards the amount of tetragonal phase increased (in the case of undoped material) but the density slightly decreased.

Fig. 3 shows the grain growth kinetics of Ce-TZP, for a constant temperature, with and without dopants, it can be seen that over the range of time considered samples without dopants had a grain size about twofold larger than those doped. For example in this log-log plot the slope for the Ce-TZP is 0.54, whereas for the Cu doped samples is 0.27 and for the Mn doped is 0.20.

The grain growth kinetics can be expressed as:

$$G^n(t) - G^n(0) = k\,t \tag{1}$$

where G(t) is the grain size after sintering for time, t; G(0) is the grain size extrapolated for time t=0; k is a proportionality constant and n is the growth exponent. For Ce-TZP n≈1.9 whereas Cu and Mn dopants rose the growth exponent, n, to ≈3.5 and ≈4.5 respectively.

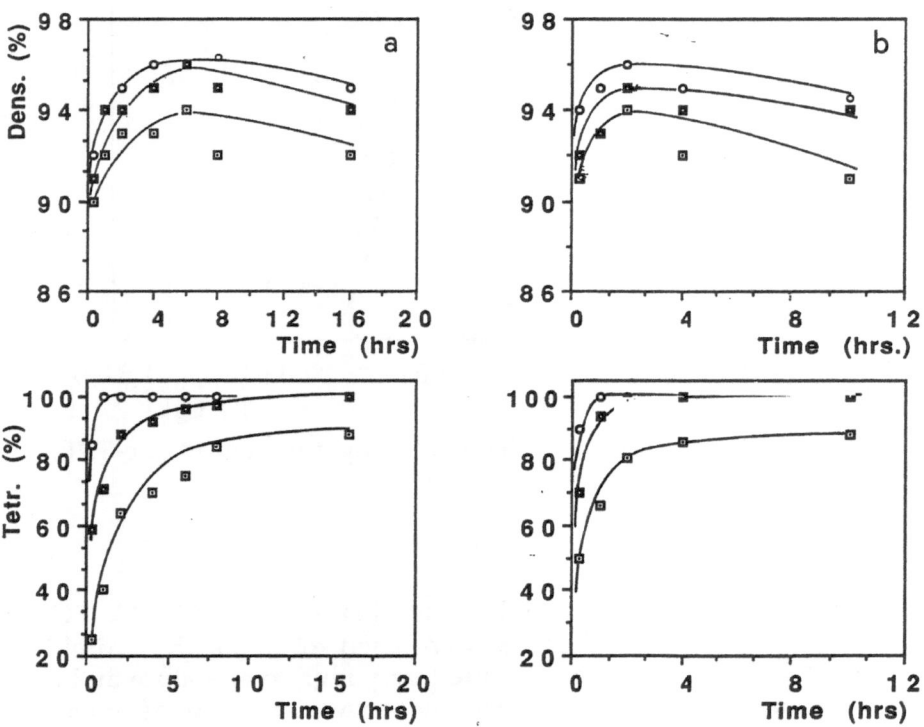

Fig. 2 Density and amount of tetragonal phase of sintered samples vs sintering time for: (a) 1450 °C, (b) 1530 °C, (□) Ce-TZP,(o)Ce-TZP+Cu, (■) Ce-TZP+Mn.

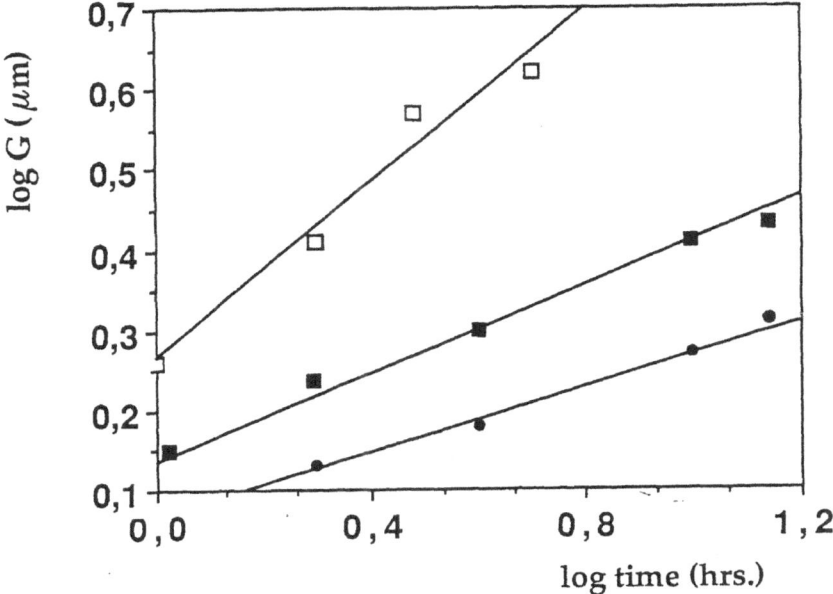

Fig. 3 Grain growth kinetics of Ce-TZP samples showed by log-log plot of grain size vs sintering time, (□) Ce-TZP, (●) Ce-TZP+Mn, (■) Ce-TZP+Cu

Fig. 4 shows the microstructure of samples sintered for 4 hours at 1530 °C, the resulting microstructure was made of fine and homogeneous grains 4-5 μm for Ce-TZP, 2 μm when CuO was used and 1.4 μm with MnO_2.

TEM analysis, Fig. 5, revealed the presence of liquid phase at the triple point grain boundaries (pockets) only when Cu was used due to the fact that CuO transforms into Cu_2O which then melts at 1236°C, whereas we were unable to spot Mn-rich areas. Therefore experimental results showed that under the same sintering conditions, small amounts of CuO could improve the sintering of Ce-TZP more effectively that MnO_2 but the latter was stronger in depleting the rate of grain growth. It is hard to explain the reasons of these different effects but the dopants influence on the densification should be discussed on the basis of the defect forming ability of the ions available after the first sintering stage common to both the pure and doped materials. Both CuO and MnO_2 undergo a high temperature reduction forming Cu_2O and $MnO_2 \cdot Mn_2O_3$ respectively. New ionic species like Cu^{+1} and Mn^{+2} become available at 1200°C to possibly interact with the sinter-reactive ceria and zirconia particles which, being both tetravalent, do not introduce oxygen vacancies in the crystal lattice.

It has been assumed that in stabilized zirconias the defect structure is fixed by the amount of the aliovalent stabilizers as Ca^{+2} and Y^{+3}. Any other dopant present in small quantity cannot possibly modify the defects' number and structure of the host matrix nor the lattice diffusion

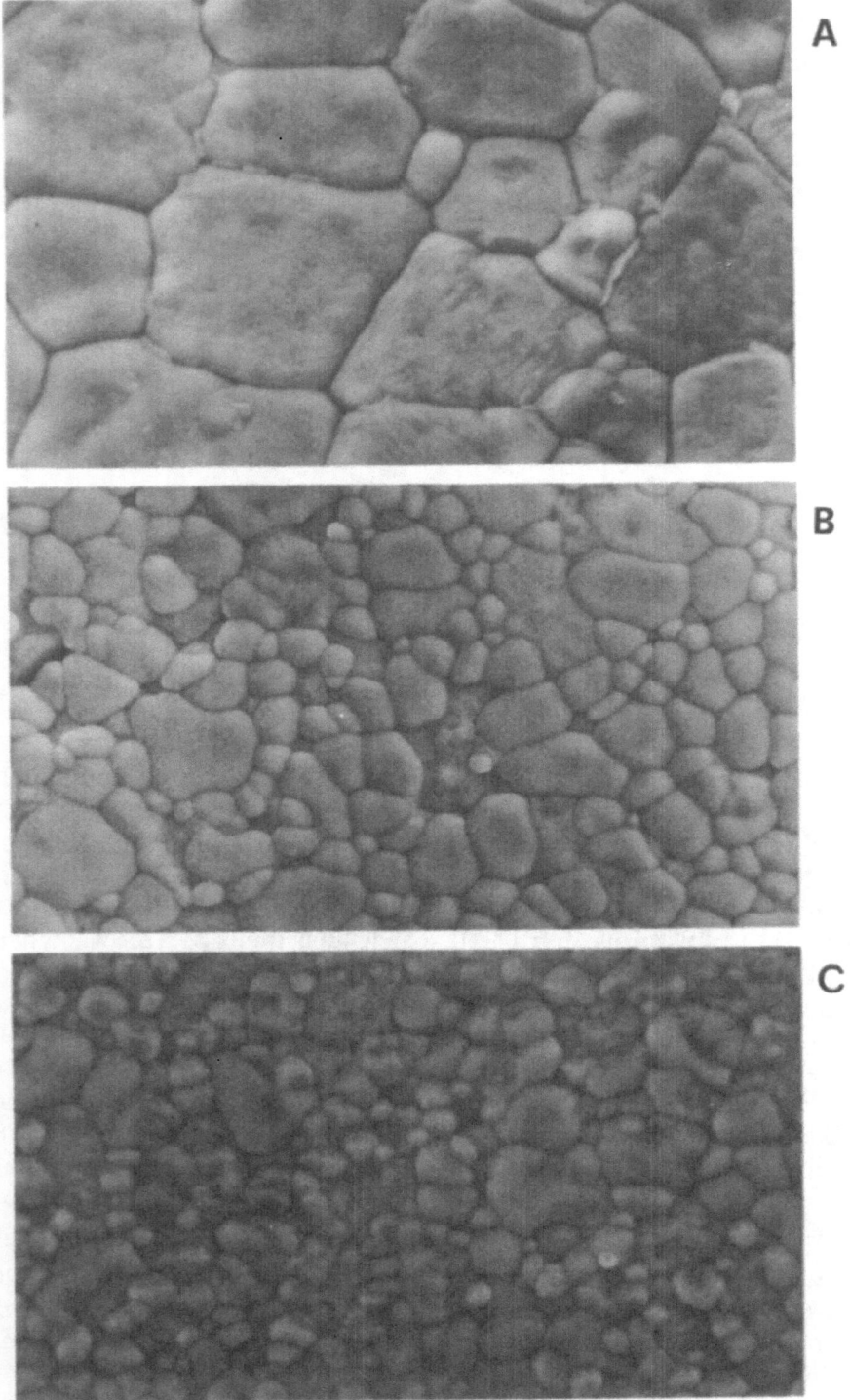

Fig. 4   SEM pictures of the Ce-TZP materials sintered at 1530 °C
for 4 hours, (A) Ce-TZP, (B) Ce-TZP+Cu, (C) Ce-TZP+Mn

coefficient of the controlling species. But this should not be the case in ceria-zirconia materials where the stabilizer CeO_2 has the same stoichiometry as the host ZrO_2 matrix at least up to very high temperature when it is reduced to Ce^{3+} introducing oxygen vacancies to balance its lower oxidation state. In this case the dopants become relevant in introducing lattice defects and enhancing cation diffusion through oxygen vacancies formation.

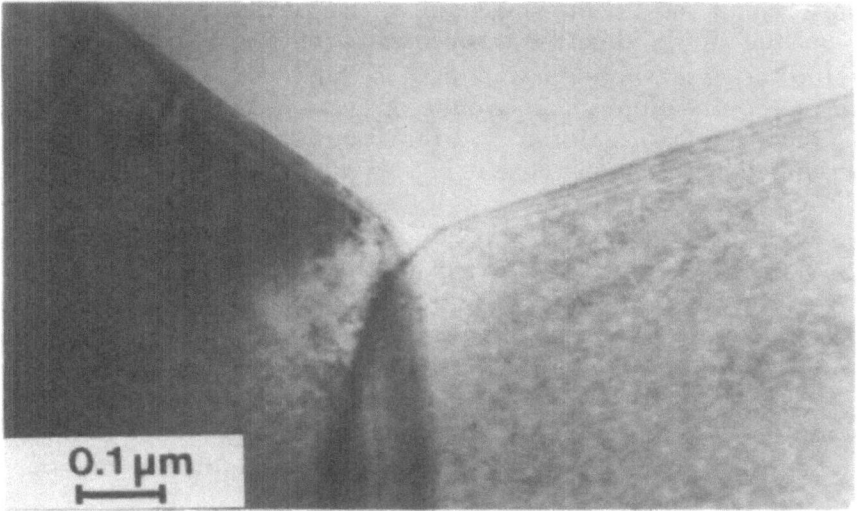

Fig. 5   TEM picture of Ce-TZP+Cu material showing a glassy phase pocket at the triple point

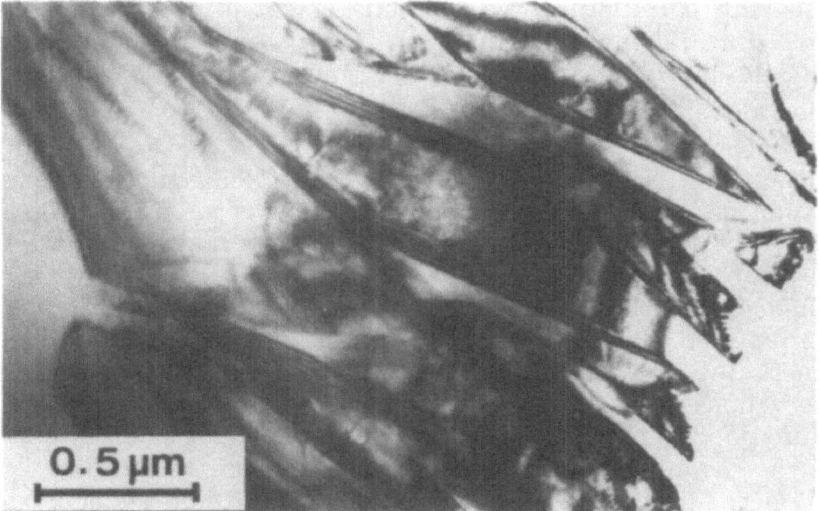

Fig. 6   Ce-TZP+Mn picture showing twins formed during the TEM analysis

From this point of view, two Cu^{+1} ions introduced three double charged oxygen vacancies whereas one Mn^{+2} or two Mn^{+3} ions form one vacancy only. Both can modify the lattice defect structure and the diffusion coefficient of the rate controlling species, namely cerium and zirconium cations[6,7]. These differences might depend upon the different dopant's ionic radii, as well as charges; in fact, in ceramics, the interaction between solutes and grain boundaries originates from (I) the electrostatic force between the charged ions and the electrical double layers in the grain boundary region and (II) the stress due to the ionic size misfit between the solute and the matrix ions. The theory most often used to explain the effect of a solute on grain-boundary motion is the Cahn solute drag model[8], which is based on solute segregation to or away from the grain boundaries, In this theory the drag force on grain boundary motion is due to an interaction energy between grain boundaries and solutes. When grain boundaries migrate, solute tend to follow the moving grain boundaries, since solutes are usually less mobile than grain boundaries, the grain boundary velocity is decreased by the solute drag force; Cu^{+1} is larger than either Mn^{+3} or Mn^{+2} and, therefore, one would expect Cu^{+1} to be a stronger segregant and perhaps a more effective grain-growth inhibitor. This simple solute segregation argument is, therefore, deficient. Therefore taking into account that in zirconia, densification is achieved by lattice diffusion $(\mathcal{D}_l)$ and coarsening by surface diffusion mechanism $(\mathcal{D}_s)$ it can be concluded that both CuO and MnO_2 affected the ratio $(\mathcal{D}_l / \mathcal{D}_s)$ but with different effectiveness.

The reason why CuO improves the sintering rate more than MnO_2 does, seems to rely upon the defects forming ability of the Cu^{+1} species more effective than Mn^{+2} besides the fact that Cu^{+1} and Ce^{+4} ionic radii are very close both contributing by the same extent to the zirconia lattice distorsion, thus raising the sintering activity and at the same time weakening the retarding action of solid solution impurities on the grain boundaries movement[9,10].

On the other hand, being Mn^{+2} and Zr^{+4} ionic radii quite similar, practically neither the distorsion is expected by the manganese ion lattice diffusion nor the stabilization of the tetragonal form can be enhanced by its addition. In fact this evidence is reflected by the mechanical properties and the materials microstructure.

The mechanical properties obtained with samples sintered at 1530 °C for 4 hours are reported in Table I. The addition of both Cu and Mn increased the hardness due to the resulting microstructure with smaller grain size. SEM observations of the surface of fractured samples showed that at room temperature doped samples were fractured mainly along the intergranular surfaces whereas in the Ce-TZP transgranular fracture was also observed. Although mechanical properties were enhanced and quite good for both doped materials, Mn appeared to be more effective than Cu in raising toughness but specially bending strength; the small difference in the average grain size, between the two doped materials, could not be responsible for the quite different mechanical properties, but a TEM analysis, Fig. 6, revealed that twins were often present in the Ce-TZP+Mn microstructure whereas they have seldom been found in the Ce-TZP+Cu materials, suggesting that the transformation toughening mechanism was more likely to occur in the former material probably due to an overstabilization of the tetragonal phase when Cu was used.

Materials	tetr. (%)	σ (MPa)	Hv (GPa)	K_{Ic} (MPa$\sqrt{m}$)
Ce-TZP.	84	150	6	8
Ce-TZP+CuO	100	350	8	11
Ce-TZP+MnO2	100	450	8	13

Table I Mechanical properties obtained after sintering at 1530 °C for 4 hours

Before any conclusion is drawn, there is one more evidence worth mentioning: the increasingly dark colour of the Cu and Mn doped materials, Fig. 7. It was previously reported that "as fired" dark coloured samples of the ceria-zirconia alloys were good conductors, through an electronic mechanism based on the presence of Ce^{3+} species, even at room temperature. However, after 300 °C in air, the samples oxidized becoming less conductive and of light colour[11].
It is therefore inferred that Cu and Mn with their variable oxidation states may contribute to maintain down to room temperature reduced Ce^{3+} specie in excess to the natural redox behaviour of the undoped material. This might be a further contribution of these additives which will be object of impedance spectroscopy investigations to find out the duration and reversibility of the phenomena.

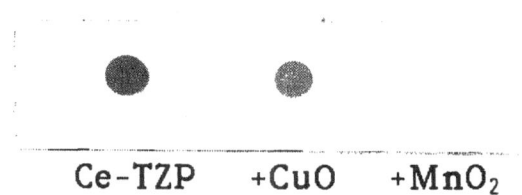

Ce-TZP   +CuO   +MnO$_2$

Fig. 7   Change of colour for the Cu and Mn doped materials due to the presence of Ce^{3+} specie

4. CONCLUSIONS

CuO and MnO_2 have shown to be good candidates as additives for the pressureless sintering of Ce-stabilized zirconia
High density, small grain size as well as almost fully tetragonal material could be achieved using short sintering time at a reasonable temperature. The temperature at which the shrinkage begun did not decrease using MnO_2 or CuO, but at the same temperature the sintering rate was higher; MnO_2 and CuO were also very effective in improving mechanical properties (hardness, bending strength and toughness)

REFERENCES

1. R.L.K. Matsumoto - "*Aging behavior of Ce-stabilized tetragonal zirconia polycrystals*" J. Amer. Ceram. Soc. 71(3) C 128-C 129 (1978)

2. K. Tsukuma, M. Shimada - "*Strength, fracture toughness and Vickers hardness of CeO-stabilized tetragonal ZrO_2 polycrystals* " J. Mater. Sci. 20, 1178, (1985)

3. S. Meriani - "*A new single-phase tetragonal CeO_2-ZrO_2 solid solution*" Mater. Sci. and Engineering 71, 369, (1985)

4. F.F. Lange, M.M. Hirlinger - "*Hyndrance of grain growth of Al_2O_3 by ZrO_2 inclusions*" J. Amer. Ceram. Soc. 67(3)164-68 (1984)

5. R.C. Garvie, P.S. Nicholson - "*Phase analysis in zirconia systems*" J. Amer. Ceram. Soc. 55(6) 303-5 (1972)

6. R.J. Brook - "*Preparation and electrical behavior of zirconia ceramics*" In Science and technology of zirconia. Ed. by H.H. Heuer, L.W. Hobbs; Amer. Ceram. Soc. Publ., Columbus, Ohio, 1981, pp.272

7. S. Wu, R.J. Brook - "*Sintering additives for zirconia ceramics*" Trans. J. Ceram. Soc., 82, 200-205, 1983

8. H.M. Cahn - "*Impurity drag effect on grain boundary motion*", Acta Metall 10(9) 789-798 (1962)

9. M.M. Gadalla, J. White - Trans. Brit. Ceram. Soc., 65(7) 387 (1966)

10 Y. Dongseng - "*The microstructure of ceramics*", J. Chinese Ceram. 10(1), 25, (1982)

11. G. Chiodelli, A. Magistris, E. Lucchini, S. Meriani - "*Microstructure-electrical properties relations in sintered ceria-zirconia*" In Science of ceramics, vol.14 Ed. by D. Taylor, The Institute of Ceramics Publ., Stoke of Trent, U.K., 1988, pp.903-908

A COMPARISON OF NUMERICAL METHODS FOR THE CALCULATION

OF STRESS INTENSITY FACTORS IN ZRO$_2$-CERAMICS

by

W.H. Müller

Hermann-Föttinger-Institut für Thermo- und Fluiddynamik
Technische Universität Berlin

ABSTRACT

The singular integral equation technique and the method of Laurent series expansion are applied in order to find exact solutions for the stress intensity factors in Zirconia toughened ceramics. The increase in toughness is calculated for various situations and the methods and results are compared with the energy approach of Claussen et al. Furthermore the case of a crack terminating at a Zirconia particle is considered where the stress singularity has no longer the form of a square root.

1. INTRODUCTION

From the experimental work of Claussen, Rühle, Heuer et al /1-4/ it is well known that the critical stress intensity factors K_{IC} of Zirconia containing ceramics are much higher than those of the original matrix materials, e.g.: $K_{IC}(Al_2O_3) = 3$ MNm$^{-3/2}$, $K_{IC}(Al_2O_3 + ZrO_2) = 6-15$ MNm$^{-3/2}$.

This is mainly due to the so-called **stress-induced transformation toughening**: Tetragonally stabilized Zirconia transforms into its monoclinic version under the influence of the enormous shear stresses in the vicinity of a crack tip /4,6/.

In fact this martensitic phase transition is accompanied by a 3-5% increase in volume. Hence the flanks of a crack which lies above a Zirconia inclusion are closed and a higher critical load can be applied externally. The same holds for a crack lying in front of an inclusion: Here the flanks are opened so that the crack is attracted and finally absorbed by the particle!

Thus in any case a **compressive process zone** is generated around the crack which explains **qualitatively** the
increase in toughness mentioned above.

Nevertheless, a verbal explanation of the Zirconia
phenomenon is not enough in order to optimize the
strength of a ZrO_2 containing ceramic since it does not
answer questions like: 'What would be the ideal matrix
combination as far as elasticity is concerned?,'How affects particle size K_{Ic}?', ' What is the influence of
the thermal mismatch?', ...,etc.

To that end **quantitative** models are necessary which
allow to check these dependencies. Several attempts have been made in this direction: Claussen and Rühle /4,
6/ arrived by means of an energy argument at the
following formula, which allows to calculate the increase in toughness K_{Ic} as function of the matrix
toughness K_0, the volume percentage v of transformed
particles, the elastic modulus E, the volume expansion
e due to the phase transition and the radius r of the
process zone:

$$K_{Ic} = \sqrt{K_0{}^2 + 2vE^2 e^2 r} \quad . \tag{1}$$

A similar result was obtained by Evans /7/:

$$\Delta K_{Ic} = 0.21Eevr/(1-\nu), \tag{2}$$

ν being Poisson's constant.

Although equations like these are undoubtedly extremely
useful for a quick and concise check of the influence
of relevant parameters they are not completely sa-
tisfying since they do not allow a minute investigation
of the "microstructure" of the ceramic, e.g. the
significance of the crack-particle distances or the
particle sizes. Furthermore they do not take into
account the influence of the particle-matrix boundary
upon the stress singularity, an aspect which is nowadays important in connection with fibre-reinforced
materials /5/.

Such problems were solved by Müller in /8-12/ upon the
basis of two numerical methods of linear elasticity,
viz. the integral equation technique and the Laurent
series expansion. The present paper presents some new

results obtained with these methods, describes their
relationship to the work of Claussen, Rühle and Evans
and concludes with an outlook on possible applications
to fibre reinforced materials.

2. MATHEMATICAL METHODS AND THEIR RESULTS

2.1 General Equations

Consider the following problem (Fig 1.): A Griffith
crack in a linear elastic medium is surrounded by seve
ral monoclinic ZrO_2 particles.

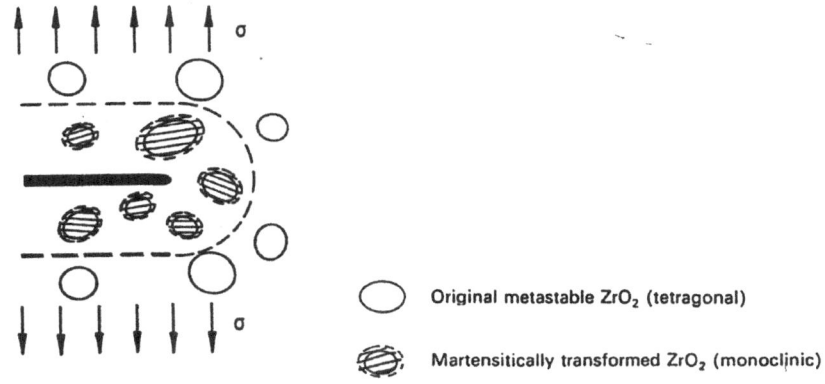

Original metastable ZrO_2 (tetragonal)

Martensitically transformed ZrO_2 (monoclinic)

Fig.1: Process zone around a Griffith crack

We ask for the critical external load which can be applied without breaking the specimen. In case of mode I
conditions this load can be calculated from the following equation:

$$K_{Ic} = K_I|_c \qquad (3)$$

Herein $K_I|_c$ denotes the stress intensity factor corres-
ponding to the situation in Fig.1, evaluated at critical conditions. K_I is a quantity which depends upon the
geometry of the specimen, its microstructure (i.e. the
particles with their different sizes and locations) and
all occurring tensions, may they be due to external

loads or to the phase transition! Thus normally K_I is not an analytical function and only in special cases given by a simple formula. E. g. for a Griffith crack of length l in an pure and infinite matrix with a load σ at infinity we have:

$$K_I = \sigma\sqrt{\pi l} \ , \tag{4}$$

an equation which allows a direct calculation of σ_c acc. to (3) provided that the K_{Ic} of the matrix is known.

In general we may write:

$$K_I = \sigma\sqrt{\pi l} \ Y(...), \tag{5}$$

where $Y(...)$ is the so-called **correction function** which subsumes all 'anomalies' like inclusions, phase transitions, etc.

From (3)and (5) it follows that compared with the critical tension of the pure matrix material (see (3) and (4)) σ_c increases if $Y(...)$ assumes values below 1 and that it decreases for $Y(...) > 1$.

Hence we expect that if the particles are above the crack $Y(...)$ will be below 1. The same must hold for cracks crossing the particle interface. But for particles which are in front of a crack $Y(...)$ should be above 1.

In the following subsections we present two techniques which allow to calculate $Y(...)$ for such situations numerically.

2.2 The Integral Equation Method - Mathematics

The integral equation approach allows the exact calculation of $Y(...)$ in the following situations (cp. Fig. 2): A Griffith crack is situated above or in a particle. The elastic constants (λ,μ) of the particle are different from those of the surrounding matrix which is under uniform tension at infinity.

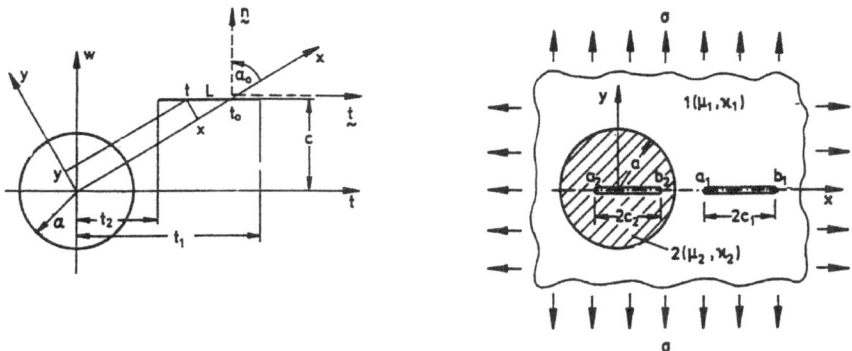

Fig.2: Griffith cracks above or in Zirconia particles

Erdogan, Gupta and Ratwani /13,14/ have shown how to tackle these problems: The crack is simulated by an un-known continuous distribution of edge dislocations. This distribution is determined by the fact that the flanks L of a crack must be free of forces. Hence the normal and tangential stresses at every point of L which are due to the dislocations have to be counterba-lanced by the external loads.

Mathematically reformulated this consideration leads to singular integral equations of the Cauchy type, e.g. in the case of a particle above the crack:

$$\int_{t_2}^{t_1} g(t_0)\frac{c}{\sqrt{c^2+t_0^2}} \frac{dt_0}{t_0-t} + \int_{t_2}^{t_1} k_{11}(t,t_0)g(t_0)dt_0 +$$

$$+ \int_{t_2}^{t_1} f(t_0)\frac{t_0}{\sqrt{c^2+t_0^2}} \frac{dt_0}{t_0-t} + \int_{t_2}^{t_1} k_{12}(t,t_0)f(t_0)dt_0 = -\frac{\pi(\kappa_1+1)}{2\mu_1} p_1(t)$$

$$\int_{t_2}^{t_1} g(t_0)\frac{t_0}{\sqrt{c^2+t_0^2}} \frac{dt_0}{t_0-t} + \int_{t_2}^{t_1} k_{21}(t,t_0)g(t_0)dt_0 -$$

(6)

$$- \int_{t_2}^{t_1} f(t_0)\frac{c}{\sqrt{c^2+t_0^2}} \frac{dt_0}{t_0-t} + \int_{t_2}^{t_1} k_{22}(t,t_0)f(t_0)dt_0 = -\frac{\pi(\kappa_1+1)}{2\mu_1} p_2(t)$$

$$\int_{t_2}^{t_1} g(t) \, dt = 0 , \qquad \int_{t_2}^{t_1} f(t) \, dt = 0 , \qquad t_2 < t < t_1$$

g and f are the components $(-b_x, -b_y)$ of infinitesimal Burger vectors. The k_{ij}'s denote nonsingular integral kernels. For their explicit forms and other details concerning the algebra the reader is referred to /12-14/.

The right sides of (6) represent the external loads in the uncracked medium, i.e. the influence of the uniform tension at infinity as well as the influence of the stresses from the phase transition. In /12/ it was shown that for a radially symmetric transformation one finds

$$p_1(t) = \sigma + \frac{a^2}{r^2} f(\sigma, e)(\sin^2 r - \cos^2 \alpha), \quad r^2 = c^2 + t^2 ,$$

$$f(\sigma, e) := \sigma \frac{\kappa_2 - 1 - (\kappa_1 - 1)m}{2m + \kappa_2 - 1} + e \frac{2m \mu_1}{2m + \kappa_2 - 1} \qquad (7)$$

The numerical treatment of (6,7) gives f and g as functions of the coordinate t from which the correction function Y(...) can be calculated:

$$Y(t_1) = - \frac{2\mu_1 \kappa}{1 + \kappa_1} \frac{1}{\sqrt{\ell(c^2 + t_1^2)}} (c\, G(t_1) + t_1\, F(t_1)) \qquad (8)$$

$$g(t) = w(t)\, G(t) , \quad f(t) = w(t)\, F(t) , w(t) = (t - t_2)^{-1/2} (t_1 - t)^{-1/2}$$

For the calculations the following material constants were used (κ: Muskhelishvili's constant for plane stress):

$$\mu_1 = 110 \text{ GPa} \qquad \kappa_1 = 2,23$$

$$\mu_2 = 66 \text{ GPa} \qquad \kappa_2 = 2,45 \qquad (9)$$

2.3 The Integral Equation Method – Results

Fig.3 shows the first iteratives $Y_{(1)}(...)$ of the correction function for various situations: If the crack is in or below the particle the values of $Y_{(1)}$ are less than 1 and this means stabilization. On the other hand the crack will be attracted by a front-particle since in this case $Y_{(1)}$ is larger than 1.

187

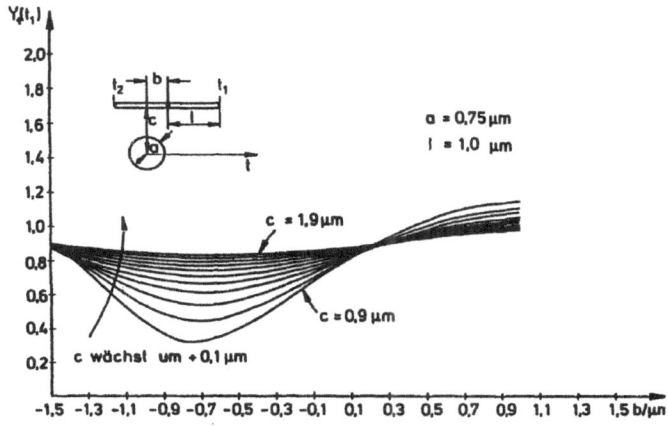

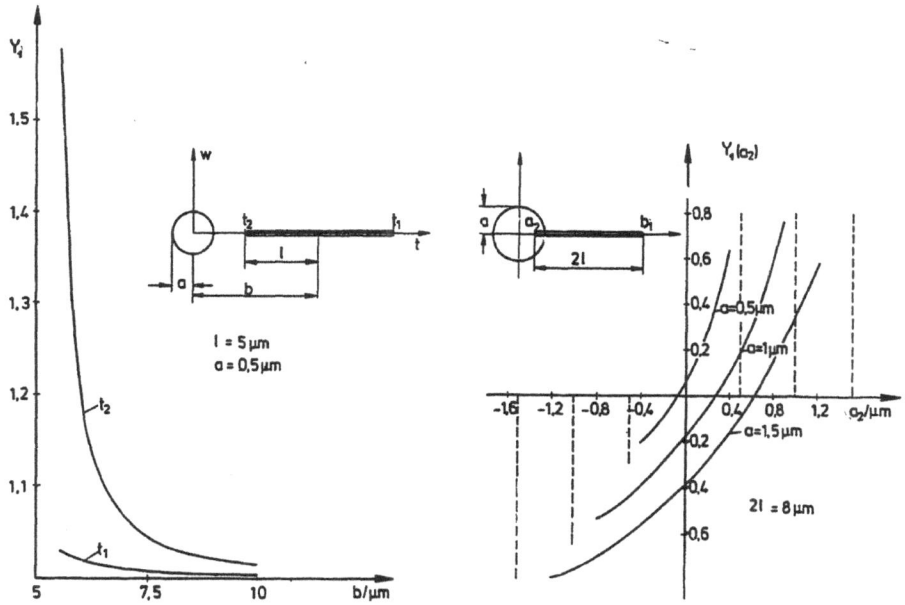

Fig.3: First iteratives of the correction function

In order to determine the exact amount of stabilization it would be necessary to calculate higher iteratives of the correction function. Since the calculation (see /12/) is a very cumbersome procedure we restrict ourselves to some selected examples:

Fig.4 presents some pictures from the film 'Zirconia in Action' /15/ which are the result of fifteen itera tions.

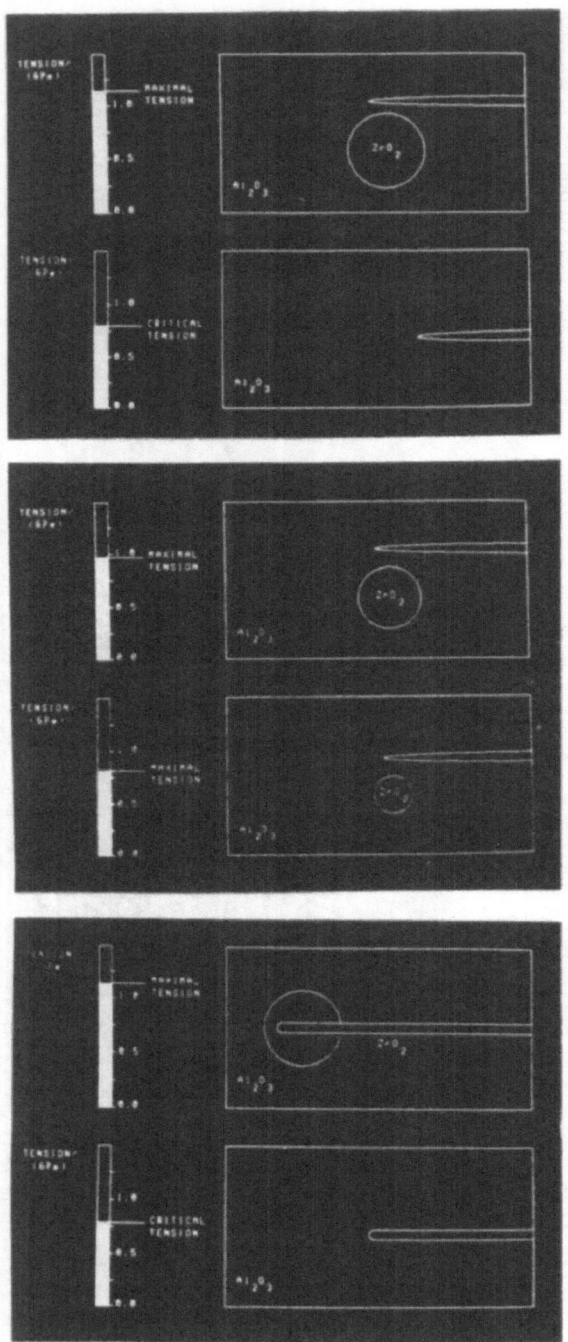

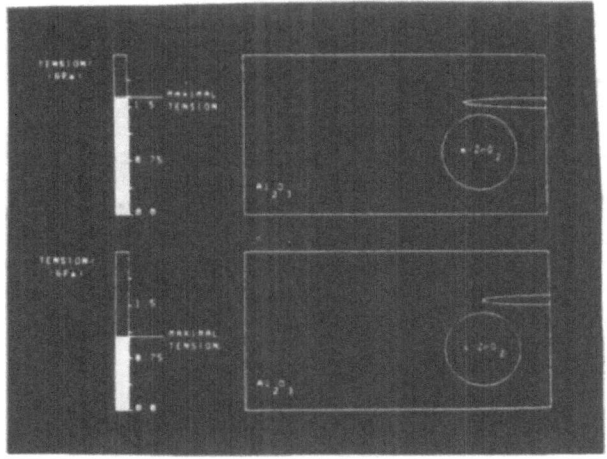

Fig.4: Sequences taken from /13/

Fig.4_1 and 4_3 demonstrate the effect of a 'single par-ticle process zone': A $2l=9\mu$m crack in an Al_2O_3 matrix ($K_{Ic}(Al_2O_3)=3MNm^{-3/2}$) becomes unstable at $\sigma_c=0.8$ GPa. Now if a 1.5 μm Zirconia particle is placed 2 microns above the crack it can be stabilized up to 1.14 GPa and $l=6.2\mu$m, which corresponds to an effective toughness $K_{Ic}(Al_2O_3+ZrO_2)=\sigma_c\sqrt{\pi l}=5.03MNm^{-3/2}$. Analogously one finds that if a $2l=16\mu$m crack is in the particle the critical tension becomes 0.89 GPa i.e. $K_{Ic}(Al_2O_3+ZrO_2)=4.46MNm^{-3/2}$.

Fig.4_2 shows the influence of the particle size: A 1.25 μm particle at a distance of 2 microns stabilizes an originally 9μm-crack up to 0.96 GPa, whereas a 0.75 μm particle at a distance of 1.5 μm (i.e. the distance between the crack and the particle surface remains constant) reaches only 0.82 GPa.

Finally Fig.4_4 proves the power of the phase transi-tion: Even if the particle does not transform (t-ZrO_2) one observes a slight increase in toughness ($\sigma_c=1.06$ GPa, $2l=5.5\mu$m, $K_{Ic}(Al_2O_3/t-ZrO_2)=3.11MNm^{-3/2}$) which is in fact rather small compared with the increase for a transformed particle (m-ZrO_2): $\sigma_c=1.63$GPa, $2l=6.9\mu$m, $K_{Ic}(Al_2O_3/m-ZrO_2)=5.37MNm^{-3/2}$. This is an important ob-servation which justifies the method of Laurent series which we are going to discuss now.

2.4 The Method of Laurent Series - Mathematics

This technique allows an exact calculation of correction functions in the following situation (Fig.5): A Griffith crack is surrounded by an arbitrary number of circular holes. The holes can be subject to an internal pressure. Since the influence of the difference between the elastic properties of the matrix and the inclusion on the toughening is neglegible to the influence of the phase transition, these pressurized holes represent the monoclinic particles fairly well provided that their internal pressure is high enough.

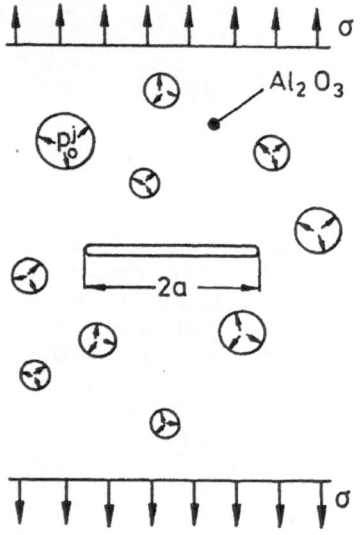

Fig.5: Pressurized holes around a Griffith crack

In order to calculate the correction function, i.e. essentially the intensity of the stresses t_{ij} at the crack tip, we start with the Muskhelishvili-Kolosov equations /16/ (d is a suitable reference length):

$$t_{xx} + t_{yy} + \frac{8i\mu}{1+\kappa}\, \omega_{xy} = 4\sigma\phi'(z)$$

$$t_{xx} - t_{yy} - 2i\, t_{xy} = -2\sigma\left\{\bar{z}\,\phi''(z) + \psi''(z)\right\} \qquad (10)$$

$$2\mu(u_x - iu_y) = \sigma d\left\{\kappa\overline{\phi}(\bar{z}) - \bar{z}\phi'(z) - \psi'(z)\right\}$$

and the problem that remains is to determine the Goursat functions Ψ and ϕ in the z-plane of Fig.5. But this is in fact an extremely difficult and wearisome analysis! Therefore we shall refrain from doing it explicitly and refer the reader for all details to /11/. Here it suffices to mention that first of all the Goursat functions have to be expanded into Laurent series, i.e. in the case of a single pressurized hole e.g.:

$$\phi(z) = \sum_{n=0}^{\infty} \left\{ (F_n^\cdot + iF_n')z^{-(n+1)} + (M_n^\cdot + iM_n')z^{n+1} \right\}$$

$$\psi(z) = - D_o^\cdot \ell nz + \sum_{n=1}^{\infty} (D_n^\cdot + iD_n')z^{-n} + \sum_{n=0}^{\infty} (K_n^\cdot + iK_n')z^{n+2} \qquad (11)$$

Then all appearing coefficients $F_n \ldots K_n$ are determined from boundary conditions and finally the correction function is calculated with Sih's formula /16/:

$$K_I - iK_{II} = 2\sigma(2\pi d)^{1/2} \ell im_{z \to \lambda} (z-\lambda)^{1/2} \phi'(z) \quad , \quad \lambda := \frac{a}{d} \qquad (12)$$

2.5 The Method of Laurent Series – Results

For the following results the size depending pressure values of /9/ were used, i.e. 2.5 GPa for a 1 micron particle.

The figure shows the first iterative of the correction function for one and two pressurized holes respectively, which are located at a fixed distance around a crack. One clearly observes that two particles lead to a stronger stabilization since their Y-minimum is deeper.

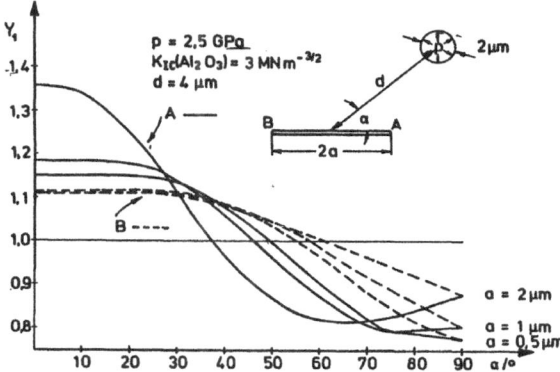

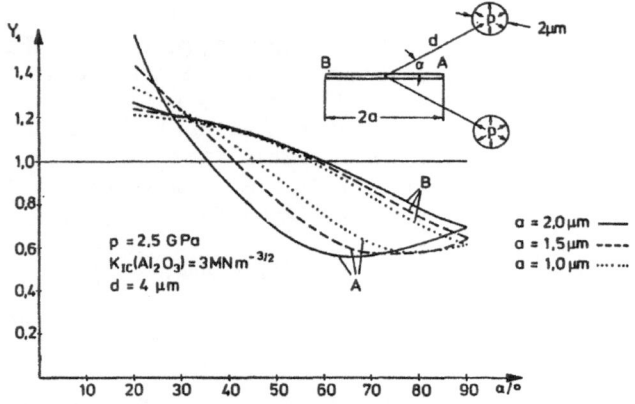

Fig.6: Correction functions for one and two particles

2.6 Analytical Methods

The foregoing chapters show drastically that one has to pay very dearly for an exact calculation of stress-intensity factors: The numerical treatment of the basic equations (6-8) or (10-11) requires high-powered computer facilities and above that several iterations are necessary to get a final result even in a comparatively simple geometric situation!

Thus from the standpoint of the practician it would be desirable to find closed form solutions which allow a quick check of the toughening effect. In the case of the integral equations it is very simple to find such a solution: As it was shown the toughening is dominated by the phase transition and not by the different elastic properties of the matrix and the inclusion. Therefore it is justified to equate the elastic constants. But in this case all integral kernels in (6) vanish and the following expression remains:

$$\int_{t_2}^{t_1} g(t_0) \cos\alpha_0 \frac{dt_0}{t_0-t} + \int_{t_2}^{t_1} f(t_0) \sin\alpha_0 \frac{dt_0}{t_0-t} = -\frac{\pi(\varkappa+1)}{2\mu} p_1(t)$$

$$\int_{t_2}^{t_1} g(t_0) \sin\alpha_0 \frac{dt_0}{t_0-t} - \int_{t_2}^{t_1} f(t_0) \cos\alpha_0 \frac{dt_0}{t_0-t} = -\frac{\pi(\varkappa+1)}{2\mu} p_2(t)$$

(13)

Now it follows from Fig.2:

$$b_t = b_x \sin \alpha_0 - b_y \cos \alpha_0$$

$$b_w = b_x \cos \alpha_0 + b_y \sin \alpha_0 \qquad (14)$$

so that (13) becomes:

$$-\frac{\pi(\kappa+1)}{2\mu} p_1(t) = \int_{t_2}^{t_1} \frac{g_2(t_0)}{t_0 - t} dt_0, \quad -\frac{\pi(\kappa+1)}{2\mu} p_2(t) = \int_{t_2}^{t_1} \frac{g_1(t_0)}{t_0 - t} dt_0, \qquad (15)$$

where (b_t, b_w) was replaced by $(-g_1, -g_2)$.

The general solution of these integral equations is well known /18/:

$$g_1(\hat{t}) = -\frac{\kappa+1}{2\pi\mu} \frac{1}{\sqrt{\ell^2 - \hat{t}^2}} \int_{-\ell}^{\ell} \frac{p_2(x) \sqrt{\ell^2 - x^2}}{\hat{t} - x} dx \qquad (16)$$

where

$$g_2(\hat{t}) = -\frac{\kappa+1}{2\pi\mu} \frac{1}{\sqrt{\ell^2 - \hat{t}^2}} \int_{-\ell}^{\ell} \frac{p_1(x) \sqrt{\ell^2 - x^2}}{\hat{t} - x} dx$$

$$t = \hat{t} + \frac{t_1 + t_2}{2}, \qquad t_0 = \hat{t}_0 + \frac{t_1 + t_2}{2}, \qquad \ell := \frac{t_1 - t_2}{2} \qquad (17)$$

was set.

Hence the stress intensity factors can be calculated via /10/:

$$K_I(1) = \lim_{t \to t_1} \frac{2\mu}{1 + \kappa} [2\pi(t_1 - t)]^{1/2} g_2(t) =$$

$$= \frac{1}{\sqrt{\pi\ell}} \int_{-\ell}^{\ell} \sqrt{\frac{\ell + x}{\ell - x}} p_1(x) dx \qquad (18)$$

Thus all that remains to be done is to solve the integral with the known function p_1 (see (7)). If we now simplify matters and assume that p_1 consists only of the external tension and a constant pressure $-p$ which shall be due to the phase transition of some Zirconia particles:

$$p = Eev \qquad (19)$$

the integral can be easily evaluated:

$$K_I = (\sigma - p)\sqrt{\pi\, l} \tag{20}$$

so that finally:

$$\frac{K_{Ic}}{K_0} = 1 + \frac{p}{\sigma_0} \tag{21}$$

This in fact is one of the basic results of /8/ and moreover it is consistent with the pioneer work of Claussen, Rühle and Evans as we shall see now.

The energy which is absorbed in the process zone can be expressed in our terms as follows:

$$U = \frac{\text{'force} * \text{distance'}}{\text{unit layer}} = pL * e\,\Delta l = Ee^2 vL\,\Delta l \tag{22}$$

L is the length of the crack tip and in the same order of magnitude as the radius of the process zone. So this expression is identical with (1) with the exception of a factor 2.

2.7 The Stress Tip Singularity

So far our cracks were either in or outside a Zirconia inclusion and their tips never terminated directly at the matrix-particle interface so that the corresponding stress-singularities were of the $r^{-1/2}$-type. However if the tips end at a boundary the singularity is of the form r^α with $-1 < \alpha < 0$, /12-13/.

The integral equation technique allows to calculate the new singularities as functions of the elastic coefficients from the following transcendental equations:

a) for a matrix crack

$$2 \cos \pi \alpha + (A_1 + A_2) - 4 A_1 (\alpha + 1)^2 = 0$$

$$A_1 := \frac{1 - m}{1 + m\kappa_1}, \qquad A_2 := \frac{\kappa_2 - m\kappa_1}{\kappa_2 + m} \qquad m := \frac{\mu_2}{\mu_1} \tag{23}$$

b) for a particle crack

$$2 \cos \pi \alpha + (A_3 + A_4) - 4 A_3 (\alpha + 1)^2 = 0$$

$$A_3 := \frac{m - 1}{m + \kappa_2} \quad , \quad A_4 := \frac{m\kappa_1 - \kappa_2}{1 + m\kappa_1} \tag{24}$$

With the data in (9) we find that:

for a): $\alpha = -0.565$ (25)

and for b): $\alpha = -0.43$ (26)

Hence the singularity at the tip of a matrix crack a) is a little stronger (< -1/2) than normal and vice versa.

Fig. 7 shows the first approximation of some stress intensity factors which correspond to the a)-singularity.

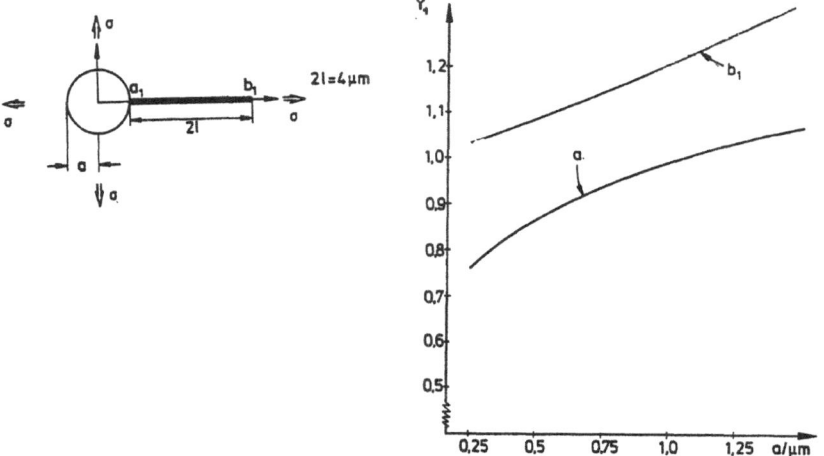

Fig.7: Stress intensities for a matrix-boundary crack

It is worth mentioning that (23-26) can be used for the calculation of the stress behaviour in Zirconia and fibre reinforced ceramics too /5/.

3. CONCLUSION AND OUTLOOK

The paper has shown that nowadays it is possible to understand the micro-mechanisms of Zirconia toughening at
various levels: from the 'spot check formula' up to a
detailed analysis based upon linear elasticity.

Above that the presented methods may serve as an orientation during the design and construction period of new
and better ceramic materials.

4. REFERENCES

/1/ N.Claussen, M.Rühle, Design of ZrO_2-Toughened
Ceramics, Adv. Cer. vol. 3, Columbus-Ohio, 1981

/2/ N.Claussen, Transformation Toughened Ceramics, in:
Cer. in Adv. Energy Tech., Petten, 1982

/3/ A.H.Heuer, N.Claussen, W.M.Kriven, M.Rühle, Stability of Tetragonal ZrO_2 Particles in Ceramic Matrices,
J. Amer. Ceram. Soc. 65 (1982) 642

/4/ N.Claussen, Umwandlungsverstärkte keramische Werk-
stoffe, Z. Werkstofftech., 13 (1982) 138

/5/ N.Claussen, K.L.Weisskopf, M.Rühle, Tetragonal Zirconia Reinforced With SiC Whiskers, J. Amer. Ceram.
Soc., 69 (1986) 288

/6/ R.Stevens, Zirconia and Zirconia Ceramics, Magnesium Electron Publication No. 113, 1986

/7/ A.G.Evans, Toughening Mechanisms in Zirconia
Alloys, Adv. Cer. vol. 12, Columbus-Ohio, 1984

/8/ W.H.Müller, Quantitative Beschreibung der Umwand-
lungsverstärkung in Al_2O_3-ZrO_2-Keramiken, Fortschritts-
ber. DKG, 1 (1985) 87

/9/ W.H.Müller, Calculation of Stress Intensity Factors
in ZrO_2-Ceramics, J. de Phys. 47 (1986) C1-607

/10/ W.H.Müller, Quantitative Modelle zur Beschreibung
der Phasenumwandlung und der Erhöhung des Bruchwider-
standes in zirkondioxidhaltigen Keramiken, thesis, TU-
Berlin, 1986

/11/ W.H.Müller, Numerical Methods for the Description of 'Ceramic Steels', Fract. Contr. Eng. Struct., vol.3, 1986, 2129

/12/ W.H.Müller, The exact Calculation of Stress-Intensity-Factors in Transformation Toughened Ceramics by Means of Integral Equations, Int. J. Fract., in print

/13/ F.Erdogan, G.D.Gupta, The Inclusion Problem With a Crack Crossing the Boundary, Int. J. Fract. 11 (1975) 13

/14/ F.Erdogan, G.D.Gupta, M.Ratwani, Interaction Between a Circular Inclusion and an Arbitrarily Oriented Crack, J. Appl. Mech. (1974) 1007

/15/ W.H.Müller, Zirconia in Action, scientific film, TU-Berlin, 1988

/16/ N.I.Muskhelishvili, Some Basic Problems of the Mathematical Theory of Elasticity, Groningen, 1953

/17/ G.C.Sih, Handbook of Stress Intensity Factors, Lehigh Univ., Bethlehem Pa., 1973

/18/ W.Schmeidler, Integralgleichungen mit Anwendungen in Physik und Technik, AVG, Leipzig, 1950

SURFACE AND INTERFACIAL ENERGIES IN ZIRCONIA -
LIQUID METAL SYSTEMS

P. Nikolopoulos, D. Sotiropoulou

Institute of Physical Metallurgy, Dpt. Chem. Engineering
University of Patras, GR.-261 10 Patras, GREECE

OBJECTIVES

The objective of the work is to determine the temperature dependence of the surface and grain-boundary energies of polycrystalline calcia-stabilized Zirconia (c-ZrO_2) ceramics as well as the interfacial energies of ZrO_2 in contact with liquid metals. This is important for the properties of composites particularly those prepared by liquid phase sintering.

EXPERIMENTAL METHOD AND RESULTS

An established method for studying the interfacial phenomena is that of the "multiphase equilibrium technique" (Fig. 1). In this method four sets of experiments are necessary to measure the equilibrium angles that develop at the interface. Combining, eqs. (1) - (4) (Fig. 1)

$$\gamma_{SV} = \gamma_{LV} \cos \Theta \; \frac{\cos\frac{\Phi}{2} \cos\frac{\Psi^*}{2}}{\cos\frac{\Psi}{2}(\cos\frac{\Phi}{2} - \cos\frac{\Psi^*}{2})} \tag{5}$$

Table I shows the groove angles (Ψ, Ψ^*) and the dihedral angle (Φ) in the system ZrO_2-Sn-Ar at 1173 and 1523 K, measured with optical interferometry. The values of the product $\gamma_{LV}\cos\Theta$ are determined by extrapolation of literature data [1].

SURFACE AND GRAIN-BOUNDARY ENERGIES

From the results shown in Table I, the surface energy (γ_{SV})
as well as the grain-boundary energy (γ_{SS}) of ZrO_2 can be
calculated using eqs. (5) and (1), respectively.
The results obtained are shown in Table II together with
data at 2123 K of Kingery [2].
Assuming that the surface and grain boundary energies (Table
II) are linear functions of temperature in the temperature
range between absolute zero and the melting point (2890 K) of
ZrO_2, one can write:

$$\gamma_{SV} = 1.496 - 0.427 \times 10^{-3} T \quad (J/m^2) \qquad (6)$$

and $\gamma_{SS} = 0.792 - 0.252 \times 10^{-3} T \quad (J/m^2) \qquad (7)$

Fig. 2 shows the temperature dependence of the surface energy
of ZrO_2 together with the available literature data [1,2,3,4].

INTERFACIAL ENERGIES

The interfacial energies in ZrO_2-liquid metal systems can be
calculated from the expression:

$$\gamma_{SL} = \gamma_{SV} - \gamma_{LV} \cos \Theta \qquad (8)$$

using the linear temperature functions, of the surface energy
for ceramic (eq. 6), of the surface energies for liquid metals
(Table III) and the measured values of contact angle Θ.
The contact angles [1,5], the interfacial energies and their
linear temperature dependence for the systems ZrO_2 in contact
with the liquid metals Sn, Bi, Pb, Cu, Ni and Co are summarize
in Table IV.
In Fig. 3 the data of interfacial energies are plotted versus
the homologous temperature (T/T_m). As the plot points out clea
ly the variation of the interfacial energies at the melting
point of the metals Bi, Pb, Cu, Ni and Co is restricted to suc

an extent, that it may be given by a mean value of

$$\overline{\gamma}_{SL} = 1.669 \pm 0.022 \ (J/m^2) \tag{9}$$

This approach seems to be independent of the type of metal and applies for those metals for which the following relation holds [6] $\gamma_{LV} \cong 3.6 \ T_m(M/\rho_L)^{-2/3}$ where γ_{LV} and (M/ρ_L) the surface energy and the molar volume at the melting points of the metals (Fig. 4).

REFERENCES

[1] P. Nikolopoulos, G. Ondracek, D. Sotiropoulou
    Ceramics International (in press)
[2] W.D. Kingery
    J. Am. Ceram. Soc. 37 (1954) 42.
[3] H.F. Holmes, E.L. Fuller, R.B. Gammage
    J. Phys. Chem. 76 (1972) 1497.
[4] J.M. Lihrmann, J.S. Haggerty
    J. Am. Ceram. Soc. 68 (1985) 81)
[5] P. Nikolopoulos, D. Sotiropoulou
    J. Mater. Sci. Letters 6 (1987) 1429
[6] B.C. Allen "Liquid Metals"
    ed. S.Z. Beer, Dekker, New York, 1972, p. 161.

TABLE I. Groove angles (Ψ, Ψ^*), dihedral angle (Φ) with median standard errors and product $\gamma_{LV}\cos\Theta$.

T (K)	Ψ (deg)	Ψ^* (deg)	Φ (deg)	$\gamma_{LV}\cos\Theta$ [1] (J.m^{-2})
1173	150.37±1.19	146.60±0.77	155.90±0.58	-0.334±0.031
1523	153.37±0.45	154.61±0.76	159.31±0.34	-0.197±0.058

TABLE II. Surface (γ_{SV}) and grain-boundary (γ_{SS}) energies of ZrO_2

T (K)	γ_{SV} (J.m^{-2})	γ_{SS} (J.m^{-2})
1173	0.997±0.145	0.510±0.077
1523	0.841±0.283	0.387±0.130
2123 [2]	0.590	0.265

TABLE III. Linear temperature functions of surface energies for liquid metals

Material	Surface energy (J.m^{-2})	
Sn	$0.544 - 0.07 \times 10^{-3}(T - T_m)$	$T \geqq T_m = 505$ K
Bi	$0.372 - 0.09 \times 10^{-3}(T - T_m)$	$T \geqq T_m = 544$ K
Pb	$0.468 - 0.13 \times 10^{-3}(T - T_m)$	$T \geqq T_m = 600$ K
Cu	$1.311 - 0.20 \times 10^{-3}(T - T_m)$	$T \geqq T_m = 1356$ K
Ni	$1.754 - 0.28 \times 10^{-3}(T - T_m)$	$T \geqq T_m = 1726$ K
Co	$1.610 - 0.29 \times 10^{-3}(T - T_m)$	$T \geqq T_m = 1768$ K

TABLE IV. Contact angles (Θ), interfacial energies and their linear temperature functions in the ZrO_2-liquid metal systems (R - correlation coefficient).

System	T (K)	Θ (deg)	γ_{SL} (J.m^{-2})	$\gamma_{SL}(T) = \gamma_{SL}(T_m) + \dfrac{d\gamma_{SL}}{dT}(T-T_m)$ (J.m^{-2})
ZrO_2-Sn	623	170.04	1.758	$1.876 - 0.820\times10^{-3}$ (T-505)
	773	169.40	1.681	(R = 0.9901)
	923	149.73	1.546	
	1073	134.92	1.393	
ZrO_2-Bi	623	162.06	1.577	$1.638 - 0.844\times10^{-3}$ (T-544)
	723	145.25	1.480	(R = 0.9986)
	823	137.77	1.401	
	923	130.77	1.322	
ZrO_2-Pb	673	154.88	1.624	$1.698 - 1.041\times10^{-3}$ (T-600)
	773	139.73	1.505	(R = 0.9931)
	873	136.36	1.436	
	973	121.55	1.300	
ZrO_2-Cu	1473	122.18	1.552	$1.664 - 0.895\times10^{-3}$ (T-1356)
	1573	121.31	1.483	(R = 0.9967)
	1673	118.38	1.374	
	1773	116.73	1.290	
ZrO_2-Ni	1740	122.33	1.688	$1.677 - 1.119\times10^{-3}$ (T-1726)
	1773	119.71	1.601	(R = 0.9774)
	1833	118.82	1.543	
	1953	117.15	1.432	
ZrO_2-Co	1823	123.21	1.590	$1.666 - 1.401\times10^{-3}$ (T-1768)
	1923	120.38	1.465	(R = 0.9851)
	1973	116.53	1.346	
	2043	116.06	1.295	

204

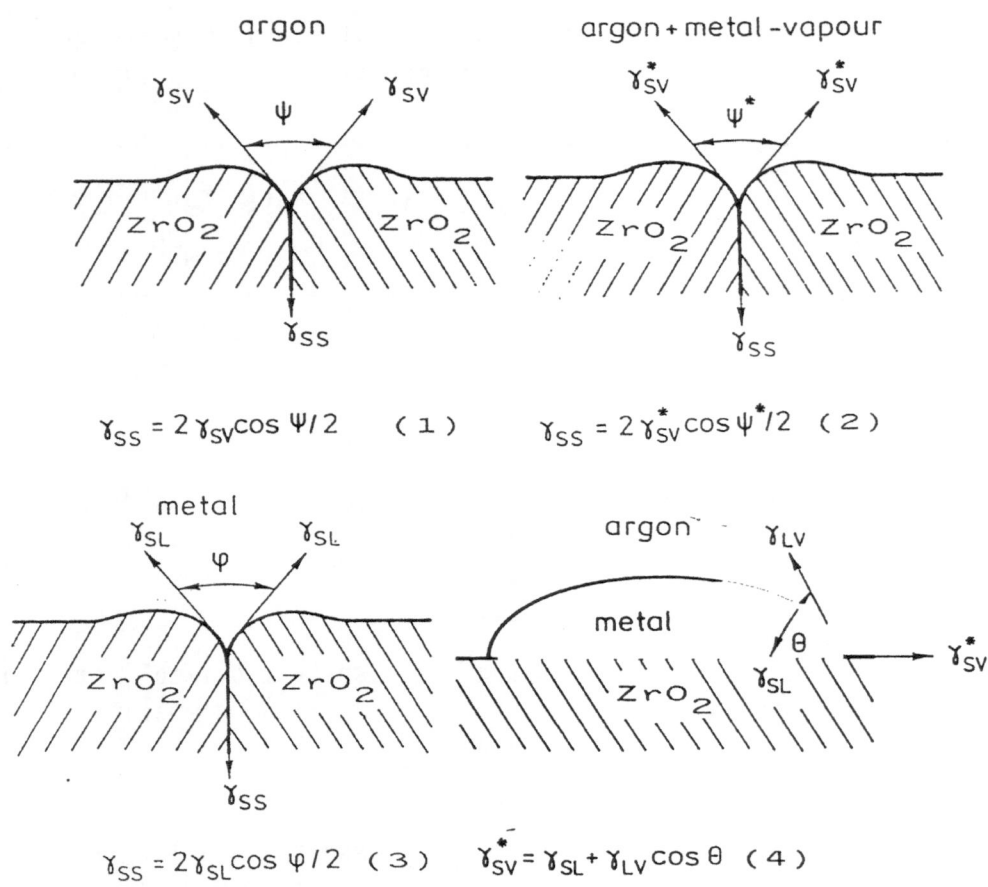

$$\gamma_{SS} = 2\gamma_{SV}\cos\Psi/2 \quad (1) \qquad \gamma_{SS} = 2\gamma_{SV}^{*}\cos\Psi^{*}/2 \quad (2)$$

$$\gamma_{SS} = 2\gamma_{SL}\cos\varphi/2 \quad (3) \qquad \gamma_{SV}^{*} = \gamma_{SL} + \gamma_{LV}\cos\theta \quad (4)$$

Fig. 1. Schematic diagramm in a solid-liquid-vapour system in
equilibrium showing groove (Ψ, Ψ^{*}) dihedral (Φ) and
contact (θ) angles (γ_{LV}, γ_{SV} = surface energies of liquid and solid, γ_{SS} = grain-boundary energy, γ_{SL} =
interfacial energy solid-liquid).

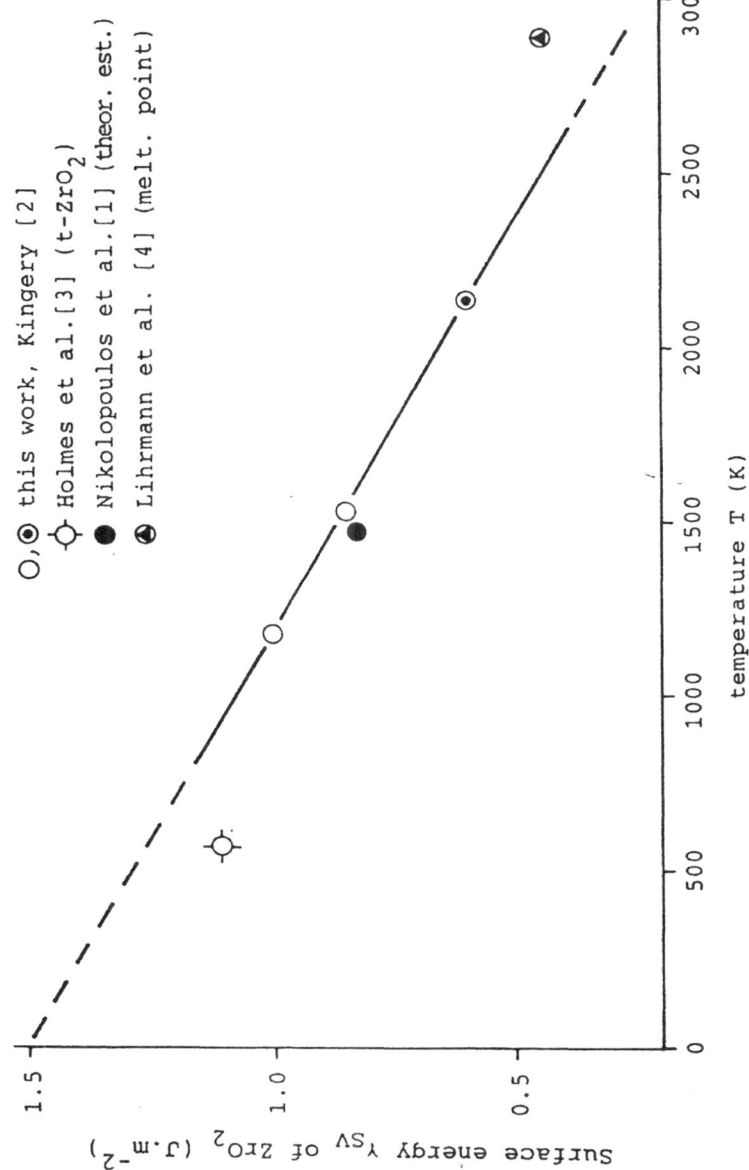

Fig. 2. The temperature dependence of the surface energy of ZrO_2.

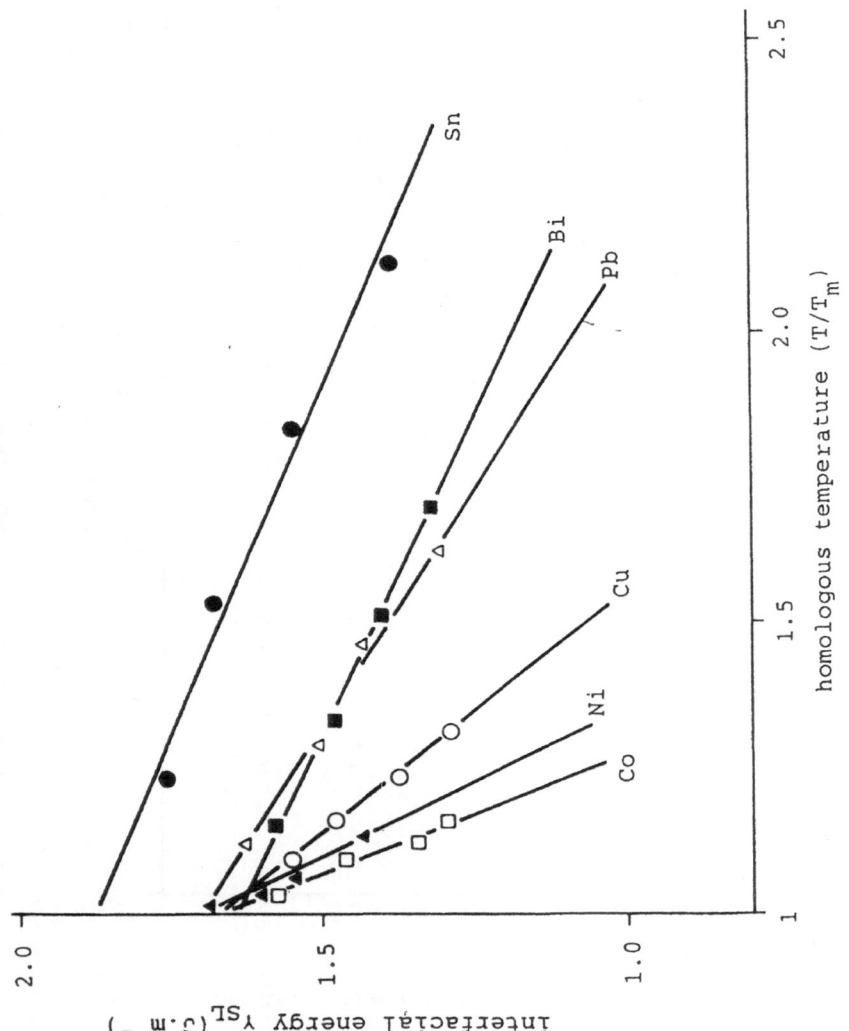

Fig. 3. Interfacial energies - temperature dependence in the ZrO_2-liquid meal systems.

207

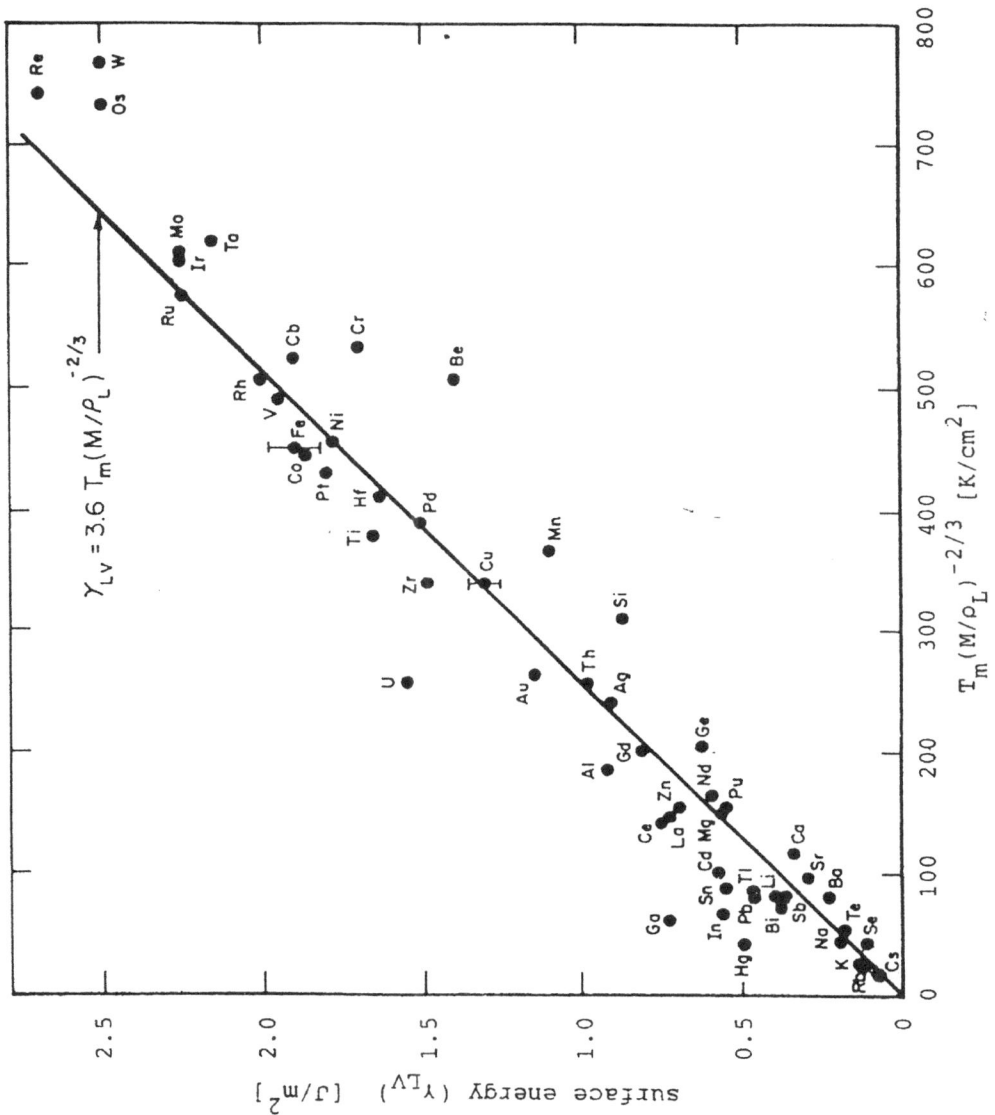

Fig. 4. Surface energy of some liquid metals at their melting
point versus (melting temperature)/(molar volume)$^{2/3}$ [6].

PROTONS IN ZrO$_2$; A SEARCH FOR EFFECTS OF WATER VAPOUR ON THE ELECTRICAL CONDUCTIVITY AND M/T PHASE TRANSFORMATION OF UNDOPED ZrO$_2$

Truls Norby

Department of Chemistry, University of Oslo, POB 1033 Blindern, N-0315 Oslo 3, Norway

Abstract
The transformation temperatures and rates of the M-T and T-M phase transformations of undoped ZrO$_2$ have been studied at high temperatures as a function of the water vapour pressure in oxygen or air. The transformation was studied by recording the electrical conductivity of a sintered, porous sample with Pt-foil electrodes. No effect of the water vapour pressure could be detected. However, the conductivity of monoclinic zirconia was dependent on the water vapour pressure. The conductivity of the tetragonal phase, on the other hand, did not depend on the water vapour pressure. It is concluded that protons are significant defects in monoclinic zirconia, while they are minority defects in tetragonal zirconia.

1. Introduction
1.1. Polymorphs of zirconia
Undoped zirconia has the monoclinic (M) structure at low temperatures. Around 1170°C it transforms to the tetragonal (T) structure. This transformation is accompanied by a decrease in volume and normally leads to cracks in dense samples. At around 2370°C the cubic (C) phase becomes stable. The presence of lower-valent cations like Mg^{2+}, Ca^{2+}, Y^{3+}, and rare-earth cations, generally stabilize the high-temperature phases to lower temperatures so that more or less metastable tetragonal or cubic phases can be obtained at lower temperatures and room temperature.

The tetragonal and cubic polymorphs obtained by the help of doping have gained much attention because of their special properties: Cubic, or fully stabilized, zirconia (FSZ) is used at 500-1000°C as a solid electrolyte with oxygen ion conduction. Metastable tetragonal zirconia polycrystals (TZP) or mixtures of tetragonal and cubic zirconia (partially stabilized zirconia, PSZ) have remarkable mechanical properties at not too high temperatures due to the so-called transformation toughening.

1.2. Protons in ZrO$_2$

In recent years it has been found that protons from hydrogen-containing gases dissolve in oxides at high temperatures and often affect or even dominate the defect structure of the oxide. This is the case, for instance, for Y$_2$O$_3$ ([1,2,3,4]), Al$_2$O$_3$ ([5,6]), TiO$_2$ ([7,8,9]), ThO$_2$ ([10]), and others ([11]). General treatments on protons in oxides at high temperatures are available ([11,12,13]).

Vest et. al. ([14]) reported that there are no effects of high hydrogen or carbon activities on the electrical conductivity of monoclinic zirconia at 1000°C.

For undoped, tetragonal zirconia some authors report evidence for effects of hydrogen defects ([15]), while others find no effects ([16]).

Wagner ([17]) predicted the possibility that protons (dissolved from water vapour) give a significant proton conductivity in zirconia and thus affect the emf of zirconia-based electrochemical cells. He determined the diffusion coefficient and concentration of protons in yttria-stabilized, cubic zirconia, and concluded that protons are minority defects and that they have no significant influence on the conductivity characteristics of the material. However, we shall show below that protons become of relatively higher importance as the dopant level is decreased, and it is thus desireable to study protons in zirconia at low dopant levels.

Water vapour is now known to play a major role in the unwanted self-triggering of the T (metastable) to M (stable) transformation of TZP and PSZ materials at intermediate temperatures ([18]). This probably has to do with the stability of yttrium hydroxide rather than with protons as defects in zirconia, but it is still a good reason to seek more information about the behaviour of hydrogen species in zirconia.

A final motivation for this work emerges from a question often raised about the possible influence of hydrogen species on the ageing process of FSZ electrolyte materials.

This paper deals with the results of a study of the influence of water vapour on the electrical conductance of undoped zirconia in the temperature region around the M/T phase transformation. Furthermore, using the conductivity, the M-T and T-M phase transformations have been followed in wet and dry atmospheres, and the characteristics of these transformations are reported.

2. Theory
2.1. Protons as defects in ZrO_2

In terms of the well-known Kröger-Vink terminology, the dissolution of protons from water vapour in zirconia (or any oxide) can be written:

$$1/2 \ H_2O \ (g) \ = \ H_i^{\cdot} \ + \ e' \ + \ 1/4 \ O_2 \ (g) \qquad (1)$$

The equation states that the protons are dissolved as interstitial, effectively positive defects, under the cogeneration of defect electrons. (It may be noted that each proton is closely bonded to an oxygen ion rather than being centered in a normal interstitial position. Many authors thus prefer to use the term "substitutional hydroxide" for the defect.)

In addition to protons introduced from the atmosphere, an excess of lower-valent metal impurities is likely to be present in undoped (as well as stabilized) zirconias. In the following discussion, this is represented by the arbitrary trivalent metal ions Ml^{3+} which are assumed to dissolve substitutionally. If we furthermore assume that Schottky-type defects (metal and oxygen vacancies) are the dominant native ionic point defects in zirconia, the electroneutrality condition can be written:

$$2[V_O^{\cdot\cdot}] \ + \ [H_i^{\cdot}] \ + \ p \ = \ 4[V_{Zr}''''] \ + \ [Ml_{Zr}'] \ + \ n \qquad (2)$$

where p and n denote the concentration of electron holes and defect electrons, respectively.

References ([11]) and ([12]) describe how the defect situation varies with the water vapour pressure under different simplified cases of Eq.(2): At low water vapour pressures, protons will be minority defects. This is illustrated in Fig.1 (left hand part), where it is furthermore assumed that the oxide is ideally pure and that the dominant defects are oxygen vacancies and defect electrons. As the water vapour pressure increases, the concentration of protons increases with the square root of the vapour pressure (cfr. Eq.(1)). Eventually, the protons may become dominating, being compensated by the dominant negative defect, i.e. defect electrons in our example. From Eq.(1) it may then be shown that

$$[H_i^{\cdot}] \ = \ n \ \propto \ p(O_2)^{-1/8} \ p(H_2O)^{1/4} \qquad (3)$$

This example of a proton-dominated defect situation is illustrated in the right-hand part of Fig.1. It is evident from the figure that a proton dominance should be seen by changes in conductivity (whether of n-, p-, or ionic type)

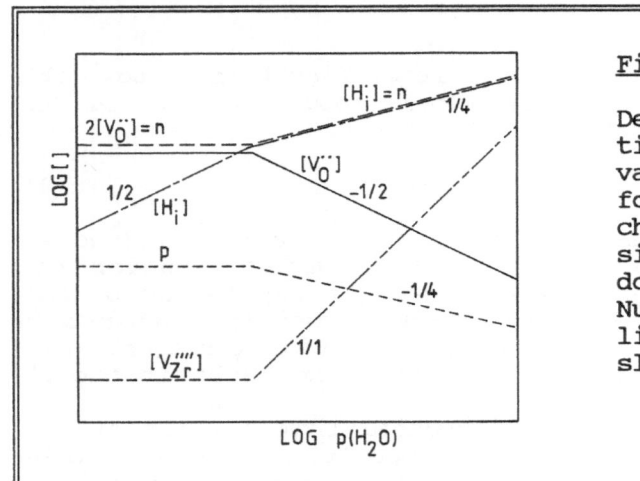

Fig.1

Defect concentra-
tions vs. water
vapour pressure
for an arbitrarily
chosen defect
situation in un-
doped ZrO_2.
Numbers along the
lines denote
slopes.

as a function of the water vapour pressure. Other negative
defects may dominate in the oxide and become the defects
compensating the protons. This would alter the slopes in
the right-hand part of Fig.1, but not the direction of
change for the different defects.

2.2. Protons and lower-valent metal dopants

Let us now look at how the defect structure at high
water vapour pressures changes when the concentration of
lower-valent metal dopants Ml is increased. A schematic

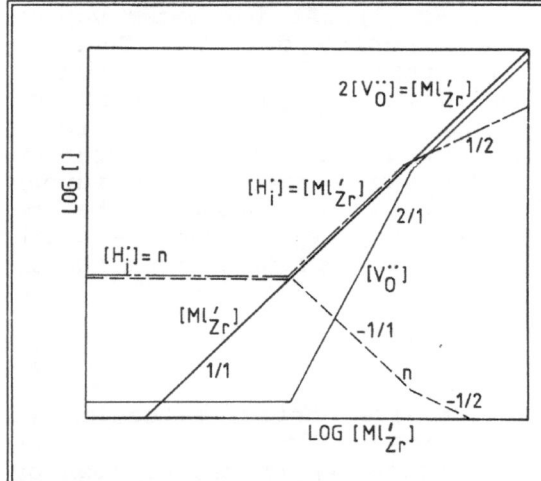

Fig.2

Schematic
illustration of
defect concentra-
tions in ZrO_2 vs.
concentration of
a substitutionally
dissolved 3+ metal
dopant Ml.

illustration of this is given in Fig.2. When the dopant becomes the dominant negative defect, it will be compensated by protons, as illustrated in the middle of Fig.2. However, the double charge of oxygen vacancies (compared to the single charge of protons) makes their concentration increase with the square of the dopant concentration. Thus, the oxygen vacancies eventually take over as the dominant positive defect at high dopant levels, as illustrated in the right-hand part of Fig.2.

Wagner [17] has shown that cubic zirconia with 8 mol-% yttria has insignificant concentrations of protons as compared to oxygen vacancies. Fig.2 shows that protons become of relatively higher importance at lower dopant levels. The aim of this work is to investigate whether proton dominance is found in undoped zirconia.

Fig.2 and the discussion of the defect relations are valid only for one structural phase at the time. We find it impossible at present to speculate about the differences in the role of protons between the phases.

2.3. Possible effects on transport properties

Dominant concentrations of protons can affect the oxide's transport properties (for instance in connection with phase transformations) simply by altering the concentrations of defects. Effects may possibly also be seen from the fact that protons decrease the radius of the oxygen ions they are bonded to [19] and locally alter the configuration of the oxygen-metal bondings. It is possible that even minority concentrations of protons play a significant role; they are generally very mobile and may have a "catalytic" effect on transport in the oxide. In this work these possibilities have been studied as regards the M-T and T-M phase transformations of zirconia.

3. Experimental

Approximately 5 g of a >99% ZrO_2 powder was cold-pressed into a tablet with diameter 22 mm and thickness 3.5 mm. The sample was hot-pressed at 1000°C for 1 hour. The degree of sintering was small, and the resulting sample had approximately 43% porosity. Sintering into a dense sample was deliberately avoided in order to have a sample which could (and in fact did) survive repeated phase transformations.

The sample was mounted between a pair of platinum foil disk electrodes. Platinum contact wires, electrodes, and sample were pressed together on top of an open alumina tube by means of a spring-loaded alumina assembly. The setup was enclosed in a larger alumina tube to allow atmosphere control. The cell was heated in a tube furnace, and the temperature was monitored by a thermocouple close to the sample. The atmosphere was air or oxygen. The water vapour level was set by drying over P_2O_5 (< 1ppm H_2O) or wetting over a saturated solution of $(NH_4)_2SO_4$ (81% RH). The dryness of the atmosphere is limited to around 30 ppm H_2O due to diffusion through the cell walls at high temperatures ([1,11]).

The ac impedance of the sample was measured with a computer-controlled Hewlett-Packard HP 4192A impedance analyzer (5Hz-13MHz) and evaluated using standard complex impedance analysis to yield the conductance of the sample.

4. Results
4.1. Complex impedance analysis
In the temperature region around the M/T phase transformation, the impedance measurements could generally be interpreted in terms of an equivalent circuit first suggested by Bauerle ([20]) and commonly found for zirconia samples: Three resistive elements in series, each in parallel with a capacitive element, are assigned to the impedance of grain interior (bulk), grain boundaries, and electrode-contacts, respectively.

The resistances found by complex impedance analysis were inverted to obtain the conductances. In the following we shall restrict the discussion to the bulk conductance around the temperatures of the M/T phase transformation. The high porosity of the sample (with the current mainly limited by the geometry of grain contact necks) makes a geometric conversion to specific conductivity very uncertain, and thus non-specific conductance is used throughout this paper.

4.2. Temperature dependence and phase transformations
At low temperatures the sample has the monoclinic (M) structure. It was heated in steps of 100°C from room temperature to 1000°C and thereafter in smaller steps as the expected phase transformation temperature was approached. Fig.3 shows the conductivity plotted versus temperature around the transformation temperature. During the measurements, the atmosphere was occasionally changed between air and oxygen and between dry and wet conditions, as shown in the plot. We shall return to the interpretation of the relatively moderate effects of these changes later.

No sign of the phase transformation was noticed up to 1152°C. Upon raising the temperature to 1157°C, however, the conductance started to increase steadily with time. This is assigned to the beginning M-T phase transformation. The increasing conductivity was recorded at 1157°C for 4

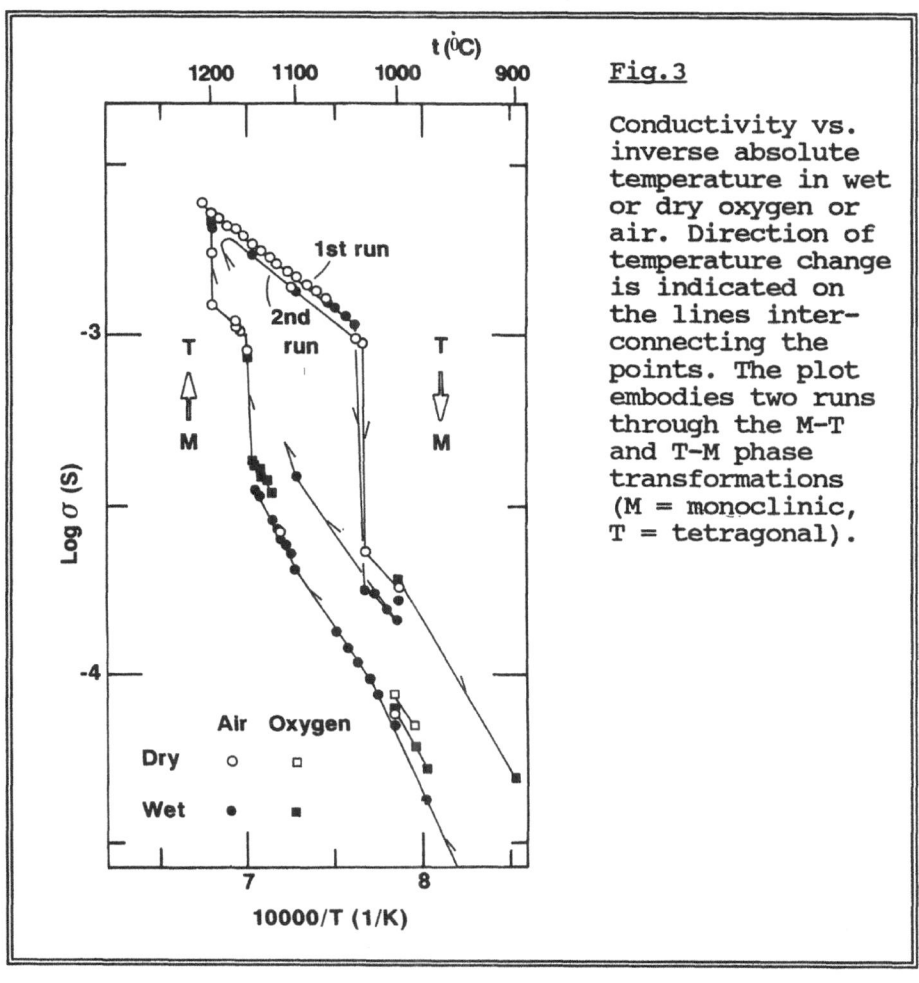

Fig.3

Conductivity vs. inverse absolute temperature in wet or dry oxygen or air. Direction of temperature change is indicated on the lines interconnecting the points. The plot embodies two runs through the M-T and T-M phase transformations (M = monoclinic, T = tetragonal).

days. During this period the atmosphere was altered between wet and dry air and oxygen. No effects could be observed on the rate of increase of the conductance. After 4 days the change in the conductance was very slow, although still noticeable. The temperature was then increased in small steps. The temperature instantly increased according to its normal temperature dependence, but then remained relatively constant. At 1200°C, however, the conductance again began to increase with time at a higher rate, before stabilizing after about 2 days (see Fig.3). Also this part of the conductance change proceeded unaffected by atmosphere variations. A further increase in temperature to 1210°C did not initiate new time-dependent conductance increases.

No attempt will be made here to discuss further the two-domain nature of the conductance increase, and the overall increase is assigned to the complete M-T phase transformation.

The temperature was now decreased in steps of 10°C. The atmosphere was varied, again using wet and dry air and oxygen, but with no significant effects. No indication of a phase transformation was noticed down to 1040°C in wet air. Upon lowering the temperature to 1030°C, however, the conductance dropped to a much lower level within a few minutes. This is interpreted as the phase transformation back to the monoclinic phase. The conductance was now higher than the original level for the monoclinic phase, as seen in Fig.3.

In order to study the last phase transformation in a second run, the M-T phase transformation was repeated (by directly heating to 1200°C to save time). Then the temperature was lowered to 1040°C, again without any sign of the phase transformation. The atmosphere changes described above were repeated to check if any of them triggered the T-M transformation, but with negative result. The tetragonal phase remained stable down to 1033°C in dry air. The T-M transformation initiated upon cooling to 1030°C and again proceeded very rapidly.

4.3. Oxygen pressure dependence of the conductance

In the monoclinic phase the conductance increased on going from air to oxygen atmospheres. This indicates a p-type conductivity, which is normally found to dominate in this phase under the conditions used ([14]).

In the tetragonal phase, on the other hand, the conductance was independent of the oxygen partial pressure. This is typical for the oxygen ion conductivity often found to dominate in this phase ([16]).

4.4. Water vapour pressure dependence of the conductance

In the monoclinic phase the conductance decreased on going from dry to wet atmospheres. This is the expected behaviour if protons from the water vapour dissolve in significant concentrations while the conductivity is p-type electronic (see Fig.1). If the protons were dominant, large effects would have been seen in the conductance over the relatively large span in water vapour pressures between dry and wet conditions. However, the observed effects were rather moderate. We thus conclude that protons are significant, but not dominant, under the conditions used.

It is likely that protons become dominant defects in monoclinic zirconia if the water vapour pressure is further increased, like in steam atmospheres. Moreover, the relative dominance of protons usually increases with decreasing temperatures.

For the tetragonal phase, no dependence on the water vapour pressure was observed. It may be concluded that protons are minority defects in tetragonal zirconia under the conditions used. When this is the case in undoped zirconia, it must also be true in doped, tetragonal zirconias under the same conditions, as discussed above.

It is possible that higher hydrogen source activities change this, and the previously mentioned effects of hydrogen atmospheres found by some investigators should be checked.

5. Summary

The M-T phase transformation was found to start at 1157°C and take several days to complete. The reverse (T-M) transformation was initiated at 1130°C and was completed within a few minutes. No effects of the water vapour pressure could be observed, neither on the transformation temperatures nor on the transformation rates.

The p-type conductivity of undoped, monoclinic zirconia in air or oxygen at high temperatures (near the M-T phase transformation) decreases moderately with increasing water vapour pressure in oxygen or air. This indicates that protons are significant, but not dominant, defects in monoclinic zirconia in wet, oxidizing atmospheres.

We find, on the other hand, no indications that protons are significant defects in tetragonal zirconia above the T-M phase transformation temperature in such atmospheres.

Acknowledgements

This work was done under partial financial support from The Royal Norwegian Council for Scientific and Industrial Research (NTNF), project MB 63.20307 "Advanced Ceramics".

References

1. T. Norby and P. Kofstad, J. Am. Ceram. Soc., **67** [12] 786-92 (1984).

2. T. Norby and P. Kofstad, J. Am. Ceram. Soc., **69** [11] 780-83 (1986).

3. T. Norby and P. Kofstad, J. Am. Ceram. Soc., **69** [11] 784-89 (1986).

4. T. Norby and P. Kofstad, Solid State Ionics, **20**, 169-84 (1986).

5. M. M. El-Aiat and F. A. Kröger, J. Appl. Phys., **53** [5] 3658-67 (1982).

6. T. Norby and P. Kofstad, High Temp. High Press., **20**, (1988), in print.

7. G. J. Hill, Br. J. Appl. Phys., Ser.2, **1**, 1151-62 (1968).

8. J. B. Bates, J. C. Wang, and R. A. Perkins, Phys. Rev. B, **19** [8] 4130-39 (1979).

9. J. C. Cathcart, R. A. Perkins, J. B. Bates, and L. C. Manley, J. Appl. Phys., **50** [6] 4110-19 (1979).

10. D. A. Shores and R. A. Rapp, J. Electrochem. Soc., **119**, 300-305 (1972).

11. T. Norby, Advances in Ceramics, **23** (1987) 107-23.

12. T. Norby and P. Kofstad, J. Phys., **47** (1986) 849-53.

13. T. Norby, Solid State Ionics, **28-30**, 1586-91 (1988).

14. R. W. Vest, N. M. Tallan, and W. C. Tripp, J. Am. Ceram. Soc., **47** [12] 635-40 (1964).

15. L. A. McLaine and C. P. Poppel, J. Electrochem. Soc., **113** [1] 80-85 (1966).

16. R. W. Vest and N. M. Tallan, J. Am. Ceram. Soc., **48** [9] 472-75 (1965).

17. C. Wagner, Ber. Bunsenges. Phys. Chem., **72** [7] 778-81 (1968).

18. F. F. Lange, G. L. Dunlop, and B. I. Davis, J. Am. Ceram. Soc., **69** [3] 237-40 (1986).

19. F. M. Ernsberger, J. Am. Ceram. Soc., **66** [11] 747-50 (1983).

20. J. E. Bauerle, J. Phys. Chem. Solids, **30**, 2657-70 (1969).

PHYSICO-CHEMICAL METHODS TO CONTROL ZIRCONIA POWDERS

P.Orlans (*), L.Montanaro (**), J.P.Lecompte (*),
B.Guilhot (*), A.Negro (**)
(*) Département de Chimie Physique des Processus
Industriels, Ecole Nationale Superieure des Mines,
St.Etienne (FRANCE)
(**) Dipartimento di Scienza dei Materiali ed
Ingegneria Chimica, Politecnico, Torino (ITALY)

1. INTRODUCTION

The development of new technologies has made it
necessary to device new materials, such as zirconium
oxide, that are capable of meeting particular needs
other solids cannot satisfy.
The main characteristics of zirconia are:
- a melting temperature of about 2700°C, i.e. higher
than that of other oxides;
- resistance to corrosion by various gases;
- ionic and electronic conductivity linked to increased
temperature or the presence of impurities;
- a degree of hardness acceptable for an oxide.
These properties enable zirconium oxide to be used in
numerous (thermal, chemical, electrical, etc)
applications at both low and high temperatures in
industry. A sound knowledge of the physico-chemical
characteristics of its initial powders is thus
essential.

2. EXPERIMENTAL

This study was conducted on three industrial zirconias:
two (A and B) were prepared by thermal decomposition,
but came from different manufacturers, and one (C) was
obtained by precipitation after chemical decomposition.
A comparison could thus be made between the physico-
chemical properties of products of different synthetic
origin. Since these properties may be influenced by the
modification of the parameters of a given synthesis, it
is important to be able to characterize the final
oxides by means of different techniques.
X-ray diffractometry (XRD) was unable to distinguish
the three oxides, since they are all monoclinic when
they crystallise at room temperature. Their analytical
characteristics are very similar, though A usually
contains fewer impurities and C has more calcium (Table
1).

TABLE 1

Analytical characteristics of the zirconium oxides

Oxide	Na_2O	SiO_2	Fe_2O_3	TiO_2	CaO	MgO	HgO_2
A	0.11	0.30	0.03	0.11	0.06	0.01	--
B	--	--	0.03	0.17	--	--	2.49
C	0.06	0.10	0.03	0.09	0.20	0.04	1.85

The specific surface area (Table 2), as measured by the
BET method, is higher for a "precipitated" as opposed
to a "thermal" zirconia. A distinction between the two
synthesis procedures can also be made from the size of
the particles, whose diameter corresponds to the 50% of
the size distribution, as measured by laser
granulometry.

TABLE 2

Morphological characteristics of the three powders

Oxide	S.S.BET (m^2/g)	$\Phi_{50\%}$ (μm)
A	2.41	4.2
B	6.68	6.6
C	25.16	1.2

Scanning electron micrographs (Figure 1 a,b,c) provide
excellent confirmation of mean particle diameters.
Zirconia A displayed small agglomerates of the order of
2 - 5 μm. Those of the other "thermal" zirconia (B)
were larger (about 10 μm) and their grains measured
nearly 1 μm.
Grains measuring > 1 μm were agglomerated in zirconia
C.
Impurities and morphology are not the only variables in

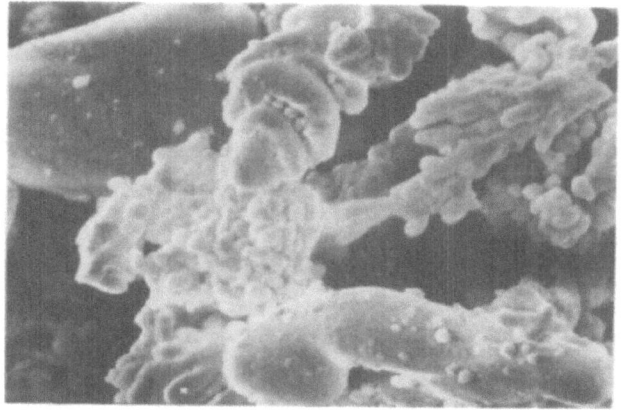

1 μm

Figure 1a - Scanning electron micrograph of zirconia A.

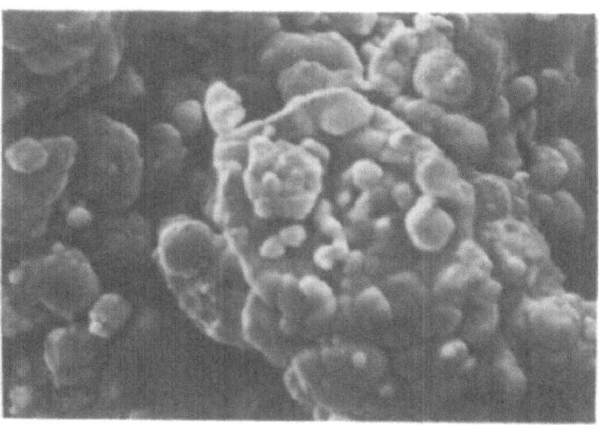

1 μm

Figure 1b - Scanning electron micrograph of zirconia B.

1μm

Figure 1c - Scanning electron micrograph of zirconia C.

zirconia synthesis, since this may either create or modify intrinsic defects in both the surface and the bulk of the material.

The nature of these defects is often difficult to determine experimentally. The energy bands theory, however, can be applied to measurements of light emission following electron transfers between the trapping centres and the emissiom centres associated with the defects.

Tribo- and thermoluminiscence are the techniques of choice. In a certain number of cases, they have made it possible to attribute defects to reactivity of the solid (2,3,4).

Triboluminescence signals are themselves enough to distinguish the three zirconias (Figure 2).

Zirconia A has two peaks at 145°C and 375°C. The surface ratio between the second and the first peak is of the order of 500.

The curve of zirconia B is more complex. There are four peaks at 75°C, 142°C, 304°C and 393°C respectively.

Their surface ratios are much smaller, the greatest being of the order of 20.

Oxide C has three peaks at 85°C, 136°C and 376°C.

Measurements of the surface of each peak will show the predominance of certain defects in an oxide.

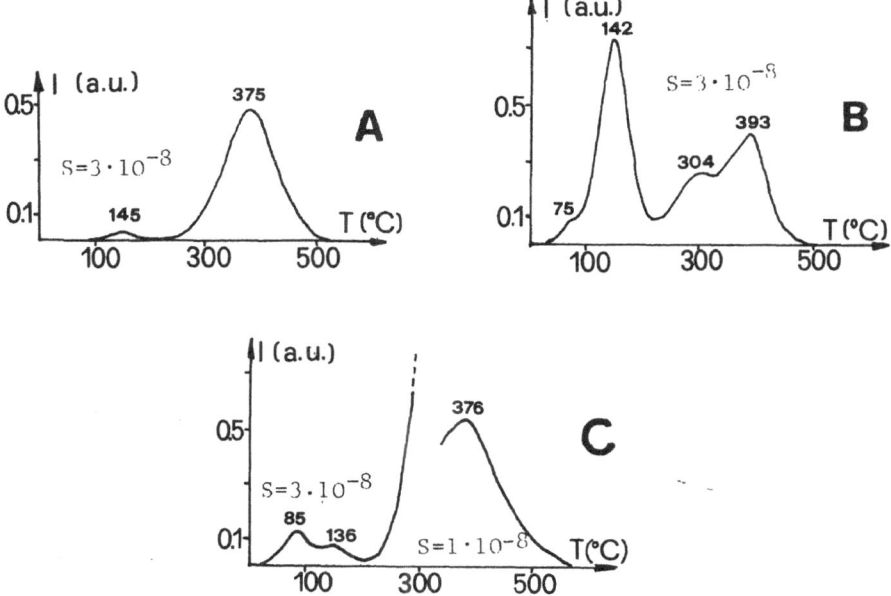

Figure 2 - Triboluminescence signals of the three zirconium oxides A,B and C.

In addition, when the peaks of several products lie at approximately the same temperature, their surface can be compared.

Peaks of this kind can be likened to constraints released by heat treatment (5). Their concentration varies from one specimen to another.

These results show that both the defects observed and their concentration are attributable to the type of synthesis.

Similar observation can be obtained using thermoluminescence, but, in this case, the defects observed are permanent in the solid.

The thermoluminescence signals for the three oxides are shown in Figure 3.

A has three peaks at 75°C, 133°C and 243°C. The second has the highest surface.

B has only one peak at 133°C. This is the most intensive of the signals for all three products.

Oxide C has two peaks of virtually the same intensity at 77°C and 138°C.

The presence of defects and differences in physico-chemical properties are responsible for changes in the pattern of the polymorphic transformation of a zirconia.

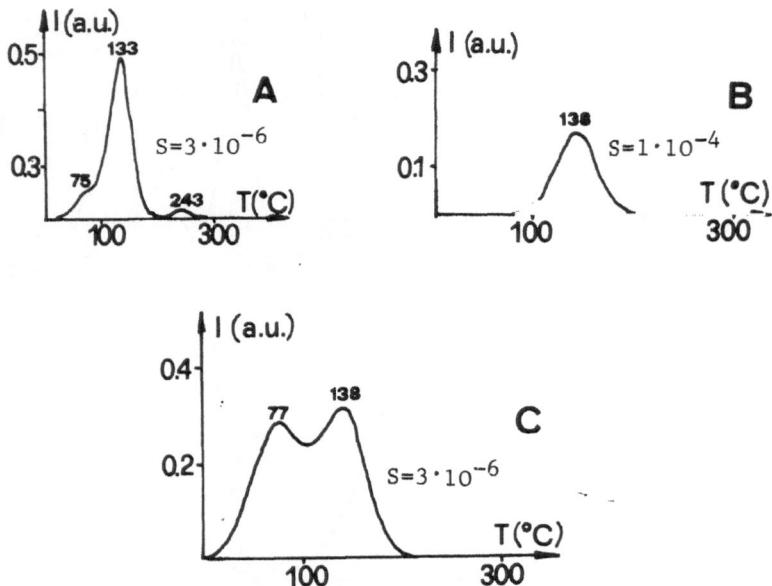

Figure 3 - Thermoluminescence signals of the three oxides A, B and C.

Differential thermal analysis (DTA) shows that an endothermic effect is associated with the monoclinic ---> tetragonal (m ---> t) transformation on heating, and an exothermic peak with the t ---> m transformation on cooling.
There are considerable differences between the DTA thermograms (Figure 4).
The m ---> t transformation begins at 1100°C and 1150°C for oxides A and B respectively.
The peak for C is weaker and no distinct start temperature can be discerned.
The t ---> m peaks begin at 967°C, 948°C and 815°C respectively.
The hysteresis associated with these transformations is thus altered, as shown by the high-temperature X-ray diffractograms (Figure 5).
Polymorphic transformation of zirconia is associated with significant changes in volume: shrinkage during m ---> t and expansion during t ---> m.
The behaviour of a zirconia during heat treatment can thus be investigated by dilatometry.
Test specimens were prepared with a 400 MPa isostatic

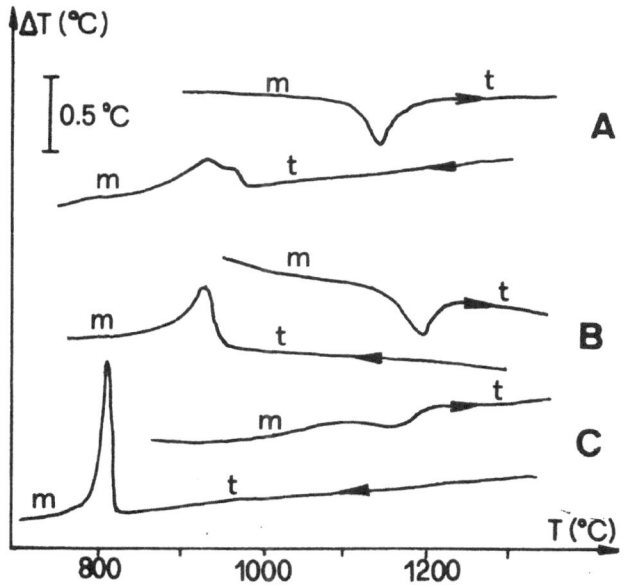

Figure 4 - DTA curves of the three oxides A, B and C.

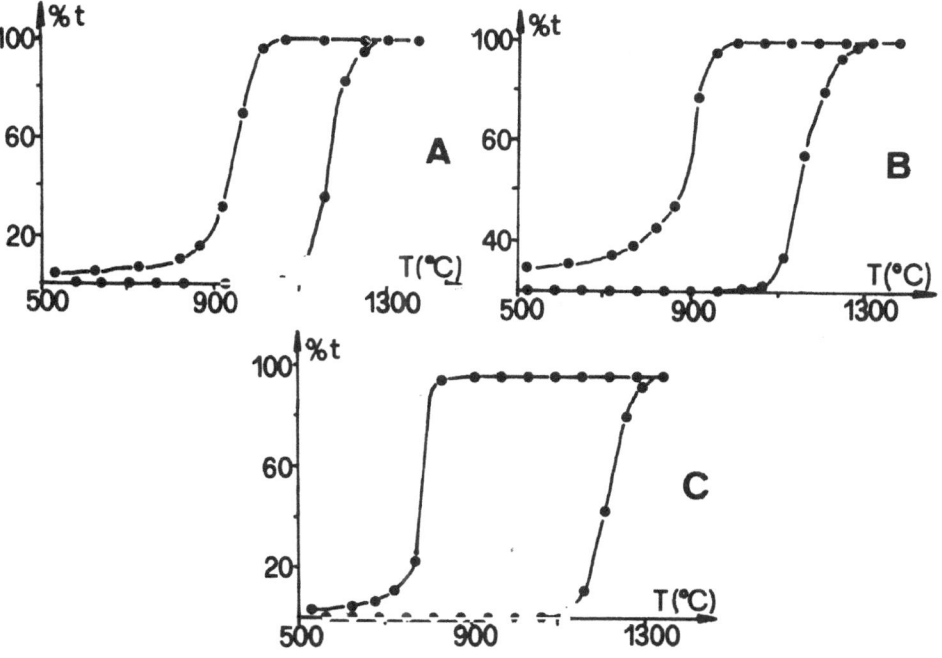

Figure 5 - Polymorphic transformations in the zirconias A, B and C followed by the high-temperature XRD.

press.

Heating and cooling rates of 20°C/min were chosen to permit comparison with the DTA thermograms and the X-ray diffractograms.

The dilatometric curves are presented in Figure 6.

An usual thermogram (expansion, shrinkage at the m ---> t transformation and expansion at the t ---> m transformation during cooling) was observed for specimen B only.

The same phenomena are apparent in curve C. Their dimensions, however, are very different. The specimen, in fact, was devoid of cohesion at the end of the thermal cycle.

Specimen A displayed uniform expansion up to 1170°C, followed by sudden expansion peaking at 1275°C. This was followed by a 1% shrinkage. During cooling, an approximately 1% expansion was noted at about 1020°C, followed by slight shrinkage. A residual expansion of about 2% was present at room temperature.

The 1020°C temperature observed for the expansion during cooling corresponds to the t ---> m transformation and is in line with the literature and the DTA and XRD findings.

On the other hand, the shrinkage beginning at 1275°C does not correspond to the m ---> t transformation (it was found that a pressure of 400 MPa had very little effect on the polymorphic transformations of the zirconia followed by DTA).

The sample is thus expanding when the transformation is over (as shown by XRD). If the thermal cycle is repeated on the same specimen, the sudden expansion observed during the first cycle does not occur (max 1275°C).

The phenomena observed are in line with the XRD data.

Sudden expansion is not observed when the temperature rise is 2°C/min instead of 20°C/min. If the powder is heated at 1450°C without pressure and then used to make a specimen, expansion appears around 1275°C.

This set of experiments suggests that the "abnormality" in specimen A is associated with the spatial arrangement of the particles in the initial powder and their elongated shape.

The influence of shape during heating can be deduced from the weaker expansion of round-particled specimens B and C.

3. CONCLUSIONS

This study has shown that the characteristics of zirconia powders can be determined.

227

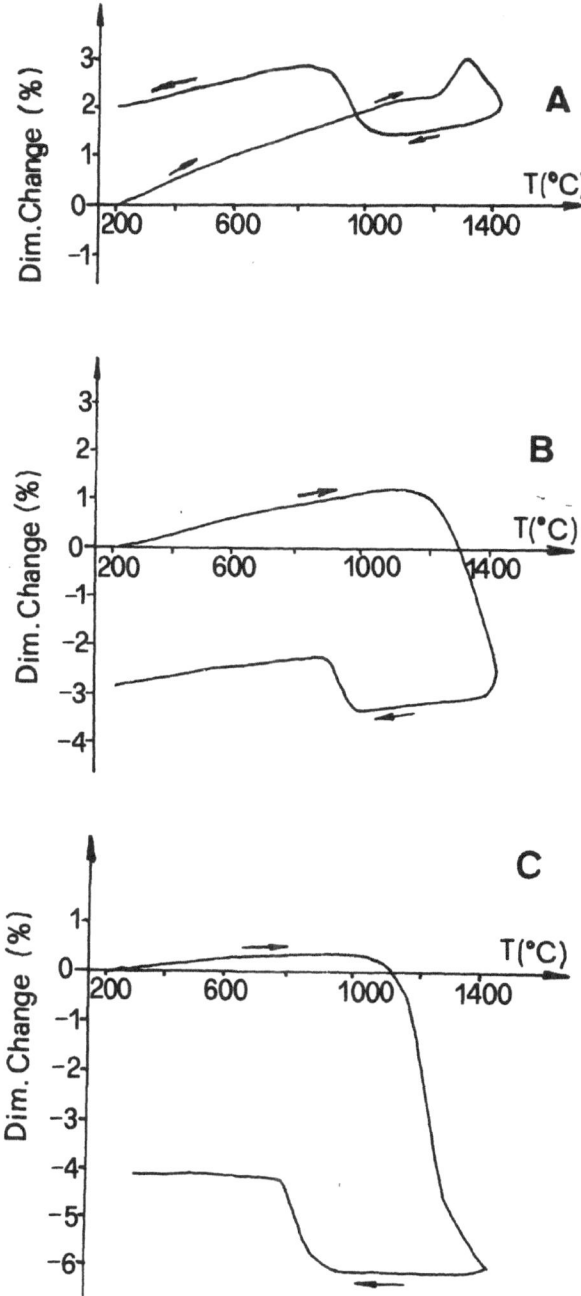

Figure 6 - Dilatometric curves of the three oxides A, B and C.

Impurities and particle morphology are not the only variables during powder synthesis, since this may generate or alter intrinsic defects in the surface or bulk of the material. These defects can be detected by tribo- and thermoluminescence.

The three zirconias investigated in this study displayed microconstraints and permanent defects resulting in significant changes in both their transformation kinetics (as shown by DTA and XRD) and their physical characteristics (as checked by dilatometry).

REFERENCES

(1) P. Iacconi, Thèse Doctorat es Science, Université de Nice (1979)

(2) P.Fierens, J.Tirlocq, J.P.Verhaegen, Cem. Concr. Res. $\underline{3}$, 549 (1973)

(3) M.Triollier, B.Guilhot, Cem. Concr. Res. $\underline{8}$, 311 (1978)

(4) A.A.Fournier, J.P.Lecompte, M.Soustelle, M.Murat, A.Bachiorrini, A.Negro, Cem.Concr.Res. $\underline{15}$,151 (1985)

(5) D.Turpin, Thèse Docteur-Ingénieur, ENSM St.Etienne (1985)

EFFECT OF DIFFERENT OXIDATION DEGREES ON THE STRUCTURE AND PROPERTIES OF STABILIZED ZIRCONIA PLASMA SPRAY COATINGS.

P.G. Orsini, R. Dal Maschio, F. Marino, P. Scardi
Dipartimento di Ingegneria, Università di Trento,
Mesiano di Povo – 38050, Trento, Italy

Introduction

The utilization of zirconia plasma spray coatings in even more severe conditions have increased in the last few years the attention to the influence of all the process parameters on the structure and properties of the coatings.

Briefly, in order to manufacture a coating, an electric arc is produced between two water-cooled electrodes, a cathode and a nozzle-shaped anode; this electric arc is compressed, constricted and well stabilized at the anode by a carrier gas. The arc heats, accelerates and ionizes the gas which becomes the plasma flame or jet.

A carrier gas is used to supply the powder to the injector where it enters the plasma. In the plasma, the powder is accelerated and heated, and becomes semi-molten or molten. The particles impact in this condition onto the surface of the work-piece producing the coating.

The plasma gases used are Ar, He, H_2, N_2, and also mixtures of these gases. Ar is very useful as it is easily ionized and forms a very stable arc at low operating voltage. The energy contained in the plasma also depends on the plasma current and the gas flow.

The aim of this work is the characterization of stabilized zirconia coatings with the composition ZrO_2-8% Y_2O_3 obtained by changing the ratio Ar/H_2 of the plasma gas. In particular samples have been prepared using H_2 of different partial pressures.

Experimental procedure

The influence of different oxidation degrees of the ceramic coatings on their microstructure has been evaluated on both as made and in air thermally treated to 500°C samples by means of X-Ray diffraction (XRD) for recognition and quantitative evaluation of the phases, X-Ray Photoelectron Spectroscopy (XPS) and Electron Paramagnetic Resonance (EPR) for the estimation of the real oxidation degree of the coatings.

The four samples (Fig.1) obtained spraying a ceramic coating 50 microns thick on an aluminum silicium alloy 2 mm thick, are characterized by different colors, changing from black (sample 1) to white (sample 4) with the increase of Ar/H_2 ratio.

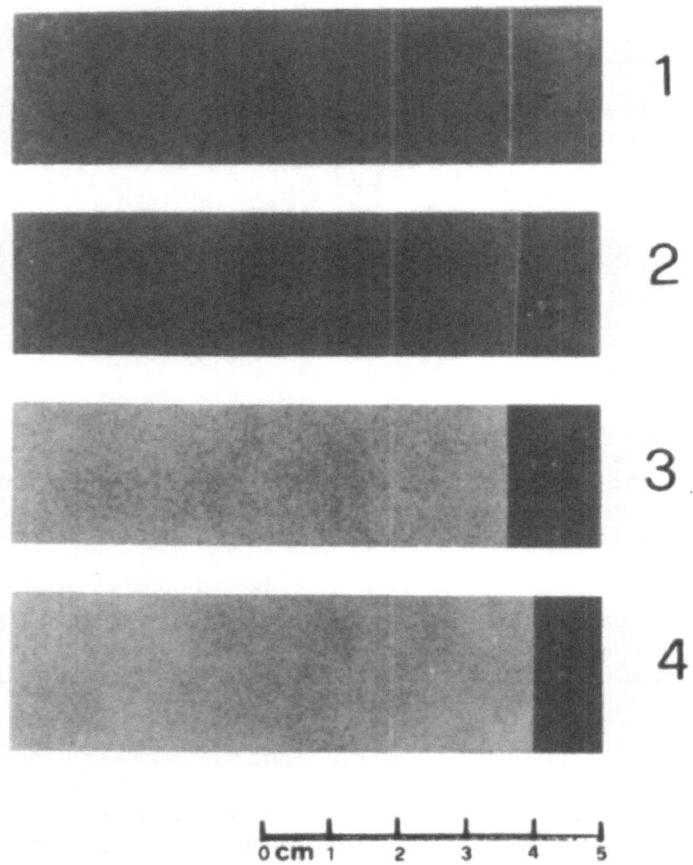

Fig. 1 : Four different samples, obtained spraying ceramic coating 50 microns thick on an aluminum silicium alloy 2 mm thick.

In Fig.2 it is possible to see an XRD spectrum relative to sample 1 as an example to show the good resolution of the instrument and the use of numerical analyses to separate the different phase contribution to the diffraction pattern [1]. All the XRD analyses were performed in a Rigaku D/max III B diffractometer adding four runs with a speed of 1 deg/min, a step of 0.02 deg and Cu Kα radiation wavelength. The peak separation method is based on a least squares non linear fitting program. In the figure are shown peaks relative to zirconia monoclinic and tetragonal phases; this information can be obtained by analyzing the 70-77 deg region where (400) peaks are.

So turning the integrated intensities into volumetric percentages by means of Toraya formulations [2,3] of (111) and (11$\bar{1}$) peaks for monoclinic and of (111) peak for tetragonal phases, it is possible to say that the monoclinic phase is about 4% and the cubic phase, if present, is in a very small quantity non observable by x-ray analysis.

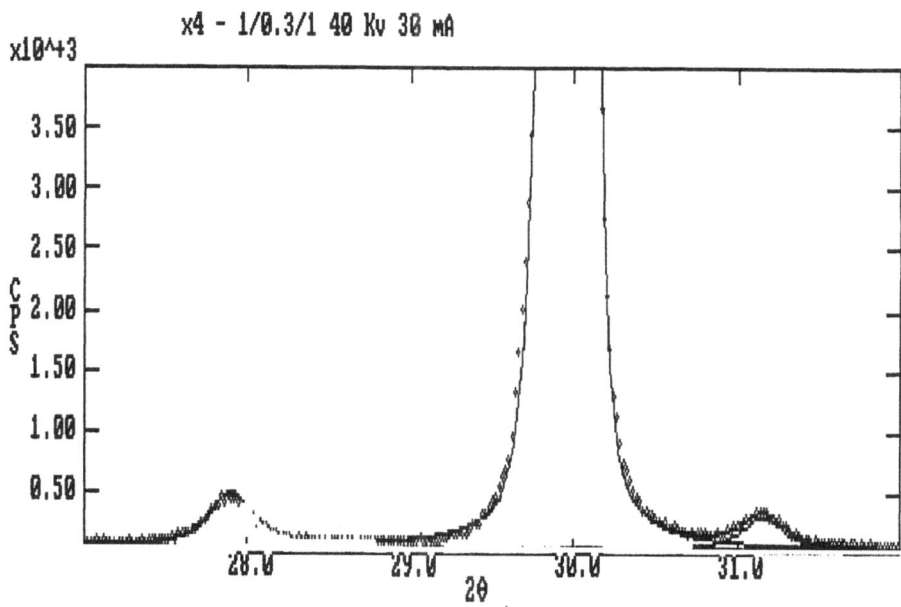

Fig. 2: XRD spectrum relative to sample 1 evidencing (111) and (11$\bar{1}$) peaks for monoclinic and (111) for tetragonal phases.

In Table 1 is a list of the percentages of monoclinic phase of the four different samples, with a difference between sample 1 and 4 of about 3%.

Fig.3 shows some representative XPS spectra in the Zr 3d region of a coating made with a very high Ar/H_2 ratio as reference and of samples 2 and 4. They were performed with a Leybold & Heraeus LHS10 spectrometer equipped with an EA11 electron analyzer using a monochromatized Al $K\alpha_{1,2}$ radiation (hv = 1486.6 eV). The electron analyzer operated in FAT mode, selecting a constant pass energy of 50 eV. Under such conditions the full width at half maximum (FWHM) of Ag3d 5/2 peak of an Ar^+ sputtered silver metal surface was 1.01 eV.

All measurements were performed at pressure lower than 10^{-10} mbar in the analysis chamber. The binding energies (BE) were referenced to the Fermi level of the electron analyzer and the BE scale was calibrated against the Au4p7/2 and Cu 2p3/2 peaks at 83.8 eV and 932.5 eV respectively.

The correction of the binding energy shift due to the steady state surface charging in insulating film was made by referring to the C_{1s} line of absorbed hydrocarbons (284.5 eV).

It is clearly possible to distinguish the multiple splitting of Zr 3d5/2 and Zr3d3/2 levels at 188.95 eV and 191.33 eV respectively.

Taking into account the charging effect and the value of C_{1s} line we obtained for the binding energies of Zr 3d 5/2 and Zr 3d 3/2 the values of 183.3 eV and 185.7 eV, suggesting the assignment of these two peaks to ZrO_2 in excellent agreement with other authors [4].

Concerning the change of the lineshapes observed for sample 2 and 4 spectra in respect to the reference sample, and more particularly the broadening of the peaks with a hump towards the lower binding energy region, we can suggest the presence of other zirconium oxides, such as ZrO_{2-x} suboxides. Therefore a second oxide component with the same lineshape as ZrO_2 was included in the curve fitting procedure and as an example, the modified data obtained for sample 2 are shown in Fig.4.

The existence of these zirconium suboxides has already been observed by means of XPS and STEM investigations [5,6], showing a degradation mechanism caused by the interaction between plasma and ceramic particles during the spraying.

As XPS is an analytical technique involving a surface layer about 50 nanometers thick, it is necessary, to confirm such results by informations concerning the bulk of the coating.

Table 1: Percentages of monoclinic phase relative to monoclinic plus tetragonal phases quantity.

SAMPLE	MONOCLINIC %
1	3.9
2	5.0
3	5.8
4	6.7

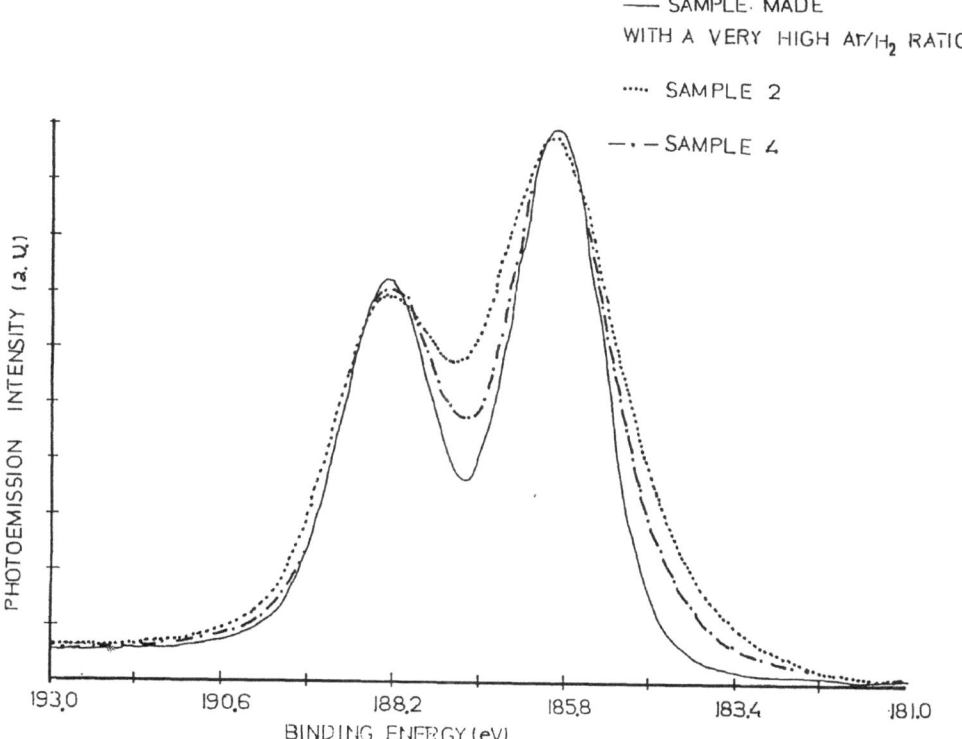

Fig. 3: XPS spectra in the Zr 3d region of the reference sample and of samples 2 and 4.

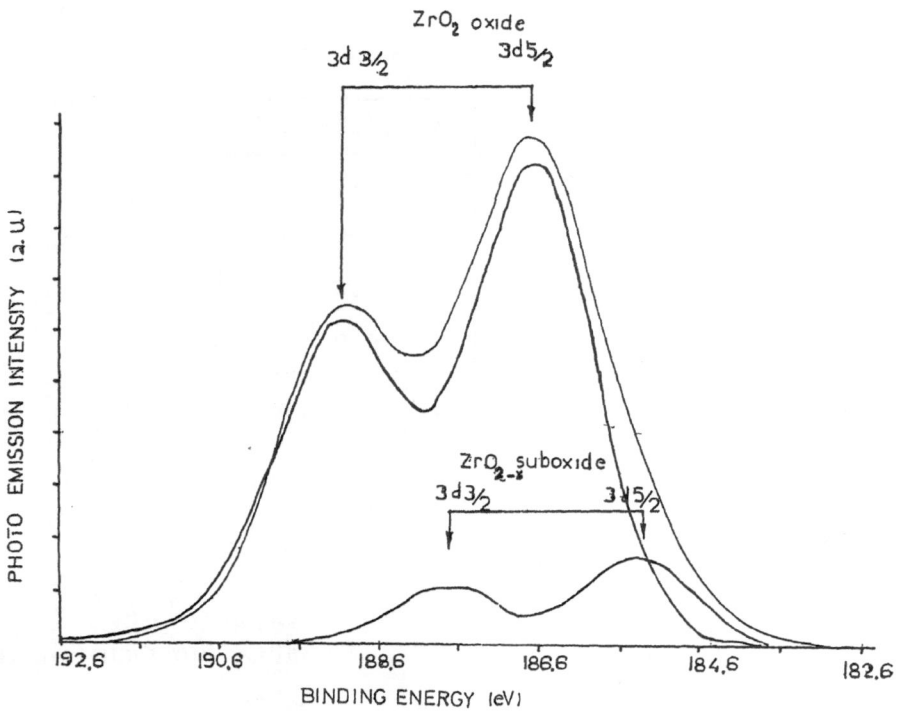

Fig. 4: XPS spectrum of sample 2 in the Zr 3d region evidencing the contributions of ZrO_2 and ZrO_{2-x}.

EPR may be a suitable technique to this goal, but it is necessary to detach the coating from the substrate, to avoid signals overlapping. In order to perform this observation, a coating 1 mm thick, made under the same experimental conditions as sample 1, was detached from the substrate, crushed and analyzed as made and after a thermal treatment at 500 °C in air. Fig. 5 shows the result of the EPR analysis, while that of the XRD-analysis is shown in Fig. 6.

EPR, a spectroscopic technique consisting in the irradiation of a solid sample put in a magnetic field of about 0.3 tesla, by means of a 9.5 GHz electromagnetic radiation, is able to detect paramagnetic centers, e.g. impaired electrons. Frequently they are ions with a particular oxidation degree. In this case the spectrum of the as made sample shows a peak with G tensor values ($g_1 = 1.98, g_2 = 1.86$) attributable to a d^1 ion, peculiar to Zr^{3+} ions. This

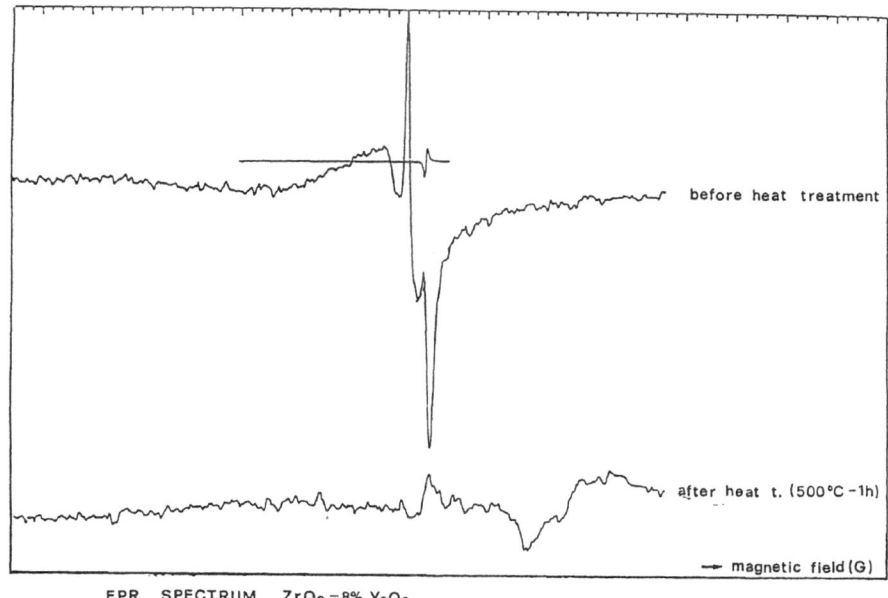

EPR SPECTRUM $ZrO_2 - 8\% \ Y_2O_3$

Fig. 5: EPR spectra on sample 1 ceramic coating detached from the substrate and crushed, as made and after a thermal treatment to 500°C for 1 h.

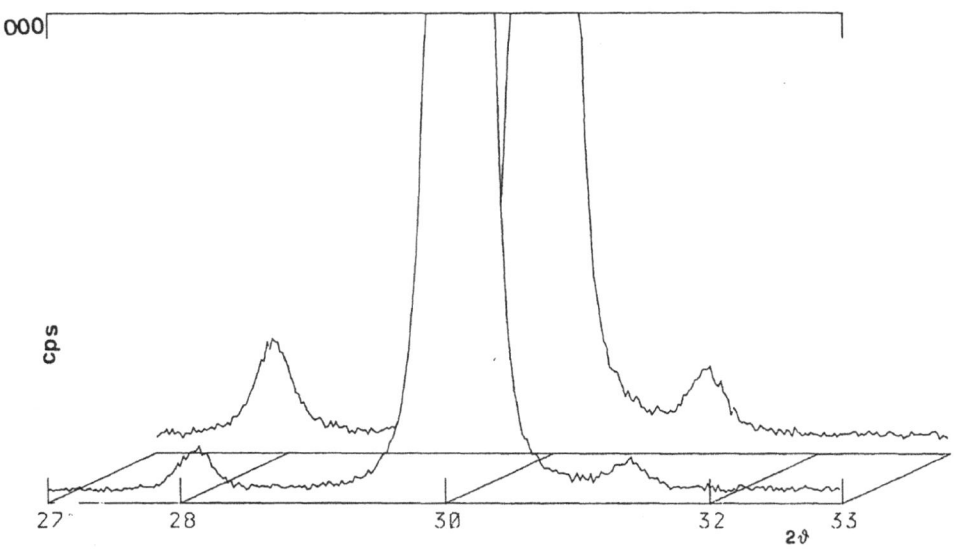

Fig. 6: XRD patterns of sample 1 ceramic coating in the some experimental conditions as in Fig. 5.

peak is conspicuously absent in the spectrum of the same sample after the thermal treatment at 500 °C.

Parallel XRD-analysis, (Fig.6) shows a very little increase in the monoclinic phase content of about 0.5% (a very little but reliable increase) as it has been inferred from the corresponding peaks integrated intensities ratio.

Discussion

Ar/H_2 ratio is one of the most important parameters in plasma spray deposition in air of ceramic coatings on metals. The effect of changing such ratio by increasing the partial pressure of H_2 is the promotion of sub stoichiometric oxides formation as evidenced by XPS and EPR spectra, both by creating more reducing conditions in the plasma and increasing its enthalpy [7] . The creation of sub stoichiometric zirconium oxides, and therefore of further anionic vacancies, in addition to vacancies due to the Y^{3+} presence stabilizes the zirconia tetragonal phase [8]. In fact, as shown in Tab.1, the ratio monoclinic/tetragonal decreases with the H_2 partial pressure increase.

Nevertheless the thermal treatment to 500°C made on the detached and crushed sample in order to restore the stoichiometry of zirconia, evidenced the disappearance in the EPR spectrum of Zr^{3+} ions signal, but a very little increase in the monoclinc content.

This fact suggests that the monoclinic/tetragonal ratio change in the four samples is mostly due to another phenomenon. The increase of partial pressure of H_2 in the plasma gas changes its enthalpy, so causing different thermal conditions during the deposition. As a consequence it induces a different residual stress field in the ceramic coating, particularly a higher compressive stress field that contributes to the tetragonal phase stabilization. Obviously in the detached and crushed sample this coupling residual stress field is negligible, so it is possible to separate the influence of the two different phenomena on the monoclinic percentage.

Conclusion

During the plasma spray deposition in air of $ZrO_2-8\%Y_2O_3$ coatings the Ar/H_2 ratio of the plasma gas has been changed, using H_2 with different partial pressures. Four samples have been prepared, in which XRD evidenced a monoclinic/tetragonal zirconia phases ratio inversely proportional to the H_2 plasma gas content. By means of XPS and EPR the presence of sub stoichiometric zirconium oxides has been shown that may contribute to the stabilization of tetragonal phase.

As the change of H_2 partial pressure causes the increase of plasma gas enthalpy, it induces different thermal conditions during the deposition, particularly different compressive residual stress field in the ceramic coating, so stabilizing the tetragonal phase. In order to separate the two different contributions a coating detached from the substrate and crushed has been analyzed before and after a thermal treatment at 500°C in air. As the stoichiometry of zirconia was restored, but with a little increase of monoclinic phase content, it can be inferred that the influence of the degree of coating oxidation on the monoclinic phase percentage is low as compared with the influence of coupling residual compressive stress field.

REFERENCES

1. S. Enzo, L. Lutterotti, P. Scardi, Proceedings of "First International Conference on Plasma Surface Engineering", September 19-23 (1988), Garmisch- Partenkirchen (FRG)

2. H. Toraya, M. Yoshimura and S. Somiya, J. Amer. Ceram. Soc. 67 (1984), C119-C121

3. Ibid., Ibid., C183-C184

4. R. Kaufmann, H. Klewe-Nebenius, H. Moers, G. Pfennig, H. Jenett and H. J. Ache, Surface and Interface Analysis, 11 (1988), pp. 502-509

5. S. Steeb and A. Riekert, Journal of the Less-Common Metals 17 (1969), pp. 429-436

6. L. Kumar, D. D. Sarma and S. Krummacher, Applied Surface Science, 32 (1988), pp. 309-319

7. E. Pfender, Surface and Coatings Technology, 34 (1988), pp. 1-14

8. P. Aldebert and J. P. Traverse, J. Amer. Ceram. Soc. 68 (1985), pp. 34-40.

AC Impedance Complex Plane Studies on
Alumina-Zirconia and Mullite-Zirconia Composites

M. I. Osendi and J. R. Jurado
Instituto de Ceramica y Vidrio,CSIC.
Arganda del Rey, Madrid. Spain.

ABSTRACT

Mullite-ZrO_2 composites with 10, 20 and 40 vol.% of ZrO_2 particles have been electrically characterized by AC impedance complex plane spectroscopy.The effect of the zirconia dispersed particles on the conductivity of mullite in the temperature range of 500-900°C has been determined. Within the temperature and frequency intervals considered only one perfect grain interior arc has been observed. On the other hand, two Al_2O_3-ZrO_2 composites have been formulated keeping constant the proportion of ZrO_2 (8 vol%), but one of them with a small amount of TiO_2 (0.5 vol%). The effects of TiO_2 doping and annealing time on the impedance semicircle evolution have been investigated on these composites. Single phase Al_2O_3 and mullite samples have also been studied for comparative purposes. The impurity controlled conduction mechanisms are also discussed in the present work.

1.INTRODUCTION

Mullite-ZrO_2 and alumina-ZrO_2 ceramics have been widely studied for structural applications due to the improved mechanical properties resulting from the presence of ZrO_2 particles[1]. Additionally, ZrO_2 ceramics are known for their properties as solid electrolytes and their use as oxygen conductors[2]. According to these statements, it seems logical to study the electrical behavior of the above composites, trying to establish the possible interest of these composites ceramics from the electrical point of view. Due probably to the presence of different phases in those composites, very few works have been done on their electrical properties.

AC impedance complex analysis has shown its usefulness in studying separately the contributions of the grain, grain boundary and electrode-ceramic interphase on the total conductivity[3]. Therefore the measurements of the electrical properties of these composites can be an additional tool to depict their microstructure. The effects of average grain size, nature of phases and heat treatments on the electrical conductivity can be an interesting exercise to characterize properly the microstructure of these materials.

The goal of the present work is to perform electrical conductivity test in order to characterize electrically Al_2O_3-ZrO_2 and mullite-ZrO_2 composites and establish the nature of the conduction mechanism in these composites.

2.EXPERIMENTAL PROCEDURE

2.1. Sample preparation

a) Mullite-ZrO_2 composites. Several compositions with increasing amounts of ZrO_2 were prepared: 10, 20 and 40 % by volume. Mullite powders were obtained by the thermochemical treatment of a halloysite by the method describe elsewhere[4].
Zirconia* powders were monoclinic in origin. Chemical analysis of both powders are collected in Table I. Materials were processed by conventional methods: attrition milling, isostatic pressing and sintering in a furnace.
b) Alumina-ZrO_2 composites. Two kind of composites were done, both with a 8 % by volume of ZrO_2 particles, but one of them with a small addition of TiO_2, 0.5 % by volume. The ZrO_2 powders were of the same type to the above mentioned. Characteristic of the Al_2O_3** and TiO_2*** powders are also listed in Table I. Powder processing was similar to the already described for mullite-ZrO_2 composites. Both composites will be named AZ and AZT, respectively. Single phase Al_2O_3 was also prepared for comparative purposes.

TABLE I.- Chemical Analysis of the Starting Powders

(wt. %)	Mullite	Zirconia	Alumina	Titania
Al_2O_3	72.2	-	99.8	-
SiO_2	26.9	0.2	0.04	-
Fe_2O_3	0.15	0.02	0.02	-
TiO_2	0.07	0.15	-	99.9
MgO	0.02	-	0.09	-
Na_2O	0.07	-	0.04	-
ZrO_2	-	99.5	-	-
Others	0.2	0.05	-	-
	(CaO)	(SO_3)		

* SC-20, Magnesium Elektron, UK.

** RC 172 DBM, Reynold USA.

***AR Grade, Merck, Darmstadt, FRG.

2.2. Ceramic Properties

The sintering temperature was 1570°C for each composite, the time ranging from 1 to 2.5 hours. Some Al_2O_3-ZrO_2 composites were also subjected to long time annealing periods (40 hours) at the same sintering temperature. In Table II and III the main ceramic properties of the both type of composites are summarized.

TABLE II.- Ceramic Properties of Mullite-ZrO_2 Composites

Sample	Density (%)	t-ZrO_2 (%)	$\bar{d}_M$ (μm)	$\bar{d}_Z$ (μm)
Mullite	97(±1)	--	0.4(±0.1)	--
M+10%ZrO_2	99(±1)	4.4(±0.1)	0.7(±0.1)	0.9(±0.1)
M+20%ZrO_2	99(±1)	2.0(±0.1)	0.8(±0.2)	0.9(±0.1)
M+40%ZrO_2	97(±1)	0	>1	~ 1.5

2.3. Electrical Properties

Bar shape specimens of dimensions 0.4cmx0.5cm and discs of 0.9 cm in diameter, with thicknesses of 0.2-0.3 cm were cut from the sintered samples for the electrical measurements. Sputtered platinum electrodes were used as electrode contacts. The two and three points methods were used indistinctly in the measurements. A computer controlled impedance analyzer[****] was used to obtain the ac impedance spectrum in the frequency interval 10-10^7 Hz. The samples were tested in the range of temperatures 500-900°C, and the spectra were taken after a period of stabilization of at least 20 min. Data were collected during both the heating and cooling regimes. Data were directly processed by the computer to obtain the impedance spectra as well as the conductivity and activation energies.

TABLE III.- Ceramic Properties of Al_2O_3-ZrO_2 Composites

Sample	Density (%)	t-ZrO_2 (%)	$\bar{d}_A$ (μm)	$\bar{d}_Z$ (μm)
Alumina	98(±1)	--	--	--
AZ	99(±1)	7.2(±0.1)	1.1(±0.1)	0.6(±0.3)
AZT	99(±1)	5.5(±0.1)	1.6(±0.1)	0.7(±0.3)
AZ(ann40h)	99(±1)	4.9(±0.1)	2.5(±0.2)	1.0(±0.5)

[****] Model 4192A, Hewlett-Packard, USA.

3. RESULTS AND DISCUSSION

A) Mullite-ZrO_2 Composites.

The impedance spectrum of the single mullite phase sample is shown in Fig. 1. It is important to note that inside the temperature range investigated only one perfect arc was observed.

This seems to indicate that the grain boundary resistivity contribution does not appear. This behavior is probably due to the nature of the grain boundary phase, formed by a glassy phase with a composition very similar to that of the mullite[5], and therefore posses the same relaxation time than mullite at those temperatures. The depression angle is very low close to zero, which involves a pure Debye dispersion process. When temperature is raised, the conductivity of the single mullite increases, but the depression angle (see Fig. 1) does not change. As a consequence, the arc intersection on the real ρ'-axis can be considered as the real resistivity of this mullite material.

When 10 % by volume ZrO_2 particles are added to the mullite, the arc behavior does not change apreciably and an impedance spectrum similar to that of the mullite is obtained. Nevertheless, in the case of the composite with 20 % by volume of ZrO_2 a certain arc evolution with the temperature is observed. Specifically an increase in the depression angle up to 19° is obtained at the temperature of 700°C. This can indicate a strong effect of zirconia particles on the dielectric dispersion process, but not on the predominant conduction mechanism. If 40% by volume of ZrO_2 particles are introduced into the mullite matrix a clear separation of arcs is obtained distinctively when temperature raises (Fig. 2). Therefore, for this proportion of ZrO_2, a mixture of conduction mechanisms is obtained corresponding to each of the present phases. Then, it can be deduced that only for ZrO_2 amounts superior to 20 % by volume the dispersed phase has an important contribution to the conduction mechanism.

From the interseption of the arc with the ρ'-axis, the conductivity has been obtained and plotted against the temperature, as depicted in Fig. 3. Table IV shows both activation enthalpy and energy for conduction in each sample. The values obtained are too low for a cationic conduction mechanism. However, the presence of impurities on the mullite starting powders, specifically Ti^{4+} and Fe^{3+} (see Table I) could induce us to consider that an electronic conduction mechanism impurity controlled takes place in these samples. Conductivity of single phase Al_2O_3 is also listed for comparison (Table IV). An important result to notice is that Al_2O_3 presents much better insulator properties than mullite.

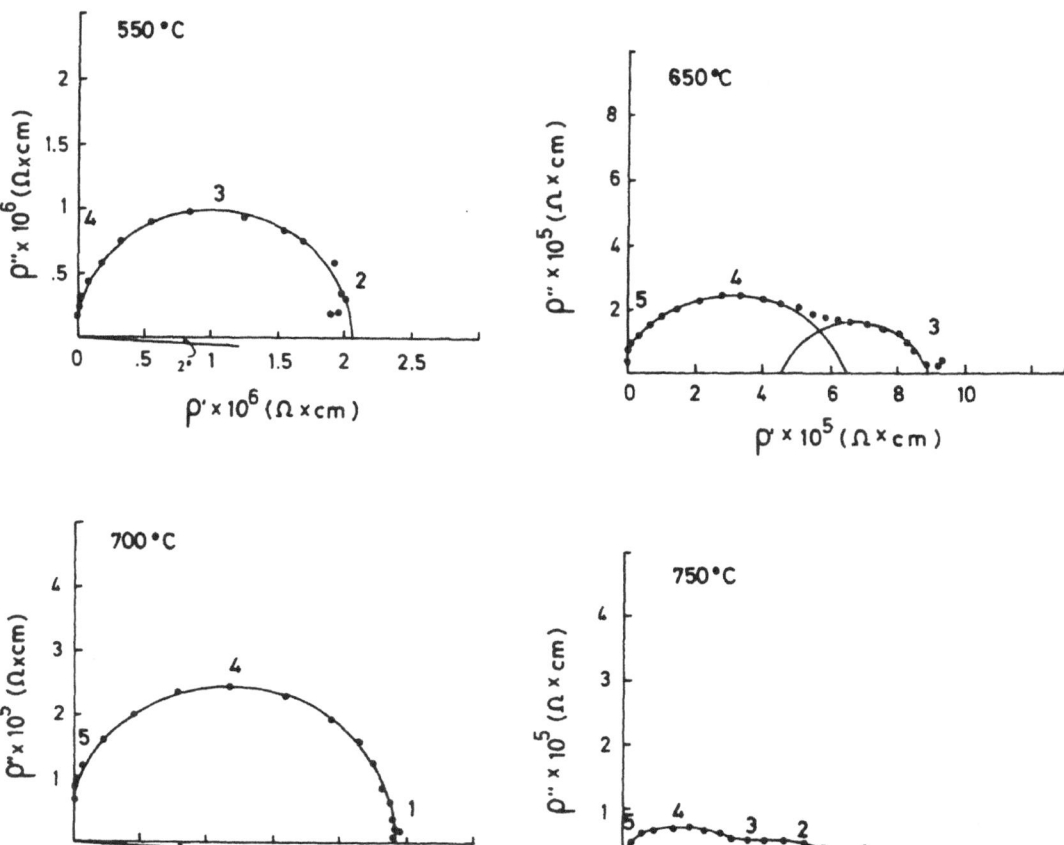

Figure 1.- AC impedance plots of Mullite at the shown temperatures.

Figure 2.- AC impedance plots of M+40vol.%ZrO_2 composite at different temperatures.

TABLE IV.- Electrical Conductivities and Activation
Energies of the Mullite-ZrO_2 and Al_2O_3-ZrO_2
Composites.

Sample	$\sigma(530°C)$ (Sxcm^{-1})		$\sigma(730°C)$ (Sxcm^{-1})		H(eV)	---E(eV)
Mullite	4.80E-7	(2°)	2.85E-6	(3°)	0.73	0.65
M+10%Z	3.66E-7	(4°)	2.47E-6	(6°)	0.78	0.69
M+20%Z	5.65E-7	(5°)	3.46E-6	(10°)	0.74	0.66
M+40%Z	3.09E-7	(21°)	4.81E-6	(19°)	1.20	1.13
Al_2O_3	n.d		1.36E-7		1.18	1.10
AZ	9.78E-8	(11°)	1.00E-6	(21°)	0.94	0.87
AZ(40h)	2.14E-8	(10°)	1.82E-7	(22°)	1.00	0.92
AZT	8.76E-7	(12°)	8.77E-6	(10°)	0.79	0.71

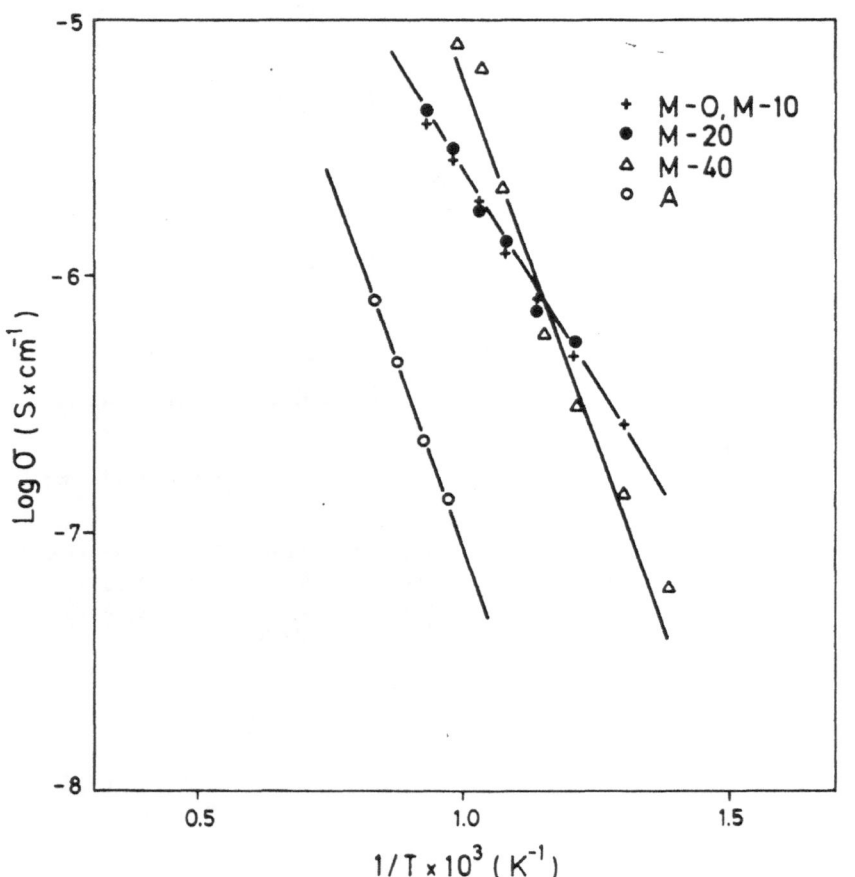

Figure 3.- Logarithm of the electrical conductivity
against temperature for the Mullite-ZrO_2
composites, results for Al_2O_3 are also
represented.

B) Alumina-ZrO$_2$ Composites.

A representative example of the impedance spectrum in the Al$_2$O$_3$-ZrO$_2$ samples is shown in Fig. 4. The impedance spectrum in throughout the temperature interval studied shows a grain interior resistivity semicircle with a depression angle of 11°, one indication that the ZrO$_2$ particles have stronger effect on the dielectric properties of Al$_2$O$_3$ matrix than have on mullite. One of the reason for this behavior is the fact that 90% of the ZrO$_2$ particles are tetragonal in this composite, while for example the mullite-40 vol.% ZrO$_2$ composite has all the particles transformed into the monoclinic phase. A small arc is evidenced in the low frequency region, which could be attributed to a small grain boundary effect or to the ZrO$_2$ particles.

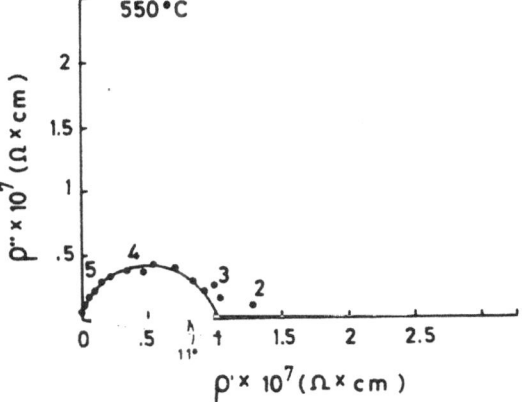

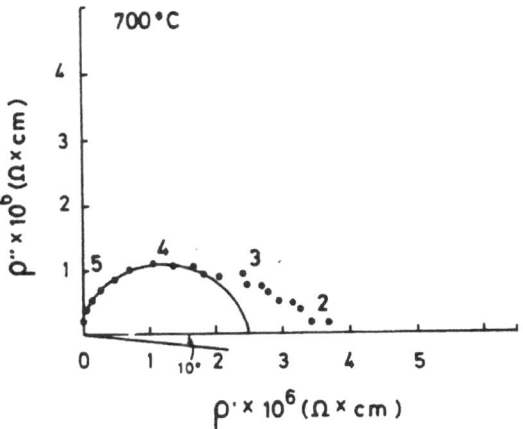

Figure 4.- AC impedance plots for the Al$_2$O$_3$-ZrO$_2$ composite at the shown temperatures.

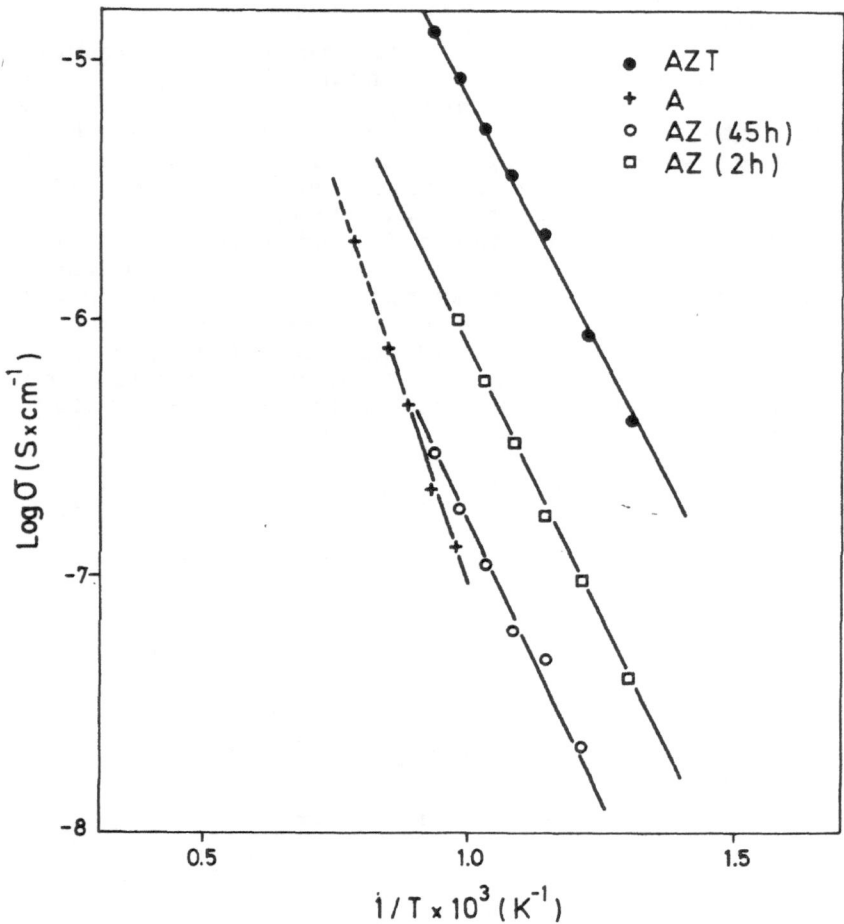

Figure 5.- Logarithm of the electrical conductivity vs.
the reciprocal temperature for the Al_2O_3-
ZrO_2 composites.

In Fig. 5 the conductivity curves are represented
as a function of the reciprocal of the temperature for
all the alumina based composites. First of all, it is
noted that when ZrO_2 is added to the Al_2O_3 provokes an
increase in conductivity, together with a decrease in
the activation energy for the conduction. As mentioned
before, the high amount of tetragonal phase in this
composite can explain the increase in conductivity
compare to the Al_2O_3 single phase. In the case of the
same composite under the annealing treatment (40 hours
at 1570°C), the conductivity decreases and becomes
similar to that of the Al_2O_3 single phase, but
activation energy stays at the same low values that the
non-annealed composite. As a consequence, the alumina

grain size growth caused by the annealing produces a strong microstructural related effect on the conductivity. As well as, the reduction on the amount of tetragonal phase in the annealed composite (see Table III) presumably influences the reduction in the conductivity of the sample.

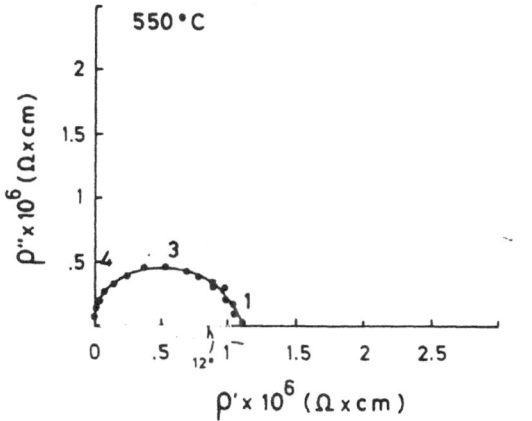

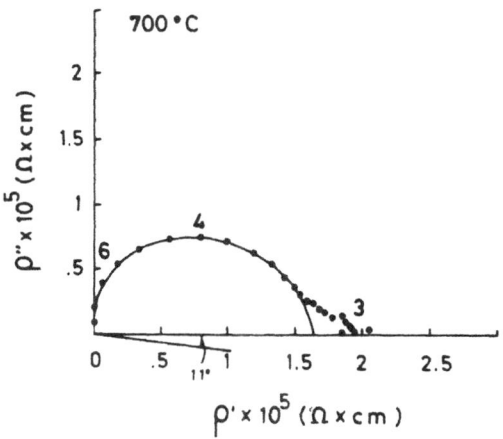

Figure 6.- AC impedance plots of the Ti-doped Al₂O₃-ZrO₂ composite.

Finally, as TiO_2 is introduced in the Al_2O_3-ZrO_2 composite its conductivity increases one order of magnitude (see Fig. 5) and the activation energy for conduction diminishes strongly (Table IV). Similar results have been obtained for Ti-doped Al_2O_3[6] and it seems clear that the Ti-induced electronic conduction is responsible for this behavior. At the same time, the existence of a grain boundary contribution to the conduction mechanism is detected in the sample (Fig. 6). This phenomenum could be favored by the Ti enrichment on the alumina grain boundaries observed on this type of composite[7]. Summarizing, the present work evidences that the incorporation of an small amount of a transition metal ion produces, via an electronic process, a strong effect on the conduction mechanism.

REFERENCES

1. A. G. Evans and R. M. Cannon; "Toughening of brittle solids by martensitic transformations", Acta Metall. **34** (5) 761-800 (1986).

2. C. Pascual, J. R. Jurado and P. Duran; "Electrical Behaviour of Doped-Ytria Stabilized Zirconia Ceramics Materials", J. Mat.Sci. **18** (5) 1315-22 (1983).

3. J. E. Bauerle; "Study of Solid Electrolyte Polarization by a Complex Admittance Method",J. Phys.Chem.Solids **30** 2657-70 (1983).

4. J. S. Moya, F. J. Valle and S. de Aza; "The sintering behavior of an active premullite obtained from kandites", Sintering-Theory and Practice, Materials Science Monographs, Vol. 14, pp 409-415, Elsevier Scientific Publishing Co., Amsterdam, 1986.

5. J. M. Rincon, G. Thomas and J. S. Moya; "Microstructural Study of sintered Mullite Obtained from Premullite", J. Am. Ceram.Soc. **69** (2) C29-C31 (1986).

6. M. M. El-Aiat, L. D. Hou, S. K. Tiku, H. A. Wang and F. A. Kröger; "High Temperature Conductivity and Creep of Polycristalline Al_2O_3 Doped with Fe and /or Ti", J. Am.Ceram. Soc. **64** (3) 174-182 (1981).

7. M. I. Osendi, B. A. Bender and D. Lewis; "Influence of TiO_2 on the Mechanical Properties at High Temperature of Zirconia-Toughened Alumina", Adv. Ceram. Mat. **3** (6) 563-68 (1988).

Precursor of zirconia based ceramics
Obtention and pyrolysis of ZrCl4 silatrane complex

G. Palavit, P. Vast, J. Ph. Rosnet

Laboratoire de Chimie Appliquée
Université des Sciences et Techniques de Lille Flandres Artois
Bât. C5 59655 Villeneuve d'Ascq Cedex   France

INTRODUCTION

Silatranes are cyclic organic compounds containing both nitrogen
and silicon - Fig 1- . The distance between these two atoms is
so short that Frye, in 1961 (1) proposed a model of pentavalent
silicon. A few years later, this idea was confirmed by Boer (2)
with radiocristallographic analysis. Even though the silicon and
the three oxygen atoms of the cage are almost coplanar, the Si-N
bond is considered as a dative bond (works of Voronkov (3) and
Frye (4)). It's only recently that Imbenotte has shown that
silatranes have a Si-N sigma bond, whose strength and length
depend upon the radical bonded with the silicon. This result
agrees with the work of Müller (6), who, by mass spectrometry,
showed that Si-N bond is more stable than Si-O.

It's interesting to note that, despite this sigma bond, the nitrogen atom keeps its basic nature. We can easily make a hydrochloride of the silatrane, but, in this case, the Si-N bond is broken. The nitrogen, which had an endo position, changes to an exo position.

As we have shown before (7), the cage structure of the silatranes allows us to obtain a complexation reaction with alkaline ions

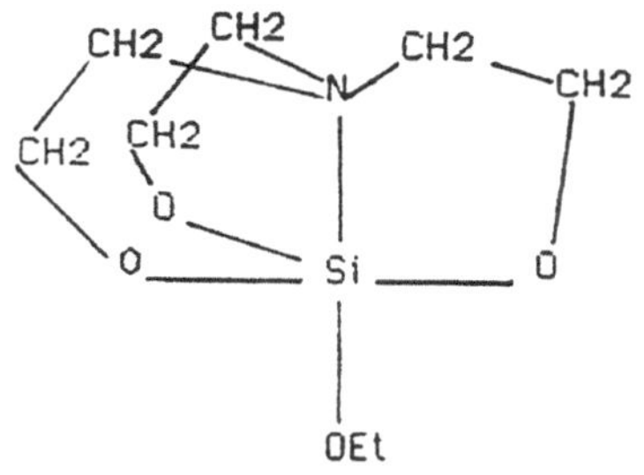

Fig 1 Ethoxyoxosilatrane
structure

Since these organo-silicious can
be considered as ceramics precur-
sors, we have tried to obtain complexes with salts of metals
giving oxygenous and nitrogenous ceramics.

We present in this paper the results we have obtained with
zirconium   tetrachloride   and   ethoxy-oxo-silatrane   and
particularly the final compounds after the pyrolysis up to
1300°C

EXPERIMENTAL DATA

The ethoxy-oxo-silatrane is obtained by heating a mixture of
tetraethoxysilane and triethanolamine at 80°C for about two

hours. After cooling, the ethoxy-oxo-silatrane crystallizes in white crystals. In order to have a high purity, it's necessary to recrystallize the product in a mixture of xylol and hexane. The zirconium tetrachloride is a commercial product of Merck-Schuchardt Company.
The apparatus used for thermal analysis was a D.S.C. 111 from SETARAM. We used sealed crucibles for having constant mass. The range of temperatures was from 20°C to 200°C, and the scanning rate was 2°C/mn.
The Infra-Red spectrometer was a IRFT 1730 from PERKIN-ELMER. We used KBr pellets, and worked in the range 1300 cm^{-1}/400 cm^{-1}.
The radiocristallographic apparatus is a PW 1010 from PHILIPS. We use copper X-Ray tube and nickel filter, with 40 kV voltage and 14 mA intensity. The goniometer was graduated in 2 θ, and the angle range was 10° - 70°.

PREPARATION OF THE PRECURSOR COMPLEX

A saturated solution of zirconium tetrachloride mixed with a solution of silatrane in diethyl ether gives a white precipitate. This complex has a formula $ZrCl_4, 2$ ethoxy-oxo-silatrane. We verify this formula by mesuring the chloride in solution, and mesuring zirconia and silica by calcination of the residue obtained after oxydation of the complex with fuming nitric and sulphuric acids.
Ethoxy-oxo-silatrane melts at 102-103°C. When melting silatrane, we have verified that the same reaction as in solution occurs.
By differential scanning calorimetry -see Fig2- it's possible to show the reaction. We mix, in differents proportions, ethoxy-oxo-silatrane and zirconium tetrachloride. This mixture is put in a sealed crucible for D.S.C., in order to avoid any loss of ethoxy-oxo-silatrane by sublimation. We observe an evolution of the endothermic melting peak of ethoxy-oxo-silatrane with the increasing of the quantity of zirconium tetrachloride. These curves can be interpreted as the addition of the endothermic phenomenon of the melting of ethoxy-oxo-silatrane and the exothermic reaction of complexation.
The Infra-Red spectrum shows clearly that the complex losses the hypervalence of the silicon atom.

PYROLYSIS
In order to study the behavior of the complex, we pyrolyse it under different atmospheres, such as argon, oxygen, and hydrogen.
We proceed in a two stages process: first, we heat the complex at a temperature of 300°C for about three hours, a sufficient temperature and length of time to destroy the complex, second we heat the products up to 1300°C for two hours, respecting the same atmospheres used in the first step. All the samples are cooled at room temperature before study.

We perform an analysis with a Fourier Transform Infra-Red spectrometer after the first and second heatings.
We only perform a radiocristallographic analysis for the samples treated up to 1300°C, because the samples obtained at 300°C are not crystalline enough.

From the Infra-Red spectros copy analysis, we obtain the following results:
At 300°C, the Infra-Red spectra of the pyrolysed complex are the same, fig 3, whatever the atmosphere used. We observe bands characteristic of the silica at 1060 cm^{-1}, 780 cm^{-1} and 454 cm^{-1}(see the reference, with the ethoxy-oxo-silatrane alone). Zirconia does not appear clearly, except in a shoulder at 600 cm^{-1}, that we perhaps could attributed to a Si-O-Zr bond, refering to the band at 615 cm^{-1} which exists in zircon $ZrSiO_4$.

At 1300°C, the gases under which the pyrolysis take places have again no effect on the spectra, but there is an evolution from the results obtained at 300°C: we conserve three bands for the silica, at 1090 cm^{-1}, 790 cm^{-1} and 470 cm^{-1}, but now, we can see three bands characteristic of zirconia at 740 cm^{-1}, 580 cm^{-1} and 410 cm^{-1} Fig 4.

From the radiocristallographic analysis, we obtain the following results (Fig 5):
a) In an argon atmosphere, we observe monoclinic zirconia and amorphous silica. The colour of the sample (black) allow us to say we have also amorphous carbon.

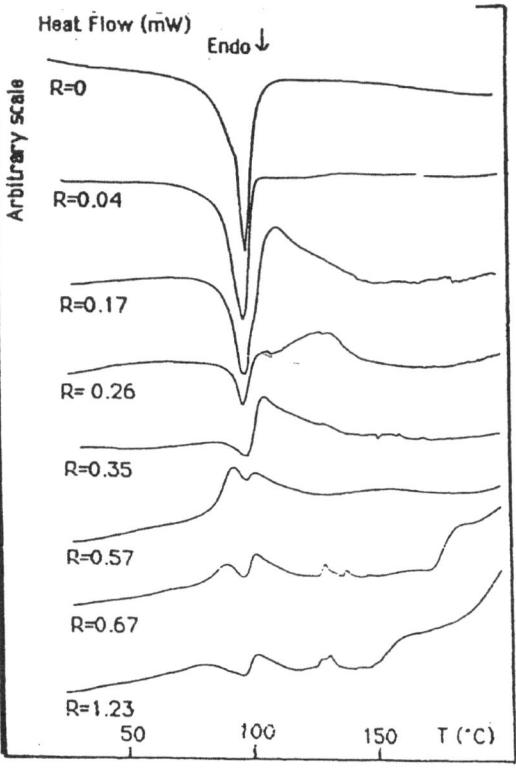

R=ZrCl$_4$/ethoxy-oxo-silatrane molar ratio

Fig2 DSC complex formation spectra

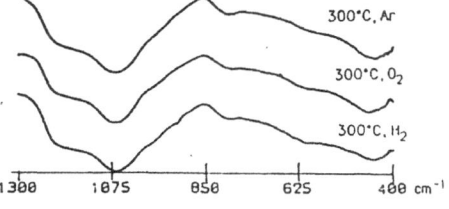

Fig3 IRTF spectra at 300°C

b) In an oxygen atmosphere, we observe monoclinic zirconia, tetragonal zirconia and α cristobalite. There is of course no carbon.

c) In an hydrogen atmosphere, we observe monoclinic zirconia, tetragonal zirconia and a little amount of αcristobalite. We note also amorphous carbon.

It's interesting to note that when we pyrolyse the ethoxy-oxo-silatrane alone in an argon atmosphere, we receive amorphous carbon and silica (α cristobalite) which is not always represented in the pyrolysis of the complex.

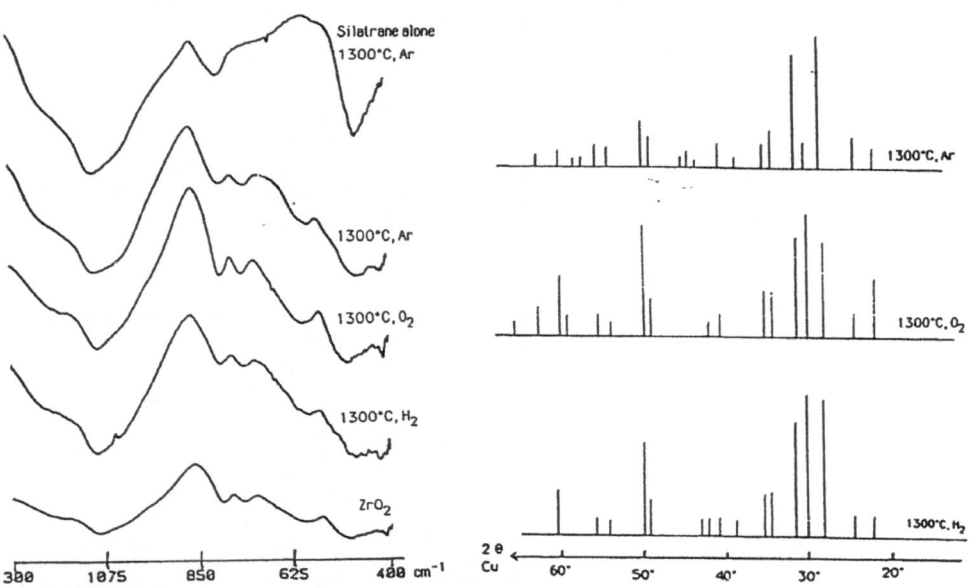

Fig 4 IRTF at 1300°C   Fig 5 X ray at 1300°C

So, we can imagine that, at 300°C, all the zirconia is bonded with a part of the silica (hence the shoulder near 600 cm^{-1}) and the excess of silica due to the composition of the complex gives the three bands of the silica.

At 1300°C, the Infra-Red spectroscopy shows the presence of zirconia and still silica. With X-Ray analysis, silica appears to be amorphous (except in an oxygen atmosphere) while zirconia appears in these two crystalline forms, monoclinic and tetragonal. We observe that an inert atmosphere enhances the monoclinic form, and oxydative or reducing atmosphere enhance the tetragonal form. But in this case, monoclinic zirconia is again present for about 50%.

CONCLUSION

This work shoows the possibility of chosing the crystalline form of the zirconia with the amosphere of pyrolysis.In a next stage, we'll study the influence of the temperature on the crystalline forms of zirconia. It could be possible, by heating at a temperature close to that of the zirconia transition, to have another factor which can help us to obtain only one crystalline form.

REFERENCES

1) C.L. Frye, G.E. Vogel and J.A. Hall; J. Am. Chem. Soc. 83, 996 (1961)

2) a) J. W. Turley and F. P. Boer; J. Am. Chem. Soc. 90, 4026 (1968)
   b) F. P. Boer, J. W. Turley and J. J. Flynn; J. Am. Chem. Soc. 90, 5102 (1968)
   c) J. W. Turley and F. P. Boer; J. Am. Chem. Soc. 91, 4129 (1969)
   d) F. P. Boer and J. W. Turley; J. Am. Chem. Soc. 91, 4134 (1969)

3) M. G. Voronkov, I. B. Mazheika, G. I. Zelchan; Khim. Geterotsikl. Soedin., Akad. Naut. Latv. SSR 1, 58 (1965)

4) C. L. Frye, G. A. Vincent and W. A. Finzel; J. Am. Chem. Soc. 93, 25 (1971)

5) M. Imbenotte, thèse de doctorat, Lille (584) 1983

6) R. Müller und H. J. Frey; Z. Anorg. Allg. Chem. 368, 113 (1969)

7) M. Imbenotte G. Palavit G. Delesalle, S. Noel VIeme International Symposium on Organosilicon chemistry Budapest 1981

MARTENSITIC TRANSFORMATION IN TZP INVESTIGATED BY A SITE SELECTIVE SPECTROSCOPY METHOD.

B. PIRIOU*, J. DEXPERT-GHYS*, B. BASTIDE** and P. ODIER***

(*) CNRS, Bellevue 92195-Meudon principal cedex (France)
(**) CRICERAM 38560-Jarrie (France)
    presently at PECHINEY 38340-Voreppe (France)
(***) CRPHT-CNRS, 45071-Orléans cedex 2 (France)

A method using the time resolved spectroscopy of europium ion under selective excitation is proposed to follow the rate of transformation in TZP below the room temperature. One ceria doped and two yttria doped samples were investigated. The best conditions of excitation and observation at 77 and 300 K are given

INTRODUCTION

A great amount of works has been devoted in the last decade to TZP (tetragonal zirconia phase) for its highly reinforced mechanical properties [1]. Reinforcement essentially comes from the transformability of the tetragonal phase which is controlled by the martensitic transformation. This transformation may be characterized by its temperature of occurrence M_s and M_f, i.e. start and finish martensitic transformation, A_s and A_f which have similar meaning for the reverse transformation. Becher et al. [2] have shown the importance to adjust M_s to the working temperature in order to have a maximal increase in toughness. Garvie et al. gave both theoretically and experimentally in a few cases the dependence of M_s versus the grain size while Anderson et al. [4] have postulated the relationship between M_s and the dopant concentration at constant grain size.

Recently [5] we have been able to test experimentally these theories on the new La-PSZ (partially stabilized zirconia), however at the present time, Y and Ce seem more appropriate for stabilizing the TZP. It has been possible on fine grains fully densified ceramics, to observe M_s at 50 °C on Y-TZP [6] and even below the room temperature [7]. Essentially three techniques have been used to follow the transformation : dilatometry, DTA [6,7] and Raman spectroscopy [8], the first two being the most currently performed although they are sometimes difficult to be carried out below the room temperature. In opposite to these macrocopic techniques, the Raman scattering has the great advantage to be performed at low temperature and this microscopic technique is sensitive to the symmetry of the unit cell . However, the spectra are generally no so well resolved due to the oxygen non-stoichiometry inherent to the stabilizing dopant. In these compounds, if the cationic sublattice is ordered, as viewed by the X-ray for example, the anionic sublattice is disordered, because of the oxygen vacancies. The phonons which involve the entire lattice lead to broadened lines even in nearly ordered zirconias [9].

Further progress in this field rely on new experimental techniques able to study with high sensitivity the small local changes in the stucture versus the temperature and especially able to work below the room temperature. The rare earth ions acting as local probes seem good candidates, and this paper presents an attempt to use the fluorescence of one of them.

The discrete energy levels of rare earth ions $4f^n$ configuration have been recorded for a large number of materials. These ions are used as optical local probe to investigate the number and symmetry of the crystallographic sites they occupy in a given structure. Among them, Eu^{3+} with $4f^6$ configuration is the favorite ion because of the simple sequence of fluorescence lines originating from the $^5D_0$ level. Moreover, the number of lines in the $^5D_0 \rightarrow {}^7F_0$ spectral range between the two non degenerate states, obviously corresponds to the number of unequivalent sites. For more details about the optical spectroscopy of Eu^{3+}, the reader may refer to general surveys [10]. The energy levels involved in this study are $^5D_2$, $^5D_1$, $^7F_3$, $^7F_2$, $^7F_1$ and obviously the $^7F_0$ ground state level.

In the case of disordered compounds like glasses, all the sites are unequivalent but they are regrouped in one or several more or less narrow distributions, also called families. This induces an inhomogeneous broadening of the lines when compared to the usual sharp lines encountered in ordered solid compounds and even in liquid solutions. For similar reasons, a broadening is observed in stabilized zirconias. Previously, one of us had determined the set of energy levels of Eu^{3+} for the typical sites in the monoclinic, tetragonal and cubic phases of yttria-doped zirconias [11,12].

In this paper the europium time-resolved site selective spectroscopy is applied to follow the martensitic transformation in TZP. Two Y-TZP and one Ce-TZP were retained as examples. For these samples, with the excitation conditions used, the fluorescence of the monoclinic phase is characterized by one family of sites while two families, labelled Q_1 and Q_2, can be attributed to the tetragonal phase.

EXPERIMENTAL PROCEDURE

1)PREPARATION AND CHARACTERIZATION

The material were prepared by using an impregnation technique of the pure zirconia. Several precursors have been used having different granulometries and chemical compositions. The sample TZ-12Ce ($ZrO_2:CeO_2:Eu_2O_3$ 87.7:12:0.3 mole per cent) was prepared by impregnation with Ce nitrate and Eu nitrate of a pure zirconia powder from Cricéram Co. The suspension was spray dryed in a Buchi atomizer, deagglomerated in isopropylic alcohol with zirconia balls and then calcined at 750 °C for 3 h in air. The resulting powder was isostatically pressed at 480 MPa and sintered at 1450 °C for 4 h in air. The sample TZ-1.5Y ($ZrO_2:Y_2O_3:Eu_2O_3$ 98.2:1.5:0.3) was prepared in a similar way but sintered at 1420 °C for 2 h, the sample TZ-2Y ($ZrO_2:Y_2O_3:Eu_2O_3$ 97.92:2:0.08) used a Toyosoda powder with 2 moles % Y_2O_3 as precursor and has been sintered at 1450 °C for 2 h.

The dilatometric anomaly at the martensitic transformation was systematically checked by the procedure described in refs [5,6] and the structure identified by X-ray diffraction analysis on the surfaces which has been studied by the spectroscopic method. It provides an estimation of the monoclinic fraction [13]. From X-ray patterns it was concluded that the sample TZ-12Ce, which was a pure tetragonal phase at room temperature just after the cooling from sintering, had 53 % of monoclinic phase after 15 min in liquid nitrogen and warming back at the room temperature. For the samples TZ-2Y and TZ-1.5Y the tranformation was absolutly reversible with respectively 5 % and 8 % of monoclinic phase at room temperature. We were unable to detect any dilatometric anomaly below or in vicinity of the room temperature for these samples. This was in fact the reason to use the optical spectroscopy which is easily carried out at the liquid nitrogen temperature.

2)FLUORESCENCE EXPERIMENTS

The fluorescence was excited either in the UV region directly by a pulsed nitrogen laser (Jobin Yvon LA 04, pulse duration 8 ns, pulse energy 5 mJ), or selectively in the $^5D_2$ sub-levels by means of a tunable pulsed dye laser (Jobin Yvon E1T). The spectral width of the line supplied by this laser was about 2 cm^{-1}, which is widely sufficient with respect to the linewidths in the disordered compounds studied here. The fluorescence was analysed through a double monochromator spectrometer (Coderg Ph 0) and detected by an Hamamatsu photomultiplier (R 1447). The signal was treated by means of a digital oscilloscope connected to a 16 bits microcomputer. Simultaneously with the scanning of the spectrometer fluorescence decay was analysed. By varying the time delay between excitation and detection, time resolved spectra for various integration gates may be recorded during one scan. A detailed description of this setup will be given elsewhere [14].

Experiments at low temperature were performed with the sample immersed in liquid nitrogen inside a fused silica dewar. To remove the disturbing effect of the bubbles, the samples were stuck on the inner wall of the dewar by means of a non fluorescent varnish.

In the following presentation a special attention will be given to the Ce-TZP sample in the preliminary experiments. Similar results were obtained with the other samples, for instance the lifetimes or adequate frequencies for the selective excitation of the Eu^{3+} sites.

RESULTS AND DISCUSSION

1)SPECTROSCOPY UNDER NON SELECTIVE EXCITATION

At first, to excite all the sites occupied by Eu^{3+} in the

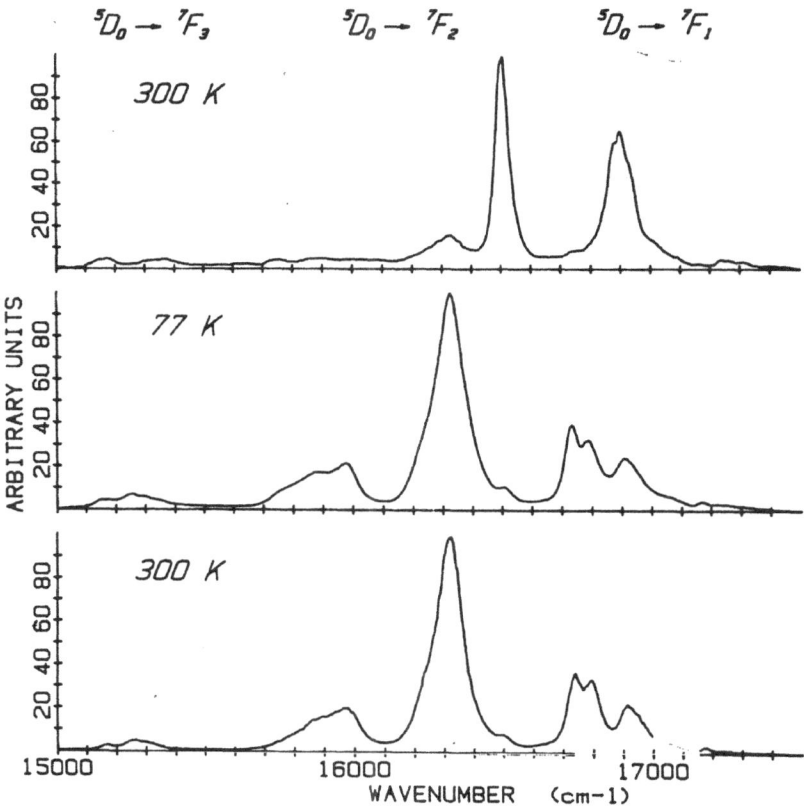

Fig.1. Emission originating from $^5D_0$ of TZ-12Ce sample under excitation at 337 nm. Upper cuve: native sample, lower curve : after cooling at 77 K.

zirconias, an UV radiation was used. One can consider as non selective the excitation at 337 nm (29665 cm^{-1}) from the nitrogen laser. Non radiative processes relax the $^5D_{3,2,1}$ upper levels on $^5D_0$, and only the transitions originating from this last level were observed, i.e. $^5D_0 \rightarrow {}^7F_0$, $^5D_0 \rightarrow {}^7F_1$, $^5D_0 \rightarrow {}^7F_2$ and $^5D_0 \rightarrow {}^7F_3$ in the spectral ranges 17200-17300 cm^{-1}, 16650-17100 cm^{-1}, 15600-16600 cm^{-1}, and 15000-15600 cm^{-1} respectively. Figure 1 shows such an emission for the TZ-12Ce sample at room temperature before and after a cooling at 77 K, and during the exposure at this low temperature. The spectra were recorded with continuous detection which integrates all the transient effects that would be able to bring some selectivity between the sites. On the upper curve one can recognize the spectum of an Eu^{3+} site in a tetragonal phase [12], whereas the two other spectra reflect that of Eu^{3+} in the monoclinic phase [11]. This means that between 300 K and 77 K the T $\rightarrow$ M transformation occurs in an irreversible way, i.e. 77 K < M_f, M_s < 300 K < A_s, A_f.

The shoulder at about 16500 cm^{-1} in the spectra of the monoclinic phase which corresponds to the main line in the upper spectrum seems to indicate that the transformation was not completely achieved. On the other hand, in the spectrum of the native zirconia (upper curve), free of the monoclinic phase as concluded from the X-ray analysis, the broad weak line at 16330 cm^{-1} could be assigned to this phase. We may also note the broadness of the lines with respect to the previous works [11,12]. For instance in the $^5D_0 \rightarrow {}^7F_2$ transitions of the monoclinic phase, a broad asymmetric line at 16320 cm^{-1} was observed instead of three expected lines at 16207, 16273 and 16311 cm^{-1}. This may confuse some assignments. As this first spectroscopic approach is insufficient to identify unambiguously the characteristics of the different phases, a more sophisticated spectroscopy is needed.

2)SITE SELECTIVE EXCITATION AND TIME RESOLVED SPECTROSCOPY

Usually, the site selective excitation is performed in the $^5D_0$ level, because the $^5D_0$ gap falls in the emission of the most efficient dye (Rhodamin 6G). We have preferred to excite in $^5D_2$ sublevels for two reasons. First, the oscillator strength of $^7F_0 \longrightarrow {}^5D_2$ is much stronger than the one of the theoretically forbidden $^7F_0 \longrightarrow {}^5D_0$ transition, thus the excitation is more efficient. The second reason is related to the unwanted excitation radiation scattered by the sample. In the observation spectral range (15000-17500 cm^{-1}) the rate of rejection by the spectrometer is better for a radiation at about 21400 cm^{-1} ($^7F_0 \longrightarrow {}^5D_2$), than at about 17250 cm^{-1} ($^7F_0 \longrightarrow {}^5D_0$).

2-1-Excitation spectra, lifetime.

The excitation spectra at 77 K are shown in figure 2. They were recorded by monitoring the main emission lines observed in figure 1 for the monoclinic and tetragonal phases respectively. Despites the narrow bandwidth used to monitor the excitation spectra, one detects in each case the response coming from a family of sites. We label M and Q_1 the families excited in each phase. In

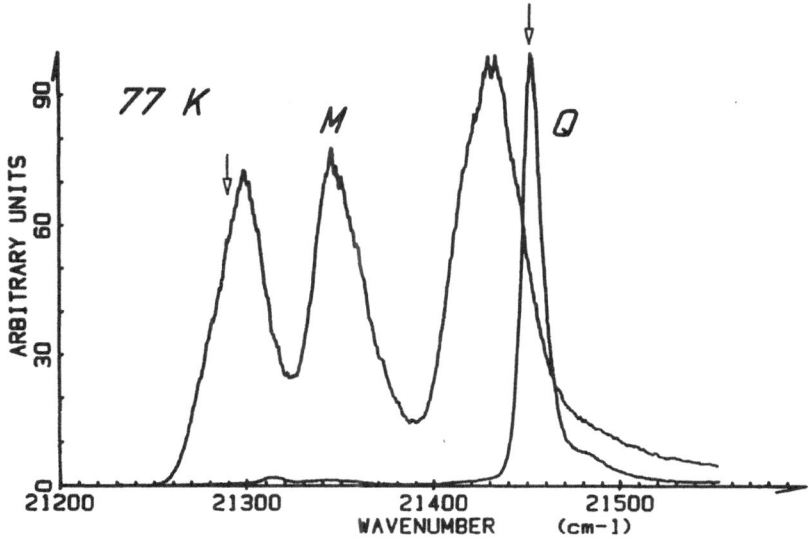

Fig.2. Excitation spectra in $^5D_2$ of M and Q_1 site at 77 K. TZ-12Ce sample.

view of the width of the $^7F_0$ --> $^5D_2$ lines, the distribution is broader for the M than for the Q_1 family. In the following we will refer to a typical M, or Q_1, site for the whole family.

From figure 2, one deduces that the monoclinic site can be excited very selectively at 21289 cm^{-1}, whereas at 21450 cm^{-1}, the frequency corresponding to the maximun excitation of the tetragonal site, M is still excited. Excepting a slight shift, at room temperature the spectra were similar and the selective frequencies were 21300 cm^{-1} and 21468 cm^{-1}. In all the forthcoming experiments, these four frequencies were retained as standard excitation lines.

The lifetime also can be an important parameter to discriminate the emissions arising from various sites. Fortunately the $^5D_0$ lifetime of the M and Q_1 sites are quite different : 0.91 ms and 1.75 ms respectively at 77 K. At 300 K the values become slightly shorter 0.87 ms and 1.61 ms. From these values we can easily calculate that 2 ms after the pulse the initial Q fluorescence is reduced by a factor 1/3 and the M fluorescence by 1/10. We

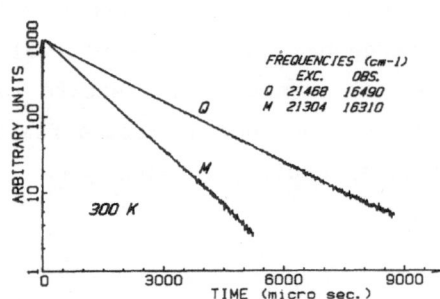

Fig.3 Fluorescence decay of $^5D_0$ level of the Q_1 site in tetragonal phase and the M site in monoclinic phase.

have turned such a favourable behaviour to account in the time resolved spectra. In the following will be presented systematically pairs of spectra. The chosen delays and gates roughly correspond to the initial emission and the emission after 2 ms. As shown on figure 3 the $^5D_0$ fluorescence decay curves for each site are remarkably exponential.

2-2-Emission of the ceria-doped zirconia

The $^5D_0$ --> $^7F_{0,1,2,3}$ emission of TZ-12Ce sample is presented on figures 4 and 5. They correspond to a tetragonal and monoclinic sites excitation respectively, the upper spectra are relevant of native sample at 300 K, the intermediate at 77 K and the lower spectra were recorded at 300 K after one cooling at 77 K. One can notice that after some thermal cycles from 300 K to 77 K the intermediate and lower spectra remain the same. In each case a pair of time resolved specta, normalized to the main line, are given. When only one site is excited no dynamical effect occurs, both spectra are fully surimposed (upper spectra fig. 4 and intermediate

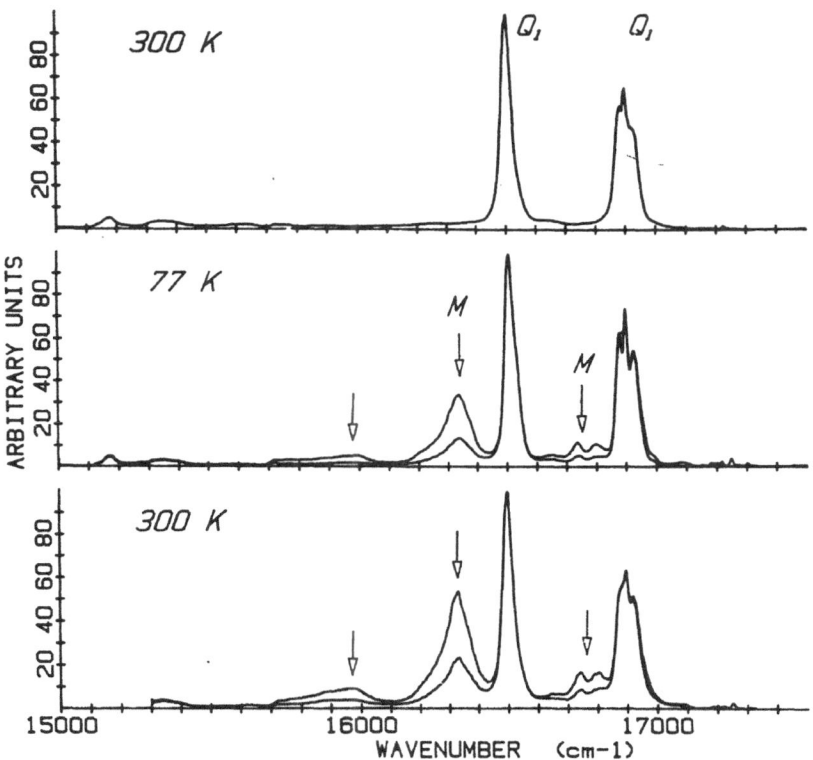

Fig.4. Time resolved emission $^5D_0 \rightarrow {}^7F_J$ of TZ-12Ce under Q_1 excitation, i.e. : 21468 cm^{-1} at 300 K and 21450 cm^{-1} at 77 K. For each pair of spectra the delays and gates are as following. First spectrum : D = 0.1 ms, G = 0.3 ms ; second spectrum : D = 1.2 ms, G = 2.1 ms. The arrows give the time evolution from the fisrt to the second spectrum.

spectra fig. 5). The comparison of the two time resolved spectra allows an unambiguous assignment of the very faint emissions (line M on native sample fig. 5 for intance). The time dependence indicated by the arrows obviously agree with the lifetimes : the fluorescence of the Q_1 site decreases less quickly than that of the M site.

According to the previous section these curves confim an irreversible transformation with the first cooling. Under the M selective excitation at 21300 cm^{-1} (fig. 5) the spectum of the

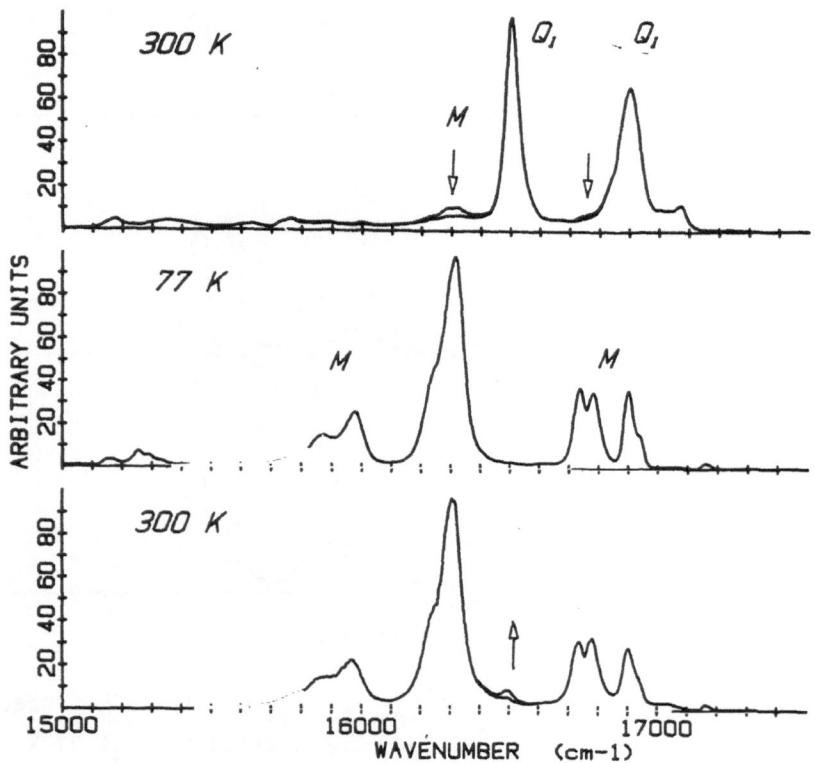

Fig.5. Time resolved emission $^5D_0 \rightarrow ^7F_J$ of TZ-12Ce under M excitation ,i.e.:21300 cm^{-1} at 300 K and 21289 cm^{-1} at 77 K. Delays and gates as in fig.4.

native ceria-doped zirconia at room temperature exhibits a low content of the monoclinic phase that was not detected on the X-ray pattern. The selectivity due to the pertinent excitation frequencies appears by comparison of the figures 4 and 5. At 77 K the whole selectivity is observed for M and partially for Q_1, as it was expected after the excitation specta (fig. 2). For sample at 77 K and reheated at 300 K the similar spectra indicate that the monoclinic content remains nearly the same , i.e. 53 % from X-ray patterns at 300 K. More quantitative measurements of the temperature dependence of this content requires a previous

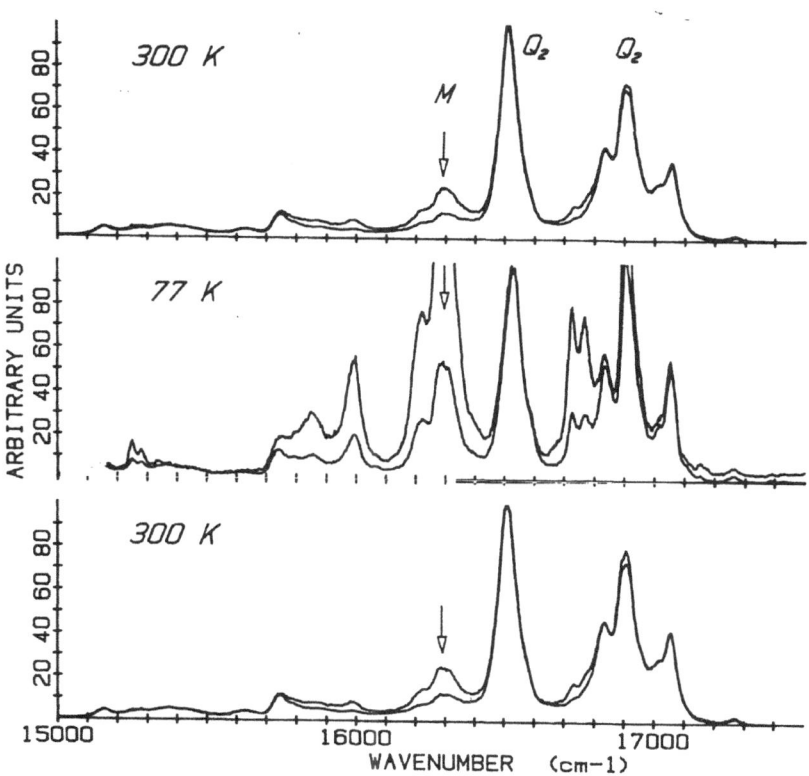

Fig.6. Time resolved emission $^5D_0 \longrightarrow{} ^7F_J$ of TZ-1.5Y under M excitation. Conditions as in fig.5.

calibration at each temperature, because the efficiencies of the selective excitations are probably temperature dependent.

One can point out that the excited tetragonal sites depend upon the frequency of the exciting line. With the Q_1 selective excitation (21468 cm^{-1} at 300 K), three $^5D_0 \to {}^7F_1$ transitions are just resolved at 16880, 16900 and 16920 cm^{-1}, whereas two lines at 16980 and 17070 cm^{-1} are observed with the excitation at 21300 cm^{-1}. In this latter case, out of the selectivity, the emission is much less intense, approximatively two hundred times less.

2-3-Emission of yttria-doped zirconias

The two europium-doped ziconias with 1.5 and 2.0 mole % of Y_2O_3, TZ-2Y and TZ-1.5Y samples, will be considered in this section. The transformation was observed to be reversible by checking with X-ray. This fact was confirmed by the spectroscopic method, the spectra at 300 K before and after one or several coolings at 77 K were always identical, as examplified for the TZ-1.5Y sample on figure 6. The time resolved emission spectra are shown in figures 7 and 8 for the TZ-2Y sample.

According to the low content of monoclinic phase at room temperature the M site appears only weakly under the Q_1 selective excitation and even under M excitation. Although having nearly the same monoclinic content at room temperature, the behaviors on cooling at 77 K of the two yttria doped zirconias are quite different. The proportion of T $\to$ M transformed phase in TZ-1.5Y (fig. 6) is much higher than in TZ-2Y (fig. 8). Such a result is in agreement whith the concentrations of yttria and it could be foreseen from the difference in monoclinic content at room temperature (8 and 5 %).

As for the ceria doped ziconia, under Q_1 excitation the fluorescence of the Q_1 tetragonal site was observed. Conversely, a new kind of quadratic sites, labelled Q_2, was excited under the M excitation line

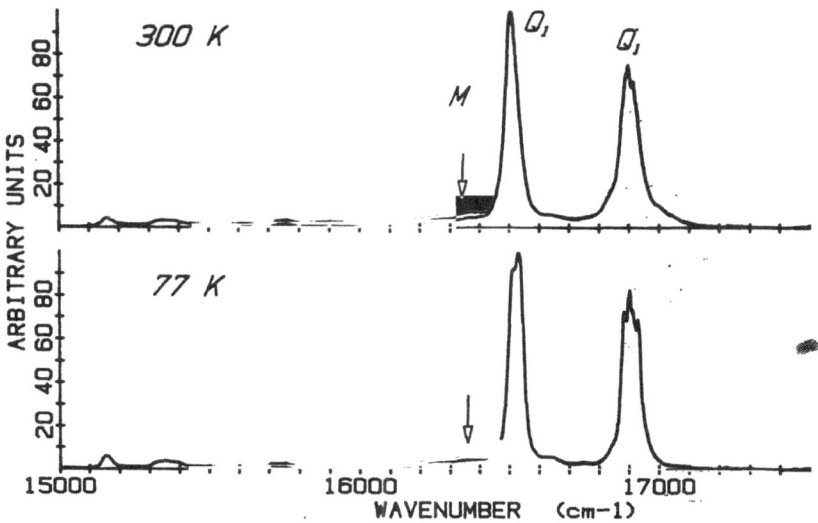

Fig.7. Time resolved emission $^5D_0 \longrightarrow {}^7F_J$ of TZ-2Y under Q excitation. Conditions as in fig.4.

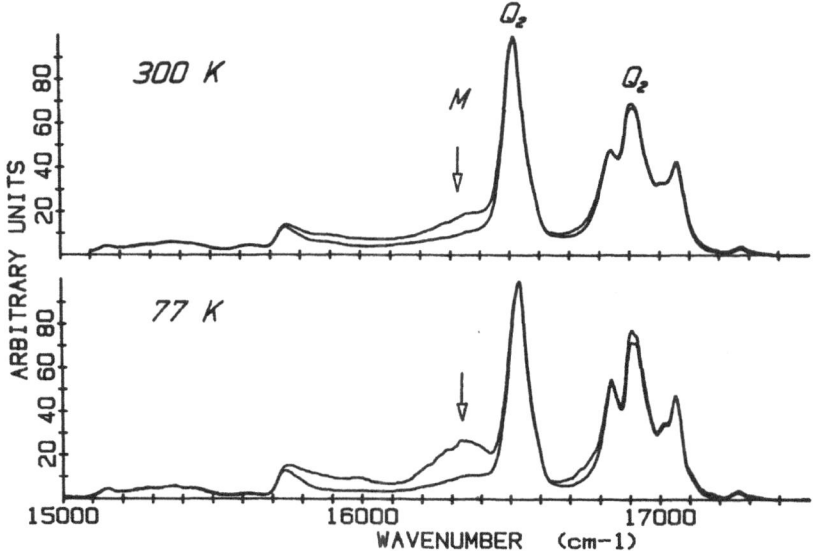

Fig.8. Time resolved emission $^5D_0 \longrightarrow {}^7F_J$ of TZ-2Y under M excitation. Conditions as in fig.5.

CONCLUSION

In conclusion we have proved that the time resolved site spectroscopy of Eu^{3+} is a technique suitable for studying the T --> M phase transformation in TZP. Its sensivity has been evidenced by the detection of small amounts of the monoclinic phase in tetragonal samples, which had not been observed by X-ray diffraction. It is very well adapted to low temperature measurements. In this work the experiments have been made at 300 and 77 K but the same investigations could be easily performed at various temperatures in the 4-300 K range. The application of this technique to a determination of the amounts of each phase in a sample would be feasible after a careful calibrations.

The interpretation of the spectroscopic characteristics of Eu^{3+} permits a microscopic approach of the stuctures involved. For instance, the lines broadness is related to the disordered nature of a sample. The monoclinic zirconia phase produced by the low temperature T --> M transformation is by far more disordered (or less well crystallysed) than the monoclinic sample synthetized at higher temperature which has been studied in a previous work [11].

On the other hand we have already mentionned the observation of a Q_1 site in the tetragonal phase of the three samples and of a Q_2 site only in the tetragonal phase of the two yttria-doped coumpounds. This clearly shows that the stabilization process does not create the same cationic point sites when the dopant ion is tetravalent (Ce) or trivalent (Y). Further work is now in progress on the detailed understanding of the spectroscopic data of Eu^{3+} in the tetragonal and monoclinic zirconia phases.

REFERENCES

1) See *Adv. Ceram*, vols 3 and 12, Edt The Am. Ceram. Soc., Columbus, 1981 and 1984.
2) P.F. BECHER, M.V. SWAIN, M.K. FERBER, J. Mat. Sci. 22, 76 (1987).
3) R.C. GARVIE, M.F. GOOS, J. Mat. Sci. 21, 1253 (1986).

4) C.A. ANDERSON, J. GREGGI Jr, T.K. GUPTA, *Adv. Ceram.*, vol.3, p 78, Edt.The   Am. Ceram. Soc., Columbus, 1981.981.

5) B. BASTIDE, P. ODIER and J.P. COUTURES, J. Am. Ceram. Soc. 71, 449 (1988).

6) B. BASTIDE, P. ODIER and J.P. COUTURES, J. Mat. Sci. Letters 7, 289 (1988)

7) P.E. REYES-MOREL and I.W. CHEN J. Am. Ceram. Soc. 71, 343 (1988).

8) C.H. PERRY and D.W. LIU and R.P. INGEL, J. Am. Ceram. Soc., 68C, 184 (1985).85).
   see also T. MAZAKI, T. TONOMURA, Y. KITANO and G. KATAGIRI, Proceeding of Sintering 87, TOKYO 1987.987.

9) B. PIRIOU and H. ARASHI, Bull. Minéral. 103, 363 (1980)

10) P. CARO, *Stucture électronique des éléments de transitions, l'atome dans le cristal,* P.U.F., Paris, (1976)976)

11) J. DEXPERT-GHYS, M. FAUCHER and P. CARO, C.R.A.S., Paris 298 (II), 621 (1984).

12) J. DEXPERT-GHYS, M. FAUCHER and P. CARO, J. Sol. State Chem. 54, 179 (1984).

13) R.C. GARVIE, P.S. NICHOLSON, J. Am. Ceram. Soc. 55, 303 (1972).

14) B. PIRIOU, J. DEXPERT-GHYS and Ph.D'ARCO, to be published.

Ce$_2$-ZrO$_2$ TOUGHENED Al$_2$O$_3$ CERAMICS OF HIGH STRENGTH AND TOUGHNESS.

G.A. ROSSI

NORTON COMPANY - Advanced Ceramics, Northborough, MA
(U.S.A.)

ABSTRACT

CeO$_2$-ZrO$_2$ toughened Al$_2$O$_3$ (Ce-ZTA) ceramics of different compositions were fabricated by pressureless sintering followed by HIPing. Chemically derived and rapidly solidified CeO$_2$-ZrO$_2$ powders were used. The Al$_2$O$_3$ powder was a commercial sub-micron a Al$_2$O$_3$ of high purity. HfO$_2$ was also used in one composition to try to enhance the high temperature mechanical properties. Strengths higher than 1 GPa and toughnness of 8 MPam$^{1/2}$ were measured at room temperature. More important, a considerable fraction of these properties was retained at 1000°C. Microstructural and fractographic analyses have shown clustering of the CeO$_2$-ZrO$_2$ grains and several types of strength limiting flaws. The minimization of such defects obtained through improved precessing is expected to result in even better properties. It is concluded that the Ce-ZTA ceramics are good potential candidates for engine components and other applications, such as wear resistant parts, where the appropriate conbination of strength, toughness and hardness are required.

INTRODUCTION

Zirconia Toughened Ceramics (ZTC) have been extensively studied as potential candidates for heat engine components and for other applications, i.e. wear and abrasion resistant parts. Three main categories of ZTC's exist, i.e. Mg-PSZ (MgO-Partially Stabilized Zirconia), Y-TZP (Y$_2$O$_3$-Tetragonal Zirconia Polycrystals) and ZTA (Zirconia Toughened Alumina). The Mg-PSZ ceramics possess excellent toughness and good strength, but suffer from high temperature instability due to the formation of m-ZrO$_2$ from the decomposition of the cubic grains, which causes severe strength degradation. The Y-TZP ceramics, although extremely strong and considerably tough at room temperature, are susceptible to degradation at low temperatures in humid environments and also rapidly lose toughness and strength at high temperature. The ZTA materials offer perhaps the highest potential, not only because Al$_2$O$_3$ is cheaper than ZrO$_2$, but also because their properties can be tailored to the application by manipulation of the microstructure and composition, i.e. ZrO$_2$/Al$_2$O$_3$ ratio, type and amount of stabilizing oxide for ZrO$_2$, grain size distribution of the ZrO$_2$ grains. Specifically, in the case of Ce-ZTA ceramics, an important advantage is expected to be the low temperature degradation resistance, due to the fact that the Ce-TZP ceramics are known to be resistant to transformation (1) in those conditions where the Y-TZP's fail.

The combination of Ce-TZP and Al_2O_3 to make the Ce-ZTA's makes sense, since the Ce-TZP's are known for their extremely high toughness, up to 20 $MPam^{1/2}$, but the toughest materials possess only moderate strength, around 350 MPa (2) . On the other hand, Al_2O_3 has low toughness, between 3 and 4 $MPam^{1/2}$ (3), but high hardness and good strength if its microstructure is sufficiently fine. Since the CeO_2-ZrO_2 grains inhibit the grain growth of the Al_2O_3 phase, optimum room temperature properties are expected if the appropriate composition and microstructure are produced.

The Ce-ZTA ceramics are also expected to retain a larger fraction of toughness and strength at thigh temperature, vis-a-vis the Y-TZP's, Ce-TZP's and Y-ZTA ceramics. One reason is better toughness retention at high temperature of the Ce-TZP's vs. the Y-TZP's (4). Furthermore, since the transformation toughening contribution is lost at high temperatures, the toughness, and consequently the strength, of the Ce-ZTA ceramics at high temperature will be dictated by the presence of other temperature independent toughening mechanisms, such as ferroelastic domain switching (5) and crack deflection by second phase particles(6).

This paper reports the temperature dependence of strength and toughness of sintered/HiPed Ce-ZTA ceramics with different CeO_2/ZrO_2 and $(CeO_2$-$ZrO_2)/Al_2O_3$ ratios in the temperature range 25°C-1000°C.

Chemically derived and rapidly solidified CeO_2-ZrO_2 powders were used, with the intention of comparing their sinterability and their influence on the microstructure and properties of the ceramics.

The partial substitution of ZrO_2 with HfO_2 in one composition was also tried in order to determine the effect of HfO_2 on the high temperature mechanical properties.

EXPERIMENTAL PROCEDURE

The CeO_2-ZrO_2 powders used were of two types, i.e. chemically derived (C/D) and rapidly solidified (R/S) from a melt. The C/D powders were prepared by co-precipitation of Ce and Zr hydroxides from a solution of reagent grade nitrates with excess NH_4OH. The gelatinous mass was dried, calcined at 1000°C and vibratory milled to produce a sub-micron sinterable powder. The R/S powders were produced by vibratory milling of a crude obtained by rapid solidification of a melt, according to a proprietary Norton technology. The C/D CeO_2-ZrO_2-HfO_2 powder was prepared as described above, from a solution of Zr, Hf and Ce nitrates. The a -Al_2O_3 powder used was a commercially available product, sub-micron in size and of high purity.

The CeO_2-ZrO_2 (or CeO_2-ZrO_2-HfO_2) and Al_2O_3 powders were mixed in a planetary mill with distilled water for 1 hr. and the slurry was poured into liquid N_2 while stirring to prevent phase segregation. The frozen mass was then freeze-dried to produce a powder free from hard agglomerates.

This powder was used to die press billets at 69 MPa, which were subsequently cold isostatically pressed at 207 MPa.

The billets were fired in air at 1500°C/3 hrs. and then HIPed in Ar at 1500°C/207 MPa/45 min. to eliminate the residual porosity.

Table 1 shows the composition of the five Ce-ZTA ceramics objects of this study.

The ceramics were characterized for phase content by XRD, for density with the Archimedes method and for microstructure by SEM.

The flexural strength was measured on 3x3x30 mm bars in 4 point, using 25 mm and 12.7 mm spans, with a cross-head speed of 0.5 mm/min. Strength values were obtained at 25°C, 600°C and 1000°C.

The fracture toughness was measured with a double torsion method at the above mentioned temperatures, using 50x25x1 mm machined samples notched and precracked.

Fractographic analysis (SEM) was done on selected broken MOR bars to identify the fracture origin.

TABLE 1 - Description of the Ce-ZTA Ceramics

MATERIAL	COMPOSITION (w/o)				POWDERS
	CeO$_2$	ZrO$_2$	HfO$_2$	Al$_2$O$_3$	
Z12CE80A-C/D	3.2	16.8	--	80.0	Gel derived CeO$_2$-ZrO$_2$+ submicron Al$_2$O$_3$
20HZ12CE80A C/D	3.2	11.8	5.0	80.0	Gel derived CeO$_2$-ZrO$_2$+ submicron Al$_2$O$_3$
Z12CE80A-R/S	3.2	16.8	--	80.0	Rapidly solidified CeO$_2$-ZrO$_2$+ submicron Al$_2$O$_3$
Z18CE80A-R/S	4.8	15.2	--	80.0	Rapidly solidified CeO$_2$-ZrO$_2$+ submicron Al$_2$O$_3$
Z12CE60A-R/S	6.4	33.8	--	60.0	Rapidly solidified CeO$_2$-ZrO$_2$+ submicron Al$_2$O$_3$

RESULTS AND DISCUSSION

Density mesurements on the billets sintered in air at 1500°C/3 hrs. have shown values between 98 and 99% of theoretical, an indication that the C/D and R/S CeO$_2$-ZrO$_2$ powders possess similar sinter-ability, when dispersed with the Al$_2$O$_3$ powder. However, shrinkage curves by dilatometry, not done in this study, could have shown different shrinkage rates. All billets had virtually theoretical density after HIPing.

The color of the sintered billets was ivory, but changed to yellow-orange after HIPing. This is believed to be caused by the reduction of CeO$_2$ to Ce$_2$O$_3$, since subsequent annealing in air at 1000°C restored the ivory color. The reduction was limited to a thin outside layer, as shown by the cross section of the billets sliced with a diamond blade.

XRD analysis performed on the as-fired surface of a billet of material Z12CE80A-C/D showed about 70% m-ZrO$_2$. However, the cross-section, off white in color, showed only about 3% m-ZrO$_2$. A similar result was obtained for a 50x25x1 mm machined (320 grit diamond wheel) sample used for K_IC measurement with the double torsion method. One face of this sample was orange-yellow and had about 15% m-ZrO_2, whereas the other face was off white and showed only about 3% m-ZrO_2.

The reason for a higher m-ZrO_2 content on the reduced material is not understood at present. Among the possible reasons are: 1) surface destabilization of the t-CeO_2-ZrO_2 grains to m-ZrO_2, caused by impurities picked up from the furnace, 2) preferential grain growth near the surface, and 3) inherent higher transformability of a reduced t-CeO_2-ZrO_2. The third hypotheses is believed to be the most probable. More work is needed to find the correct answer.

The microstructures of the Ce-ZTA ceramics have shown the presence of clusters of CeO_2-ZrO_2 grains. Figure 1 shows that of materal Z12CE80A-R/S as an example. This indicates that the method used for dispersing the CeO_2-ZrO_2 phase in the Al_2O_3 matrix is not very effective and could be improved, for example by improving the chemical homogeneity of the aqueous slurry of the two powders, perhaps by optimizing the pH after measuring the zeta potential of the two powders or by adding water soluble polymers to impart steric stabilization to the particles.

Table 2 shows the temperature dependence of strength for five Ce-ZTA ceramics made with different CeO_2-ZrO_2 powders, CeO_2/ZrO_2 and (CeO_2-ZrO_2)/Al_2O_3 ratios. For the sake of comparison, the data for a Y-TZP ceramic made with a Norton R/S powder is also included. Table 3 shows the temperature dependence of fracture toughness for the same ceramics.

TABLE 2 - Temperature Dependance of Strength for Ce-ZTA and Y-TZP Ceramics

MATERIAL	25 °C	600 °C	1000 °C	% Ret.
Z12CE80A-C/D	741 (7)	489 (3)	508 (4)	69
20HZ12CE80A C/D	703 (6)	505 (5)	462 (5)	66
Z12CE80A-R/S	889 (10)	--	544 (10)	61
Z18CE80A-R/S	840 (10)	--	461 (10)	55
Z12CE60A-R/S	1020 (10)	--	537 (7)	53
Y-TZP-S/H (2.5 m/o Y_2O_3)	1474 (10)	--	544 (10)	37

NOTE: Strength values in MPa; number of bars tested are in parenthesis; % Ret. = percentage of room temperature strength retained at 1000 °C.

TABLE 3 - Temperature Dependance of Toughness for Ce-ZTA and Y-TZP Ceramics

MATERIAL	25 °C	600 °C	1000 °C	% Ret.
Z12CE80A-C/D	8.2 (2)	5.9 (3)	4.5 (3)	55
20HZ12CE80A C/D	6.3 (3)	5.8 (4)	4.7 (3)	75
Z12CE80A-R/S	6.1 (5)	--	3.0 (5)	49
Z18CE80A-R/S	5.2 (5)	--	2.5 (5)	48
Z12CE60A-R/S	7.5 (2)	--	--	--
Y-TZP-S/H (2.5 m/o Y_2O_3)	8.8 (5)	--	2.2 (5)	25

NOTE: Toughness values in MPa . $\sqrt{m}$; number of bars tested are in parenthesis; % Ret. = percentage of room temperature toughness retained at 1000 °C.

Several intersting results are found in Tables 2 and3. First, the strength retention at 1000°C for the Ce-ZTA ceramics is much higher than that of the Y-TZP material, which is much stronger at room temperature. Since it is known that the Ce-TZP's retain at 1000°C about 40% of their room tmperature strength (4), and since the strength of polycrystalline alumina is relatively temperature independent up to 1000°C (3), the high percentages

278

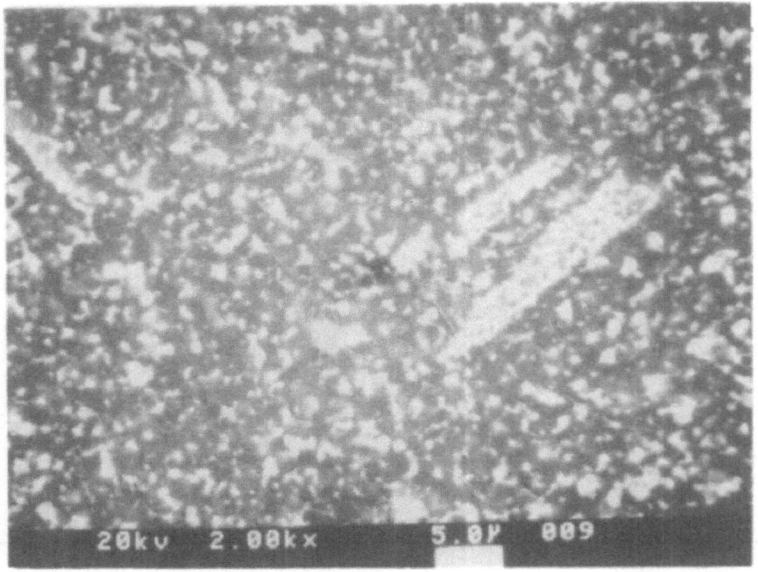

Fig. 1: Microstructure (SEM) of material Z12CE80A-R/S showing clusteringof the CeO_2-ZrO_2 grains.

Fig . 2: Fractograph (SEM) of material Z18CE80A   (24 w/o CeO_2) showing ZrO_2-rich inclusion as fracture origin.

of retained strength of Table 2, from 53% to 69%, seem to indicate that in these Ce-ZTA composites toughening mechanisms other than transformation toughening exist at high temperature, such as crack deflection by second phase particles. Secondly, for the same composition (Z12CE80A) the ceramic made with the R/S CeO_2-ZrO_2 powder exhibits higher strength, but lower toughness, than, that made with the C/D CeO_2-ZrO_2 powder (Table 3). The lower strength of the ceramic made with the C/D powder was explained by fractographic analysis of bend bars, which showed a large fraction of the breaks at the surface or at the bar chamfers, resulting from machining damage. The fracture origins in the bend bars of the ceramics made with the R/S powder, on the other hand, appeared to be in most cases internal inclusions rich in ZrO_2, as shown by Figure 2. These materials turned out to be less susceptible to machining damage, which is not understood given their lower toughness, as shown in Table 3. The difference in toughness for the same composition could be due to microstructure differences and a careful quantitative analysis shoulld provide the answer. A larger grain size of the t-CeO_2-ZrO_2 grains in the Z12CE80A-C/D ceramic could explain its higher toughness, since a larger grain is more easily transformable under stress (7). However, the XRD patterns of the fracture surfaces of the bend bars broken at 25°C, 600°C and 1000°C have revealed no m-ZrO_2. Therefore, one must explain the high toughness (8.2 MPam$^{1/2}$) of material Z12CE80A-C/D either by postulating a <u>reversible</u> t-m transformation, as suggested by Coyle et al. (8), or by assuming that other toughening mechanisms are active in these ceramics. The drop in toughness with temperature suggests that transformation toughening exists at low temperature, as pointed out by Matsumoto et al. (4). These authors state that, in Ce-TZP ceramics, "the contribution of transformation toughening to the overall toughness is negligible above 750°C".

Three other results are shown by Tables 2 and 3. First, the partial substitution of ZrO_2 with HfO_2 (20m/o) in material 20HZCE80A-C:D has not resulted in better high temperature properties. Since HfO_2, compared to ZrO_2, has a higher t-m transformation temperature, higher toughness was expected at 1000°C. This is not shown by the data of Table 2. Second, lower toughness and strength are exhibited by the Z18CE80A-R/S, in comparison with the Z12CE80A-R/S ceramic. This result agrees with the prediction, since it is known that a larger concentration of stabilizer (in this case 18 m/o vs. 12 m/o CeO_2) in ZrO_2 lowers the M_s temperature. Therefore, the t-CeO_2-ZrO_2 grains become thermodynamically more stable and more difficult to transform to m-ZrO_2 under stress. Third, the ceramic with higher (CeO_2-ZrO_2)/Al_2O_3 ratio (Z12CE60A-R/S) exhibits higher room temperature strength in comparison to that with lower ratio (Z12CE80A-R/S). This result can be explained by its higher toughness (7.5 vs. 6.1 $MPam^{1/2}$). However, as Table 2 shows, material Z12CE60A-R/S retains a lower fraction of its strength at 1000°C, which agrees with the hypotheses that the relative contribution of transformation toughening to the total toughness decreases with increasing temperature.

CONCLUSIONS

The main conclusions from this study are as follows:
1) No appreciable difference in sinterability was observed between chemically derived and rapidly solidified CeO_2-ZrO_2 powders, when mixed with Al_2O_3 powder.
2) Sintered/HIPed Ce-ZTA ceramics with different CeO_2/ZrO_2 and $(CeO_2$-$ZrO_2)/Al_2O_3$ ratios have shown good room temperature strength (700-1000 MPa) and toughness (5-8 $MPam^{1/2}$), a substantial improvement vis a vis polycrystalline Al_2O_3. They have also exhibited better retention of these properties at high temperature, in comparison with Y-TZP's and Ce-TZP's. This result seems to indicate that other toughening mechanisms operate in these composites, besides transformation toughening.
3) More m-ZrO_2 was found by XRD on reduced than on oxidized ground surfaces. Additional work is needed to explain this result.
4) Ceramics with the same nominal composition (Z12CE80A), but made with different powders, have exhibited different toughness. Quantitative microstructural analysis is needed to find an explanation to this discrepancy.
5) No m-ZrO_2 was detected on fracture surfaces of bend bars. Therefore, a reversible t-m transformation is postulated in order to explain the high toughness observed.
6) Partial substitution of ZrO_2 with HfO_2 has not resulted in better high temperature properties, a result in contrast with the expectation.
7) Microstructural examination has shown clustering of the CeO_2-ZrO_2 grains. Improved processing is needed to obtain a more homogeneous dispersion.
8) More basic scientific understanding is needed in these composites to establish a correlation between chemistry, microstructure and mechanical properties.

REFERENCES

1) R.L.K. Matsumoto: "Aging behavior of Ceria-Stabilized Tetragonal Zirconia Polycrystals", J.Am.Ceram.Soc.,71, (3), C128-C129 (1988).
2) S. Blackburn et al., "Toughened Zirconia Ceramic from Electro-Refined PSZ Powders", paper presented at 89th ACERS Meeting, Pittsburgh, PA, (1987).
3) K. Niihara et al., "High Temperature Mechanical Properties of Al_2O_3-SiC Composites",Fracture Mechanics of Ceramics, Vol.7, p.

103-116, Ed. by R.C. Bradt, A.G.Evans, D.P.H.Hasselman and F.F.Lange, Plenum Press Pub., NY, (1986).
4) R.L.K. Matsumoto and R.J.Mayhew, "High Temperature Evaluation of New Ferroelastic Toughened Ceramic Materials", Final Report, Dept. of the Army, Contract N.DAAL04-87-C-0058.
5) A.V.Virkar and R.L.K. Matsumoto, "Ferroelastic Domain Switching as a Toughening Mechanism in Tetragonal Zirconia", J.Am.Ceram.Soc., 69, (10), C-224-C-226, (1986).
6) A.G.Evans, "Toughening Mechanisms in Zirconia Alloys", Fracture in Ceramic Materials, p. 16-36, Ed. by A.G. Evans, Noyes Pub., (1984).
7) K. Tsukuma and M; Shimada, "Strength Fracture Toughness and Vickers Hardness of CeO_2-Stabilized Tetragonal ZrO_2 Polycrystals (Ce-TZP)", J.Mater.Sci., 20, 1178-1184, (1985).
8) T.W.Coyle, et al., "Transformation Toughening in Large Grain-Size CeO_2 Doped ZrO_2 Polycrystals", J.Am.Ceram.Soc., 71, (2), C88-C92, (1988).

ACKNOWLEDGMENTS

The author thanks Dr.E.Lilley for helpful discussions. This work was part of an ORNL/DOE Contract (No. 86X-22031C) entitled "Zirconia Toughened Ceramics for Heat Engine Applications".

THE DOMAIN STRUCTURE OF TETRAGONAL ZIRCONIA IN ZrO_2-Y_2O_3 ALLOYS

T. SAKUMA and H. HATA*

The University of Tokyo
Department of Materials Science
Faculty of Engineering
Tokyo 113
Japan.

*Graduate Student
 The University of Tokyo.

ABSTRACT

The microstructural features formed by the diffusionless cubic-to-tetragonal(c-t) transition in ZrO_2-Y_2O_3 alloys were examined by electron microscopy with a special interest in domain structure. The diffusionless c-t transition accompanies two types of microstructures, thin plates or lenticular features and domain structure. These microstructures were found in ZrO_2-(3-7)mol%Y_2O_3 alloys rapidly-cooled from single cubic phase region, but not in ZrO_2-8 mol%Y_2O_3 alloy. The domain structure was always found in a whole region of the sample, and the domain size decreased with an increase of yttria content. On the basis of microstructural examinations, it was concluded that the diffusionless c-t transition has a nature of second order phase transition.

Key words ; partially-stabilized zirconia,
phase transformation, martensitic transformation,
microstructure, domain structure

INTRODUCTION

The cubic-to-tetragonal(c-t) transition in pure zirconia was first found by Smith and Cline[1], and later reconfirmed by several workers[2,3]. Cohen and Shaner have reported that the c-t diffusionless transition induced in ZrO_2-UO_2 is of martensitic type[4]. Andersson and Gupta have shown that the martensitic c-t transformation can be expected from ZrO_2-Y_2O_3 phase diagram, if an alloy was cooled rapidly from single cubic phase region to a temperature sufficiently below T_0[5]. Andersson et al. have reported that the plates of tetragonal zirconia(t-ZrO_2) formed in ZrO_2-4.5 mol%Y_2O_3 alloy are the product of martensitic transformation[6]. Plates or lenticular features have commonly been found in alloys rapidly-cooled from single cubic phase region in zirconia alloys[7-9]. These features resemble the microstructures formed by martensitic transformation in metallic alloys[8]. On the other hand, Heuer and his colleagues have observed another microstructure, the domain structure showing the contrast of anti-phase domain boundaries(APB's), which was also formed by the diffusionless c-t transition[10,11]. They have insisted that the domain structure is t-ZrO$_2$* and plates or lenticular features are twins which were induced for accommodating the transformation strain. One of the present authors has proposed that the diffusionless c-t transition has a nature of second order phase transition[13,14], if the domain structure itself is t-ZrO$_2$. This proposal is based on the fact that the domain structure always appears in a whole region of the sample[10,11,13]. The diffusionless c-t transition and the resultant microstructure are thus complicated. It is necessary to make further detailed examinations to elucidate the nature of this transition. In the present study, the microstructures formed by the diffusionless c-t transition were examined in ZrO$_2$-Y$_2$O$_3$ alloys for making clear its nature.

EXPERIMENTAL PROCEDURE

Zirconia powders containing various yttria contents supplied by TOSOH Co. Ltd. were used for starting materials. They were pressed into a size of about 5x7x10 mm^3 and arc-melted in an argon atmosphere on a water-cooled copper hearth. They were cooled from melting temperature to 1200°C by 10 s. Six alloys containing between 3 and 8 mol%Y$_2$O$_3$ were prepared by this procedure. Metallographic examinations were mainly carried out in these alloys without any heat treatments after melting. X-ray diffraction experiments were made by CuKα radiation in RIGAKU RAD II-A. Thin foils prepared by ion-milling were examined by electron microscope(EM) HITACHI H-800 operated at 200 kV.

RESULTS

Fig. 1 shows the XRD profiles in these alloys. The intensity profiles of 2θ angles between 72° and 75.5° are shown in Fig. 1, because 400 type reflections are sensitive to distinguish t-ZrO$_2$ from cubic phase(c-ZrO$_2$). The peaks in ZrO$_2$-4 mol% and 6 mol%Y$_2$O$_3$

---

* The t-ZrO$_2$ containing a sufficient amount of cubic-stabilizing oxides is fairly stable and is not transformed into monoclinic phase(m-ZrO$_2$). Such t-ZrO$_2$ is often referred to be t'-ZrO$_2$ for distinguishing from unstable and transformable t-ZrO$_2$[12]. However, t-ZrO$_2$ and t'-ZrO$_2$ are essentially of the same crystal symmetry and differ only chemical composition. In this paper, the terminology t-ZrO$_2$ is used to express tetragonal zirconia.

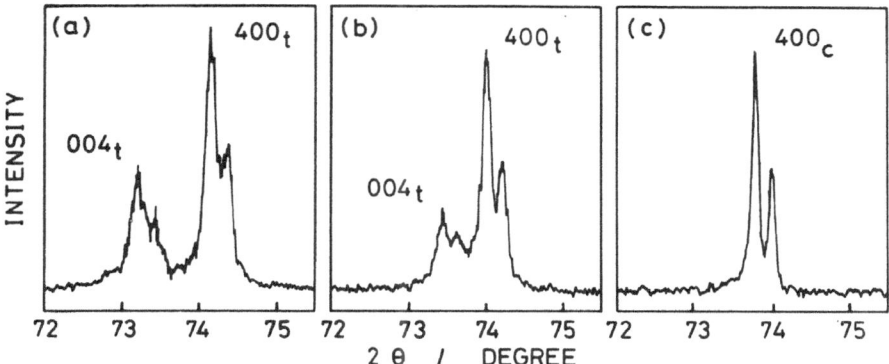

Fig. 1 X-ray intensity profiles in ZrO_2-(a) 4 mol%,
(b) 6 mol% and (c) 8 mol%Y_2O_3 alloys. Peaks
of each reflection appears as doublets from
Kα1 and Kα2 radiations.

alloys are regarded to be from single phase t-ZrO_2, while those
in ZrO_2-8 mol%Y_2O_3 alloy are from c-ZrO_2. The peaks only from
single phase, either t-ZrO_2 or c-ZrO_2, were found and peaks from
these two phases did not appear simultaneously in present alloys.
This result is characteristic of alloys cooled rapidly from
single c-ZrO_2 region.
    Fig. 2 is the electron micrographs in ZrO_2-4 mol%Y_2O_3 alloy.
Bright field image (a) and dark field image using a 112
reflection (b) were taken in the same area. Different
microstructural features are revealed in the bright and dark
field images; lenticular features are seen in Fig. 2(a), while
the domain structure is observed in Fig. 2(b).

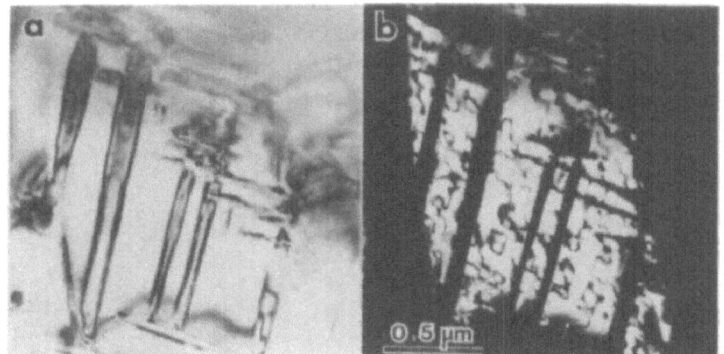

Fig. 2 Microstructures in ZrO_2-4 mol%Y_2O_3 alloys;
bright field image (a) and dark field image
taken by a 112 reflection (b).

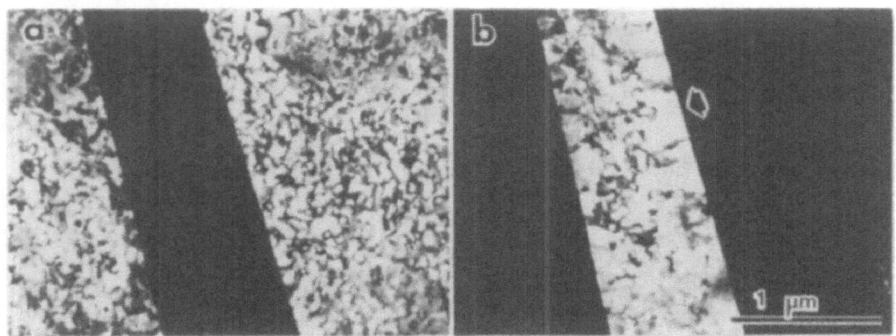

Fig. 3 Dark field images taken by two different 112
reflections. Two micrographs show the same
area, which consists of two variants of t-ZrO$_2$.

Fig. 3 shows the dark field images of the same area taken by
different 112 reflections. This area consists of two variants of
t-ZrO$_2$. The domain structure is observed in both variants. It was
found that the domain structure was always formed in a whole
region of sample in alloys containing 3-7 mol%Y$_2$O$_3$. Since
the domain structure is regarded to be a microstructure of
t-ZrO$_2$, then the above result means that the alloys with 3-7
mol%Y$_2$O$_3$ are fully tetragonal at room temperature in accordance
with the result of XRD analysis.

In Fig. 3(b), some large domains in contact with the
interface between two variants are seen as shown by an arrow.
Such coarse domains were sometimes found in each alloy
particularly at the interface. Some domain boundaries were found
to be facetted and to run nearly normal to the interface. The
facetted boundaries were often formed in alloys annealed at low
temperatures after melting. Fig. 4 is an example of such
boundaries in ZrO$_2$-6 mol%Y$_2$O$_3$ alloy, which was annealed at 300°C
for 100 h in air. The nonuniformity of domain size and facetting

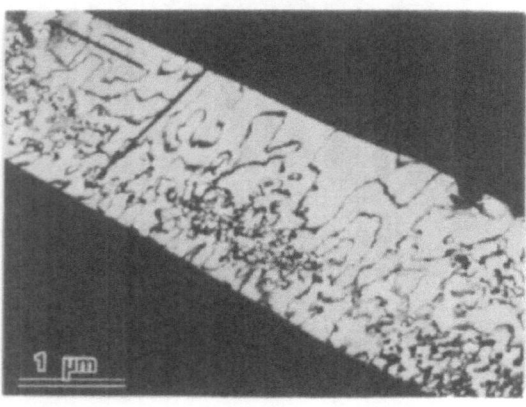

Fig. 4 The domain
structure in ZrO$_2$-
6 mol%Y$_2$O$_3$ alloy
isothermally-annealed
at 300 °C for 100 h in
air after arc melting.

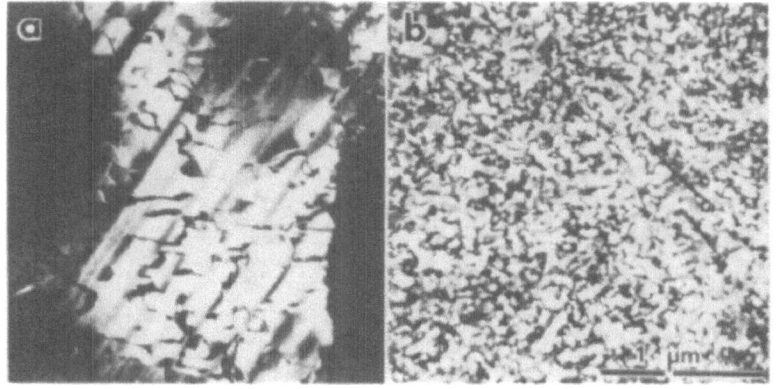

Fig. 5   The domain structure of ZrO_2-(a) 3 mol% and
(b) 6 mol%Y_2O_3 alloys.

of boundaries are clearly seen in this micrograph.
The domain structure in two alloys is shown in Fig. 5. The domains in ZrO_2-3 mol%Y_2O_3 alloy are much larger than those in ZrO_2-6 mol%Y_2O_3 alloy. Although the domain size varies locally in each alloy, the average size was smaller in alloys with higher yttria content. The domain size is plotted against yttria content in Fig. 6. The change in domain size is marked in the composition range between 3 and 5 mol%Y_2O_3.

DISCUSSION

The XRD and EM analyses have shown that the room temperature phase consists only of t-ZrO_2 in ZrO_2-3 mol% to 7 mol%Y_2O_3 alloys rapidly-cooled from single cubic phase region, but of c-ZrO_2 in ZrO_2-8 mol%Y_2O_3 alloys. The mixed structure composed of t-ZrO_2

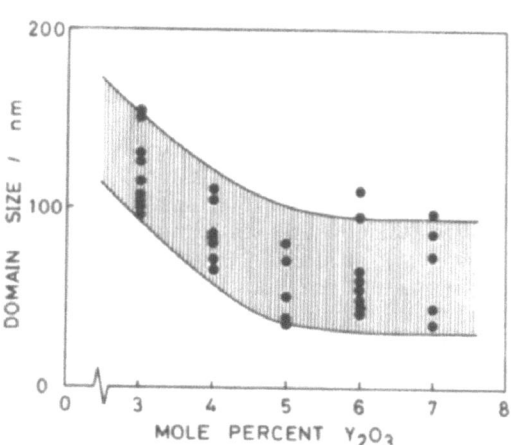

Fig. 6   The domain
size as a function
of yttria content.

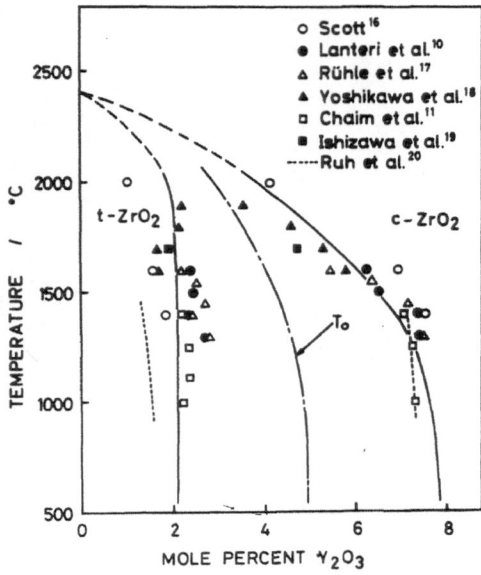

Fig. 7 The data of c-t
equilibrium in
ZrO$_2$-Y$_2$O$_3$ system.

and c-ZrO$_2$ was never found in these alloys. This result may
indicate that the diffusionless c-t transition is always
completed [10,11,13]. The generation of t-ZrO$_2$ by the
diffusionless transition is likely to be characterized by the
initial development of domain structure, which is followed by
the formation of plate-like or lenticular features[13]. These
features are twins or variants of t-ZrO$_2$, which are probably
induced for accommodating the transformation strain[10,11]. These
evidences suggest that the nature of this transition is not the
first order phase transition as has generally been accepted
[6,7,10,11,15]. This nature seems to be of second order type as
discussed below.

The two phase equilibrium between t-ZrO$_2$ and c-ZrO$_2$ is
compiled in Fig. 7[10,11,16-20]. The T_0 line in Fig. 7 is drawn
between the two phase boundaries, t-ZrO$_2$ side and c-ZrO$_2$ side.
This is a common way to decide T_0 line. Andersson and Gupta have
argued that the Ms line may be defined somewhat below the T_0
line, and the c-t transition will be induced by martensitic
mechanism during rapid cooling from single cubic phase region to
a temperature below Ms. However, Fig. 7 shows that some present
alloys, ZrO$_2$-6 and 7 mol%Y$_2$O$_3$, have lower T_0 than room
temperature. That is, the diffusionless c-t transition takes
place even "above" T_0 in ZrO$_2$-6 mol% and 7 mol%Y$_2$O$_3$ alloys. This
fact undoubtedly indicates that this transition is not of simple
martensitic type.

For discussing the fact in more detail, let us consider the
free energy-composition(G-x) diagram in this system. Fig. 8 shows

the c-t equilibrium(a), the generally-accepted G-x diagram (b) and the newly-proposed G-x diagram(c). In general, the diffusionless c-t transition and the precipitation of t-ZrO_2 in c-ZrO_2 matrix have been discussed from the G-x diagram such as Fig. 8(b), in which t-ZrO_2 and c-ZrO_2 have different G-x curves[5,15]. These two phases take the same free energy at a composition x_1 in Fig. 8(b). If this is the case, only an alloy with less yttria content than x_1 has a driving force for the diffusionless c-t transition. The diffusionless c-t transition in ZrO_2-6 and 7 mol%Y_2O_3 alloys cannot be explained from Fig. 8 (b). Someone may insist that T_0 is not necessarily at the center between the two phase boundaries and the above discussion is based on such unreliable determination of T_0. The T_0 temperature is not at the center, when a relevant phase has very low solid solubility or a stoichiometric compound is participated in equilibrium. In such a case, T_0 is often shifted from the center towards the side of the low solid solubility phase or of the stoichiometric compound. However, it is not possible to expect that T_0 is shifted towards c-ZrO_2 side rather than depicted in Fig. 8(b), because c-ZrO_2 has a wide solid solution range. It is, therefore concluded that the construction of G-x diagram in

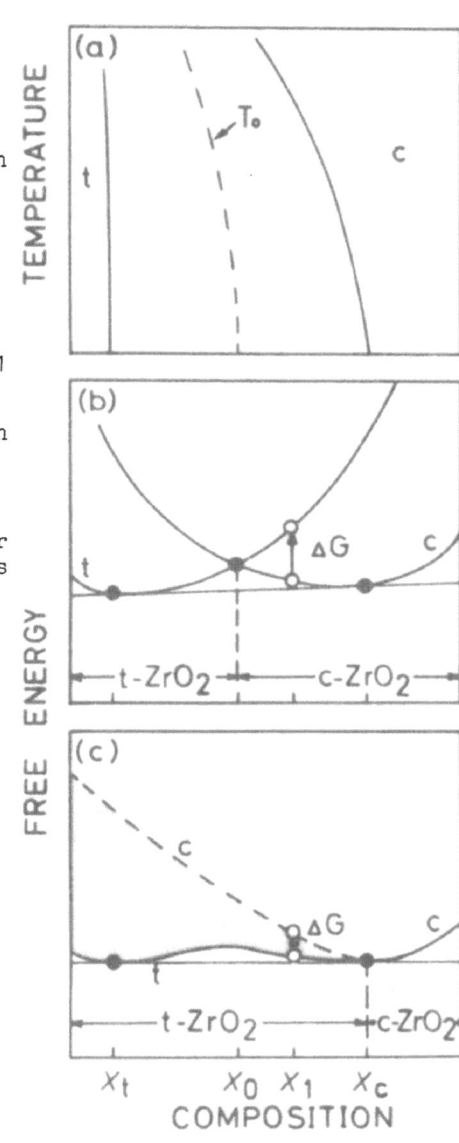

Fig. 8 The c-t equilibrium (a) and the corresponding free energy-composition diagrams (b) and (c). The diagram (b) has commonly been used to discuss the c-t equilibrium and phase transformation or precipitation, while (c) is the proposed one to explain the nature of the diffusionless c-t transition(see text).

Fig. 8(b) predicts a negative driving force for the diffusionless
c-t transition in the alloy with a composition x_1, which is
between x_0 and x_c. This is not consistent with the present
experimental results. The inconsistent prediction arises from the
improper estimation of driving force ΔG. It is necessary to
identify the tetragonality of t-ZrO$_2$ for estimating ΔG. It has
experimentally been clarified that the room-temperature
tetragonality of t-ZrO$_2$ decreases with an increase of Y$_2$O$_3$
content[16,21]. Writing the room-temperature tetragonality of
t-ZrO$_2$ with compositions x_t and x_1 as $(c/a)_t$ and $(c/a)_1$, the
t-ZrO$_2$ formed by the diffusionless c-t transition in alloy x_1 has
a tetragonality $(c/a)_1$, not $(c/a)_t$. Whereas, the G-x curve for
the t-ZrO$_2$ in Fig. 8(b) is considered to be of t-ZrO$_2$ with
$(c/a)_t$. This is the reason why a negative ΔG is expected for the
diffusionless c-t transition in the alloy x_1. The free energy for
t-ZrO$_2$ with $(c/a)_1$ must be lower than that for c-ZrO$_2$ at this
composition x_1 so that a positive driving force acts for inducing
the diffusionless c-t transition.

It may be reasonable to assume that the t-ZrO$_2$ can take
various tetragonalities at each composition, but the most stable
one appears at each temperature. The tetragonality of t-ZrO$_2$
depends on composition and temperature. The temperature
dependence of tetragonality has been examined in pure zirconia,
whose tetragonality increases sharply with a decrease of
temperature just below the c-t transition temperature[22]. Here,
an important question arises whether or not the tetragonality
changes continuously from zero to a certain value, and the t-ZrO$_2$
with zero tetragonality is regarded to be c-ZrO$_2$. The
experimental data has shown that the extrapolation of lattice
parameter of t-ZrO$_2$ into zero tetragonality falls very close to
that of c-ZrO$_2$, as if the lattice parameter changes continuously
from c-ZrO$_2$ to t-ZrO$_2$[16,21]. The XRD measurements are not so
accurate enough to confirm the continuous lattice parameter
change between the two phases. However, the present results can
reasonably be explained by assuming this continuous change.

A possible G-x diagram when the diffusionless transition
takes place continuously from c-ZrO$_2$ to t-ZrO$_2$ is depicted in
Fig. 8(c). Here, the t-ZrO$_2$ with zero tetragonality is assumed to
be c-ZrO$_2$. The t-ZrO$_2$ with a particular tetragonality must have a
single G-x curve. When the G-x curves of t-ZrO$_2$ for various
tetragonalities are drawn, they will become a free energy surface
such as the hatched area in Fig. 8(c), provided that the
tetragonality changes continuously from zero to various values.
The base line of this free energy surface is a locus of minimum
free energy of t-ZrO$_2$ at each composition. Judging from the
lattice parameter data, the tetragonality of t-ZrO$_2$ with the
minimum free energy must change continuously with yttria content.
The base line must be convex upward in the central part, because
t-ZrO$_2$ and c-ZrO$_2$ coexist in a certain composition range at
equilibrium. The G-x surface and its base line in Fig. 8(c) are

constructed so as to be consistent with the present results and also the reported c-t equilibrium. The diffusionless c-t transition can be induced even in alloys with higher composition than x_0 when the G-x diagram is such as Fig. 8(c). For example, there is a driving force ΔG for this transition in an alloy with a composition x_1. Note that a driving force for the reverse t-c transition is expected in this alloy for Fig. 8(b). It may be reasonable that the driving force ΔG in Fig. 8(c) will not be worked instantaneously. The diffusionless c-t transition will take place in a whole region under an extremely small driving force, and further cooling results in an increase of tetragonality. In such a case, the diffusionless c-t transition always goes to completion in alloys with composition less than x_c, while this transition does not occur in alloys with composition higher than x_c. Actually, this is in accord with the present experimental result. The diffusionless c-t transition has a similar nature with the order-disorder transition in β-brass [14,19], which is believed to be the second order phase transition.

Finally, it may be necessary to point out that T_0 line is only regarded to be middle composition between t-ZrO$_2$ and c-ZrO$_2$ in equilibrium for Fig. 8(c). This is different from the definition of T_0 for Fig. 8(b), at which t-ZrO$_2$ and c-ZrO$_2$ have an equal free energy. The Ms line cannot be defined in Fig. 8(c). Then, it is concluded that plates or lenticular features are not formed by martensitic transformation[14], but are regarded to be twins or variants of t-ZrO$_2$ induced for the strain accommodation [10,11].

CONCLUSION

Rapid cooling from high-temperature cubic phase region in ZrO$_2$-Y$_2$O$_3$ alloys resulted always in single phase structure, either t-ZrO$_2$ or c-ZrO$_2$, at room temperature. Alloys containing 3-7 mol%Y$_2$O$_3$ were fully tetragonal, while ZrO$_2$-8 mol%Y$_2$O$_3$ alloy was cubic phase only. Two-phase structure composed of t-ZrO$_2$ and c-ZrO$_2$ was not found in the rapidly-cooled alloys. The room-temperature t-ZrO$_2$ was characterized by two types of microstructures, domain structure and thin plates or lenticular features. The domain structure was formed initially and thin plates or lenticular features later during the diffusionless c-t transition. The diffusionless c-t transition was considered to have a nature of second order phase transition such as order-disorder transition of β-brass, which is always completed, because two-phase structure was never detected by XRD analysis, and the domain structure was always found in entire region of sample by EM observation. Generally-accepted G-x diagram in ZrO$_2$-Y$_2$O$_3$ system was contradictory to present experimental results. A new construction of G-x diagram consistent with experimental results obtained so far was proposed.

ACKNOWLEDGEMENTS

Financial support by Grant-in-Aid Shiken-kenkyu(2) 62856132 for Fundamental Research form the Ministry of Education in Japan is gratefully acknowledged.

REFERENCES

[1] D. K. Smith and C. F. Cline, J. Am. Ceram. Soc., 45(1962)249.
[2] G. W. Wolten, *ibid.*, 46(1963)418.
[3] A. G. Boganov, V. S. Rudenko and L. P. Makarov, Dokl. Akad. Nauk SSSR, 160(1965)1065.
[4] I. Cohen and B. E. Shaner, J. Nuclear Mater., 9(1963)18.
[5] C. A. Andersson and T. K. Gupta, Advances in Ceramics, Vol. 3, Science and Technology of Zirconia, ed. by A. H. Heuer and L. W. Hobbs, The Am. Ceram. Soc., OH,(1981) p. 184.
[6] C. A. Andersson, J. Greggi Jr. and T. K. Gupta, Advances in Ceramics, Vol. 12, Science and Technology of Zirconia II, ed. by N. Claussen, M. Rühle and A. H. Heuer, The Am. Ceram. Soc., OH, (1984) p. 78.
[7] T. Sakuma, Y. Yoshizawa and H. Suto, J. Mater. Sci., 20(1985)2339.
[8] T. Sakuma, H. Eda and H. Suto, Proc. Int. Conf. on Martensitic Trnsformations, The Jpn. Inst. Met., (1986) p. 1149.
[9] R. P. Ingel, D. Lewis III, B. A. Bender, and S. C. Semken, Advances in Ceramics, Science and Technology of Zirconia III, in Press.
[10] V. Lanteri, A. H. Heuer and T. E. Mitchel, Ref.[6], p. 118.
[11] R. Chaim, M. Rühle and A. H. Heuer, J. Am. Ceram. Soc., 68(1985)427.
[12] R. A. Miller, J. L. Simalek and R. G. Garlik, Ref.[5], p. 241.
[13] T. Sakuma, J. Mater. Sci., 22(1987)4470.
[14] T. Sakuma, A paper presented at MRS meeting, Tokyo(1988).
[15] A. H. Heuer and M. Rühle, Ref.[6], p. 1
[16] H. G. Scott, J. Mater. Sci., 10(1975)1527.
[17] M. Rühle, N. Claussen and A. H. Heuer, Ref. [6], p. 352.
[18] N. Yoshikawa, H. Eda and H. Suto, J. Jpn. Inst. Met., 50(1986)113
[19] N. Ishizawa, A. Saiki, T. Yagi, N. Mizutani and M. Kato, J. Am. Ceram. Soc., 62(1986)c-18.
[20] R. Ruh, K. S. Mazdiyasni, P. G. Valentine and H. O. Bielstein, *ibid.*, 67(1984)c-190.
[21] H. Suto, T. Sakuma and N. Yoshikawa, Trans. Jpn. Inst. Met., 28(1987)623.
[22] P. Aldebert and J. P. Traverse, J. Am. Ceram. Soc., 68(1985)34.

Post-Sintering Hot Isostatic Pressing of Ceria-Doped Tetragonal
Zirconia/Alumina Composites

Tsugio Sato, Tadashi Endo and Masahiko Shimada
Department of Applied Chemistry, Faculty of Engineering,
Tohoku University, Sendai 980, Japan

Abstract

Ceria-doped tetragonal zirconia/alumina composites (Ce-TZP/Al$_2$O$_3$) and yttria and ceria-doped teragonal zirconia/alumina composites ((Y,Ce)-TZP/Al$_2$O$_3$) were fabricated by sintering at 1450°C for 10 hr in air, followed by hot isostatic pressing (Post-sintering HIPing) at 1450°C and 100 MPa for 1 hr in an 80 vol% Ar and 20 vol% O$_2$ gas atmosphere. Dispersion of Al$_2$O$_3$ particles into TZP was useful in supressing the grain growth of zirconia and increasing the relative density of TZP prior to HIPing. Post-sintering HIPing was useful to densify Ce-TZP/Al$_2$O$_3$ composites to almost fully theoretical density without grain growth and to improve the fracture strength and thermal shock resistance.

Introduction

Although yttria-doped tetragonal zirconia polycrystals (Y-TZP) show high fracture strength and fracture toughness, the mechanical properties of them are sometimes greatly degraded by low temperature annealing at 200-300°C in air due to the formation of microcracks accompanied with the tetragonal (t)-to-monoclinic (m) phase transformation (t$\rightarrow$m) during annealing.[1-3] The thermal stability of TZP was depended on the stability of the tetragonal zirconia and could be improved by microstructural modification such as doping with stabilizers, dispersing the particles with high Young's modulus and decreasing the grain size of zirconia.[4] Doping with ceria, in which ionic radius of Ce^{4+} is larger than that of Zr^{4+}, is one of the best method to improve the thermal stability of TZP because it can greatly decrease the critical transformation temperature below which t$\rightarrow$m phase transformation occur.[4] It was noted that Ce-TZP shows excellent fracture toughness (above 15 MPa$\cdot$m$^{1/2}$), but the sinterability and the fracture strength are modest.[5] It was demonstrated that the fracture strength of Y-TZP could be greatly improved by dispersing Al$_2$O$_3$ particles into TZP matrix. Therefore, it can be expected that the fracture strength of Ce-TZP may be improved by dispersing Al$_2$O$_3$ particles if the composites are densified well. Hot isostatic pressing is often applied to densify the poor sinterable materials, but this conventional technique using Ar gas was not applicable to the fabrication of Ce-TZP ceramics, because CeO$_2$ is reduced to Ce$_2$O$_3$ and the tetragonal zirconia transformed to monoclinic phase. In the present study, (Y,Ce)-TZP/Al$_2$O$_3$ and Ce-TZP/Al$_2$O$_3$ composites were fabricated by normal sintering in air and post-sintering hot isostatic pressing in an Ar-O$_2$ gas atmosphere in order to evaluate the thermal stability and the mechanical properties of the composites.

Experimental

The starting materials of Y-TZP powders containing 2 mol% Y_2O_3 (2Y-TZP) and Ce-TZP powders containing 12 mol% CeO_2 (12Ce-TZP) fabricated by a hydrolysis method[5] and Al_2O_3 powders were supplied by Toso Co. Ltd. and Sumitomo Chemical Co. Ltd., respectively. CeO_2 powders were of analytical grade. 2Y-TZP, CeO_2, 12Ce-TZP and Al_2O_3 powders were mixed by ball milling with ethyl alcohol and zirconia balls in a polyethylene container, followed by drying at 80°C for 5 hr in air. The powders were uniaxially pressed at 50 MPa to form plates, 5 mm x 30 mm x 50 mm prior to cold isostatic pressing at 200 MPa. The samples were sintered at 1450-1600°C for 10 hr in air. The sintered bodies obtained were hot isostatically pressed at 1450°C and 100 MPa for 1 hr in a gas atmosphere containing 80 vol% Ar and 20 vol% O_2. The sample plates were cut into rectangular coupons, 2 mm x 4 mm x 15 mm and polished successively with 10-, 3- and 0.3-µm diamond paste to mirror-like finish. The edges of the samples were beveled with 600-grit diamond grinding wheel. In order to evaluate the resistance to the t→m phase transformation during low temperature annealing, the specimens obtained in this way were annealed in water at 120°C for 100 hr. Phase identification was carried out by powder x-ray diffraction analysis on the plate surface. The bulk density was determined by Archimedes' technique. The average grain size was determined by the intercept method.[7] Fracture toughness was determined by indentation fracture method.[9] Fracture strength was determined by three-point bending tests with a cross head speed of 0.5 mm/min and a span width of 10 mm. Five specimens were fractured at each test condition.

Results and Discussion

Since it was reported that the tetragonal to monoclinic phase transformation of TZP ceramics by low temperature annealing could be controlled by doping more than 12 mol% CeO2, (2Y,12Ce)-TZP/Al_2O_3 and 12Ce-TZP/Al_2O_3 were fabricated by sintering at 1450°C for 10 hr in air and followed HIPing at 100MPa and 1450°C for 1hr in 80vol% Ar-20vol% O_2 gas atmosphere. No phase transformation from tetragonal to monoclinic phase occurred after post-sintering HIPing under the present gas atmosphere condition. Scanning electron micrographs of the fracture surface of 12Ce-TZP and 12Ce-TZP/20wt% Al_2O_3 composites sintered at 1450°C for 10 hr in air are shown in Fig. 1. It can be seen that Al_2O_3 grains of about 0.5 µm are located at the grain boundaries of zirconia grains and that the grain growth of the zirconia matrix was greatly supressed by dispersing Al_2O_3.

Fig. 1 Scanning electron micrographs of the fracture surface of 12Ce-TZP (A) and 12Ce-TZP/20 wt% Al_2O_3 (B) sintered at 1450°C for 10 hr in air. Arrows indicate Al_2O_3 grains.

Relative density, ρ_r, grain size of zirconia, d, fracture toughness, K_{IC}, and three point bending strength, σ_{3b}, of (2Y,12Ce)-TZP/Al_2O_3 are shown in Fig. 2. Dispersion of Al_2O_3 resulted in the decrease of the grain size of zirconia. The relative density of as-sintered bodies in air was almost constant at about 92% up to 15wt% of Al_2O_3 content and then decreased. The fracture toughness of as-sintered bodies in air increased with increasing Al_2O_3 content, while fracture strength increased from 300MPa to 450 MPa by dispersing 5 wt% of Al_2O_3, and then was almost constant up to 20 wt%. On the other hand, relative density increased to about 97.5% without grain growth by post-sintering HIPing, but no noticeable improvemet was found for the fracture strength. These results indicated that the amount of small pore in the pre-sintered sample in air could be reduced by post-sintering HIPing, but the large pores still remained in the specimens.

In order to decrease the crack size of sample by post-sintering HIPing, it is required to use the densified pre-sintered body without open-pore as post-HIPing material. The grain size of the zirconia, relative density, fracture strength and fracture toughness of 12Ce-TZP/Al_2O_3 are shown in Fig. 3. The grain size of the zirconia in the sintered bodies greatly decreased from 3.1 um to 1.5 um and the relative density increased from 95.5 to 98 % with increasing Al_2O_3 contents from 0 to 20 wt%. As expected, all samples of 12Ce-TZP and 12Ce-TZP/Al_2O_3 composites were densified to almost fully theoretical density by post-sintering HIPing treatment without grain growth of zirconia. These results

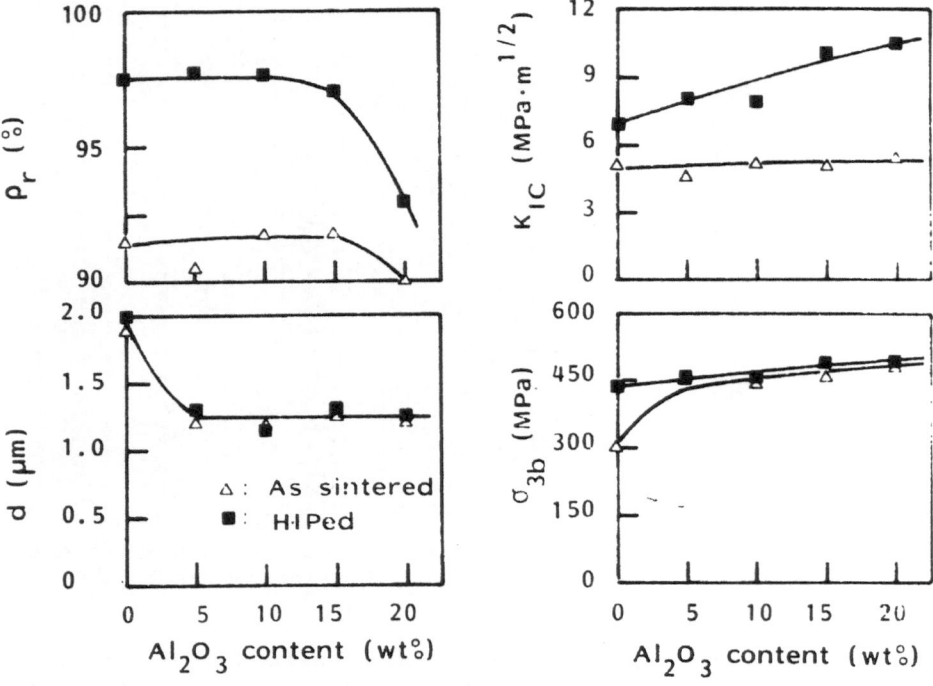

Fig. 2 Relative density, grain size of zirconia, fracture toughness and bending strength of (2Y,12Ce)-TZP/Al$_2$O$_3$ as-sintered at 1450°C for 1 hr in air and post-HIPed at 1450°C and 100 MPa for 1 hr in 80 vol% Ar and 20 vol% O$_2$ gas atmosphere

indicate that the pressure increase up to 100 MPa enhanced the densification process of these materials. The fracture strength of as-sintered body was greatly increased by the addition of Al$_2$O$_3$, while the fracture toughness decreased a little. No noticeable difference in the fracture toughness was observed between as-sintered 12Ce-TZP/Al$_2$O$_3$ composites and post-sintering HIPed samples, but the fracture strength increased from 500-580 MPa to 620-760 MPa following the post-sintering HIPing treatment. These results indicated that the critical flaw size in post-sintering HIPed samples decreased to about half of that in as-sintered bodies. On the other hand, the fracture toughness of 12Ce-TZP increased from 15 to 22 MPa·m$^{1/2}$ following the post-sintering HIP treatment, whereas the fracture strength was almost the same value before and after post-sintering HIPing treatment. Swain[10] reported a relationship between fracture strength and fracture toughness of various zirconia ceramics, and concluded that failure of zirconia ceramics could be attributed to the intrinsic flaws or to transformation generated flaws.

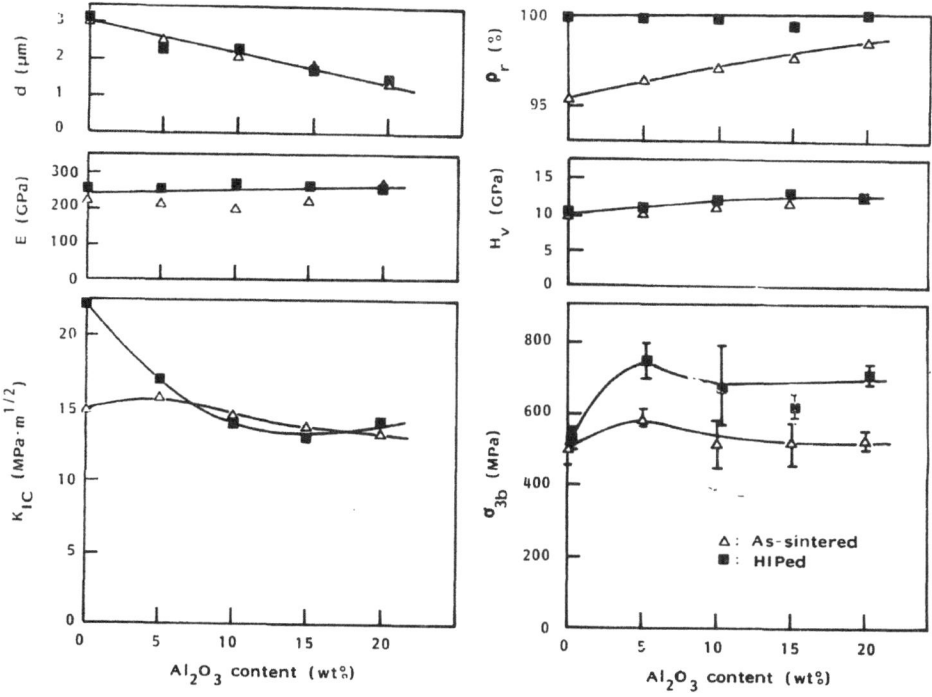

Fig. 3 Grain size of zirconia, relative density, fracture strength and fracture toughness of 12Ce-TZP and 12Ce-TZP/Al$_2$O$_3$ composites sintered at 1450°C for 10 hr in air and post-sintering HIPed at 1450°C and 100 MPa for 1 hr in an 80 vol% Ar-20 vol% O$_2$ gas atmosphere.

The relationship between fracture strength and fracture toughness of (2Y,12Ce)-TZP/Al$_2$O$_3$, 12Ce-TZP and 12Ce-TZP/Al$_2$O$_3$ samples before and after post-sintering HIPing treatment is summarized in Fig. 4. As seen in this figure, the fracture strength of these zirconia ceramics linearly increased with increasing the fracture toughness at first, and then decreased. It was considered that fracture in the former region was attributed to intrinsic flaws and flaw size affected to the slope of the straight line, and in the latter region fracture was attributed to transformation generated flaws.[10] That is to say, fracture of 12Ce-TZP as-sintered and post-sintering HIPed materials might be attributed to intrinsic flaws and transformation generated flaws, respectively.

Since t → m phase transformation of TZP ceramics during low temperature annealing was greatly enhanced in water,[3] the thermal stability of 2Y-TZP, (2Y,12Ce)-TZP/Al$_2$O$_3$, 12Ce-TZP and 12Ce-TZP/20 wt% Al$_2$O$_3$ composites as-sintered and post-sintering HIPed was

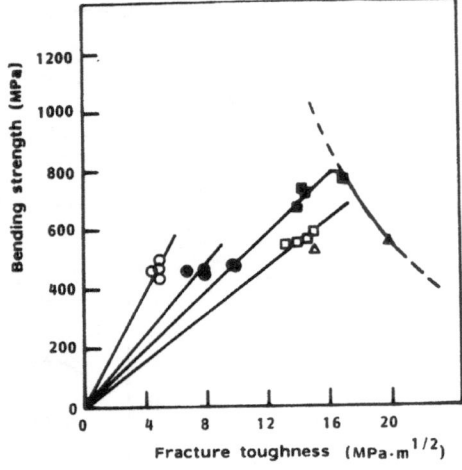

O : (2Y,12Ce)-TZP/Al$_2$O$_3$ as-sintered
● : (2Y,12Ce)-TZP/Al$_2$O$_3$ HIPed
△ : 12Ce-TZP as-sintered
▲ : 12Ce-TZP HIPed
□ : 12Ce-TZP/Al$_2$O$_3$ as-sintered
■ : 12Ce-TZP/Al$_2$O$_3$ HIPed

Fig. 4 Relationship between fracture strength and fracture toughness of (2Y,12Ce)-TZP/Al$_2$O$_3$ and 12Ce-TZP/Al$_2$O$_3$ composites.

evaluated by annealing in water at 120°C for 100hr. No t→m phase transformation occurred during annealing. The fracture strengths of the specimens obtained before and after annealing are shown in Fig. 5. The fracture strength of 2Y-TZP decreased from 1200 MPa to 100 MPa due to the formation of microcracks accompanied with the t→m phase transformation during annealing, but no degradation was observed for zirconia ceramics doped with 12 mol% of CeO$_2$. The fracture strength after annealing incresed in the order 12Ce-TZP/20 wt% Al$_2$O$_3$ post-sintering HIPed > 12Ce-TZP/20 wt% Al$_2$O$_3$ as-sintered > 12Ce-TZP > (2Y,12Ce)-TZP/15 wt% Al$_2$O$_3$ as-sintered > (2Y,12Ce)-TZP as-sintered >> 2Y-TZP.

The thermal shock resistance of 12Ce-TZP/20 wt% Al$_2$O$_3$ as sintered and post-sintering HIPed was evaluated by a quenching method using water at 0°C as a quenching medium. The results are shown in Fig. 6. The critical quenching temperature difference, ΔT_c, at which the fracture strength greatly degraded, increased from 240°C to 280°C following the post-sintering HIPing treatment. Generally, the critical quenching temperature difference can be written by :[11,12)

$$\Delta T_c = [\sigma_t(1-V)/\alpha E][1.451(1+3.42k/r_o h)]$$

where σ_t is the tensile stress, V is Poisson's ratio, is the thermal expansion coefficient, E is Young's modulus, k is the thermal conductivity of specimen, r_o is half thickness of the

299

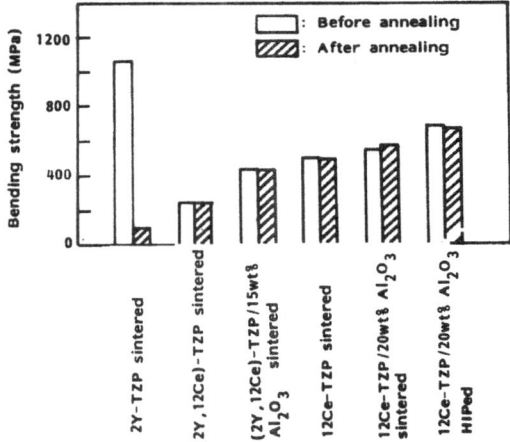

Fig. 5 Fracture strength of various zirconia ceramics as-prepared and annealed in water at 120°C for 100 hr.

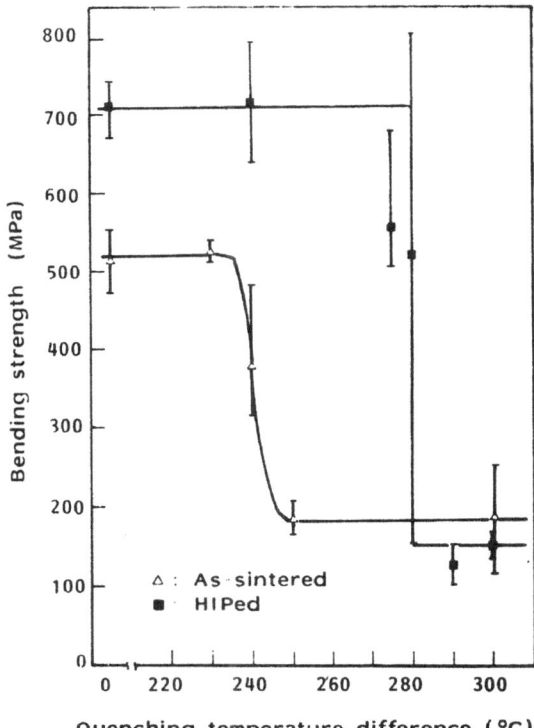

Quenching temperature difference (°C)

Fig. 6 Thermal shock resistance of 12Ce-TZP/20 wt% Al₂O₃ composites sintered at 1450°C for 10 hr in air and post-sintering HIPed at 1450°C and 100 MPa for 1 hr in an 80 vol% Ar-20 vol% O₂ gas atmosphere.

sample plate and h is heat transfer coefficient. Since the post-sintering HIPing of 12Ce-TZP/20 wt% Al_2O_3 caused only limited increase in relative density from 98 % to almost full theoretical density as shown in Fig. 3, no significant change can be expected for α, V, E and k. Therefore, the improvement of the thermal shock resistance of this material by post-sintering HIPing might be due to the increase in the fracture strength.

Conclusions

The thermal stability of TZP ceramics could be improved by alloying CeO_2, but the fracture strength of ceria-doped tetragonal zirconia was modest. Dispersion of Al_2O_3 into both (2Y,12Ce)-TZP and 12Ce-TZP was useful to depress the grain growth of zirconia and to increase the fracture strength. Post-sintering HIPing was useful to densify the sample without grain growth and to improve the fracture strength and thermal shock resistance of 12Ce-TZP/Al_2O_3 composites, but only limited improvement of the fracture strength was observed for (2Y,12Ce)-TZP/Al_2O_3 composites.

Acknowledgement

The authors wish to thank Dr. Fujikawa, IP Center of Kobe Steel, Ltd. for technical services of HIPing. This work was partly supported by a Grant-in-Aid for Science Research from the Ministry of Education, Science and Culture and by a Grant-in-Aid for Developmental Scientific Research from the Ministry of Education, Science and Culture.

References

1) K. Kobayashi, H. Kuwajima and T. Tsukidate, Solid State Ionics, 3-4, 489 (1981).
2) T. Sato and M. Shimada, J. Am. Ceram. Soc., 67, C212 (1984).
3) T. Sato, S. Ohtakia and M. Shimada, J. Am. Ceram. Soc., 68, 356 (1985).
4) T. Sato, S. Ohtaki, T. Endo and M. Shimada, Int. J. High Technol. Ceram., 2, 167 (1986).
5) K. Tsukuma and M. Shimada, J. Mater. Sci., 20, 1178 (1985).
6) T. Sato, T. Fukushima T. Endo and M. Shimada, Proceeding of International Conference Science of Ceramics 14, pp. 843 (1988).
7) R.L. Fullman, J. Metal Trans AIME, 197, 447 (1953).
8) D.B. Marshall, T. Noma and A.G. Evans, J. Am. Ceram. Soc., 65, C175 (1982).
9) K. Niihara, R. Morena and D.P.H. Hasselman, J. Mater. Sci., 1, 13 (1982).
10) M. Swain, Acta Metall., 33, 2083 (1985).
11) J.C. Jaeger, Philos. Mag., 36, 418 (1945).
12) K. Satyamurthy, J.P. Singh, D.P.H. Hasselman and M.P. Kamat, J. Am. Ceram. Soc., 63, 694 (1980).

REDOX BEHAVIOUR OF A CERIA ZIRCONIA ALLOY

V. Sergo, C. Schmid and S. Meriani
Istituto di Chimica Applicata e Industriale,
Università di Trieste, ITALY

ABSTRACT

The thermal history of ceria zirconia solid solutions greatly influences their structure and properties, particularly their electric conductivity. This in turn depends on a reversible redox reaction here investigated by means of TG, DTA and TDM measurements.

INTRODUCTION

As part of our research on the electrical and mechanical properties of ceria-zirconia sintered materials, the relation between their microstructure and electrical properties was investigated by impedance spectroscopy [1,2]. Results indicated that the bulk conductivities of ceria-zirconia solid solutions are very much dependent on their previous thermal histories. Samples with the same composition, in fact, displayed conductivities differing by several orders of magnitude, according to whether they were "as fired" or "annealed". Fig.1, from reference 2, gives the conductivity vs. temperature Arrhenius'plots.

Since ceria has long been known [3] to undergo a reduction at high temperature, the present paper describes thermoanalysis (TG + DTA =STA) of the 50 mol% composition in an investigation of the phenomena giving rise to these differences in conductivity. Furthermore XRD data are recorded for materials both in the oxidized and in the reduced conditions.

EXPERIMENTAL

Commercial ceria-zirconia powders of 99.0 wt.% purity were used; mixing was obtained by using agate balls and mortar for 24 hours. The suspension was dried in a microwave oven, pressed into flat disks and fired as follows: heating to 1873 K at 10 K·min^{-1}, soaking for 4 hours, coolin at about 30 K·min^{-1} to 1273 K and then to room temperature in a further 3 hours.

Previous investigations [4] showed that 40-60 % samples yield a single phase with tetragonal symmetry on cooling from 1873 K. Conversely, 25-40 % samples present a second tetragonal phase. The phase common to the whole range has been defined CeTZP' (ceria tetragonal zirconia polycrystal prime); the other has a lower ceria content and has been defined CeTZP°. For our STA investigation, all samples were with 50 mol% ceria; they were slightly grey and by

SEM microanalysis they averaged about 2 mol% of silica (high silica content). After annealing, i.e. holding the sample at 1173 K for at least 4 hours and letting it cool down for a further 8 hours, the conductivity dropped by about 6 orders of magnitude at the same measuring temperature, as illustrated in Fig. 1. the colour then became pale yellow.

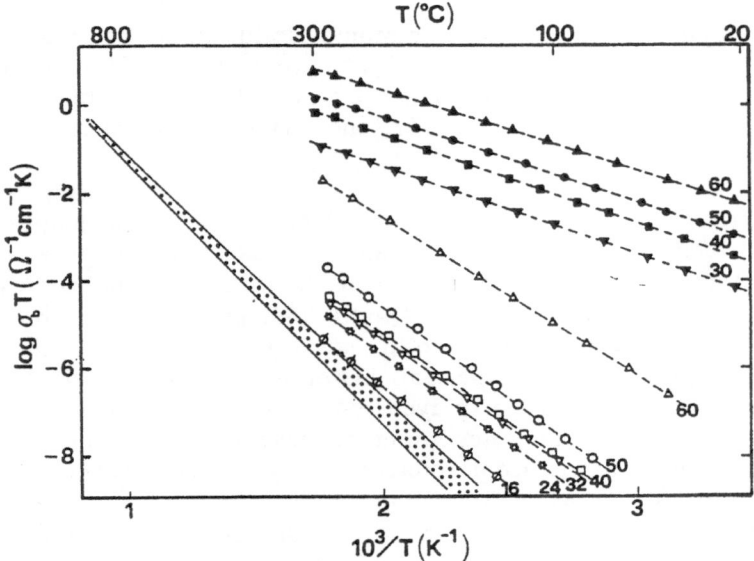

Fig. 1: Bulk conductivity vs temperature of as-fired samples at indicated composition (in mol% of CeO_2) with high-SiO_2 (closed points) and low-SiO_2 (open points) contents. All sigma values of annealed samples are enclosed within the continuous lines.

The TG investigations were performed on flat disks laid on the top of the crucibles of the thermobalance. The initial weight of the oxidized samples was in the range 0.8-1.2 g. The DTA measurements were performed on powders obtained by grinding the sintered pellets and using alumina as reference material. The TDM measurements were performed on small bars cut from a flat disk 3mm thick and 20mm in diameter. Typical dimensions were 10x5x3mm.

Different gases were introduced at 15 ml·min^{-1} into the measurement chamber after it had been emptied with a vacuum system before the beginning of the experiment. Since TG measurements are very sensitive to the gas flow, the balance was zeroed when the system was in dynamical equilibrium between incoming and outgoing gas. In addition, buoyancy effects were also taken into account by measurements with inert powders. The oxidizing gases were O_2 and air, while the reducing gas was Ar + H_2

The apparatus was a Netzsch 409 Thermal Analyzer equipped on line with a Balzers Quadrupole Mass Spectrometer (QMS) mod. QMG 420 with Netzsch software used on a HP 86 PC.

XRD data were collected with a CuKα (Ni) radiation on powders which were exposed to the redox cycles as above.

RESULTS AND DISCUSSION

The annealed samples which exhibited a very low conductivity [2] were regarded as fully oxidized because their weight was stable after further oxygen firing. X-ray diffraction confirmed their tetragonal symmetry. TG and DTA results are illustrated in Fig.2. There is an increasing weight loss beginning at about 500 K. The full weight loss is accomplished after a few hours, it closely corresponds to the theoretical 2.71 loss% expected with complete reduction, according to the reaction:

$$2 \ (CeO_2 \cdot ZrO_2) \ ---> \ Ce_2O_3 \cdot 2ZrO_2 + 1/2 \ O_2 \qquad (1)$$

The reduced, black material, displayed a cubic XRD pattern reflecting the CeCZP (ceria cubic zirconia polycrystal) previously reported [5]. The reduction mechanism necessarily relies upon the presence of H_2 since under a simply neutral atmosphere, no appreciable weight loss could be measured up to 1473 K.

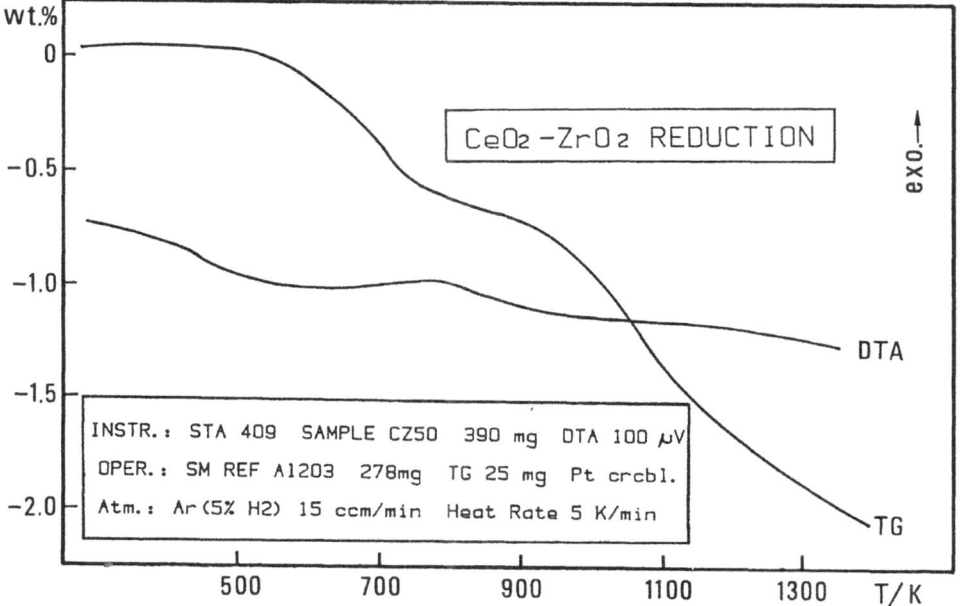

Fig. 2: TG + DTA curves of the reduction reaction.

The reduced sample was recycled in the STA apparatus under oxygen or air. In both cases, with minor differences in the rate of weight gain, a very rapid oxidation was recorded at temperatures close to 500 K. The DTA exotherm was very large as expected for an oxidation reaction. Most of the weight gain indicates rapid oxidation accompanied by a large exothermic effect that was clearly recorded by the DTA peak trace (Fig.3). By contrast, reduction of the sample was accompanied by a poorly defined DTA curve with a slightly endothermic trend during the whole weight loss period.

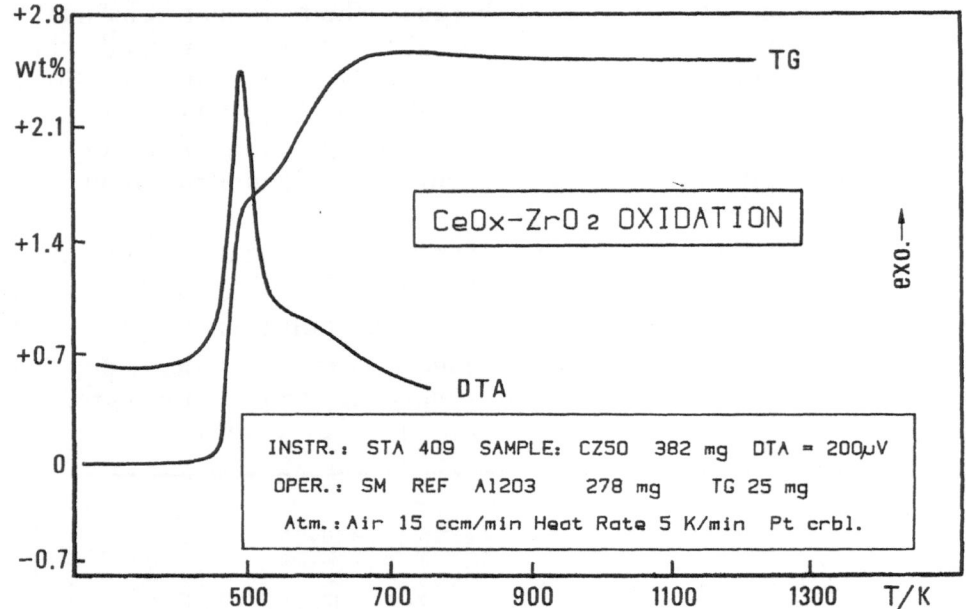

Fig. 3: TG + DTA curves of the oxidation reaction

The lack of an evident DTA effect can be attributed to both the departure of an oxygen atom from the crystal lattice and the formation of a water molecule with the hydrogen in the atmosphere surrounding the crystal surface. The two events, endo- and exothermic respectively, are most probably of very similar energy content and their overall effect is very much balanced.

The experiments indicate that the reduction and oxidation steps have a different kinetics; although exhaustive evaluation of these phenomena has not yet been completed, a "rapid" oxidation and a comparatively "slow" reduction were observed. The reduction mechanism is connected with the formation of a water molecule at the crystal surface; its release rate may hinder the overall reaction kinetics. since different reduction rates were noted under constant experimental conditions. Further investigations will be necessary to pin down the rate-determining parameters, e.g. the bulk porosity or the number of cycles the material has been exposed to. The latter seems to effectively influence the promptness of both the reducing and the oxidizing processes. At this stage, we can only suppose a relation between the redox cycles and the porosity.

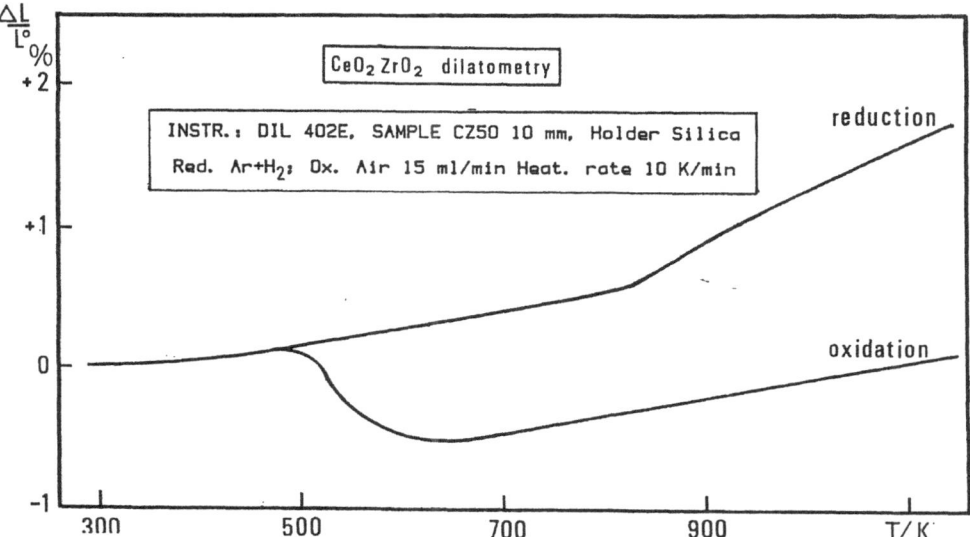

Fig. 4: TDM curves of both the oxidation and the reduction reaction.

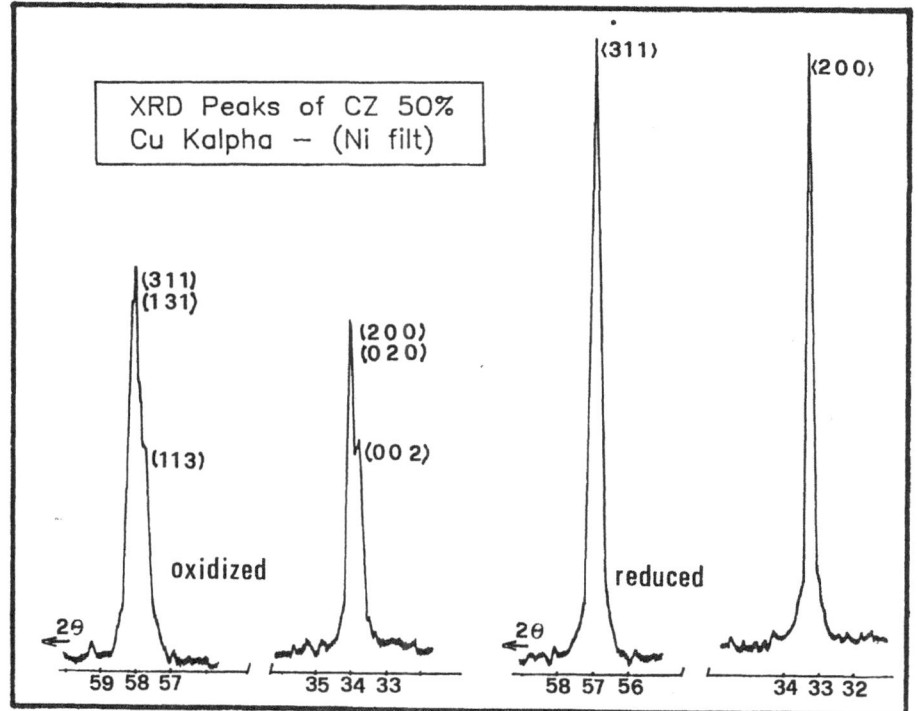

The reversible redox behaviour of the material was also investigated by TDM under the same conditions. Fig.4 shows the TDM trace of a reduction and an oxidation; it clearly indicates that the former is accompanied by a volume increase, whereas the latter leads to a decrease.

XRD evidence (Fig.5) is in agreement with this phenomenon; the reduced material is cubic with spacings shifted to lower 2Θ angles that correspond to a larger unit cell. The opposite reasoning applies to the oxidized, tetragonal material. Comparison of cubic and tetragonal structures is quite easy since the c/a ratio is very close to unity [6].

CONCLUSIONS

We have shown that cera-zirconia solid solutions undergo prompt, reversible and practically quantitative redox reactions when treated in an appropriate atmosphere. Although these enhance the phenomena, they explain previous observations pointing to a cubic symmetry of samples "quenched", i.e. taken directly from 1873 K to room temperature, indicating that a reduced state, stable at high temperature, can be frozen down.

By contrast the furnace-cooled samples displayed the tetragonal symmetry of the CeTZP' phase, because on cooling it became almost completely oxidized. The extent and promptness of the oxidation reaction is a matter of compromise between oxygen availability and its diffusion in the bulk of the sintered body. Therefore, by cooling sintered samples from 1873 K, more or less reduced material can be found at room temperature. These will display a very different electrical response because the Ce^{+3} greatly contributes to the overall conductivity with an electron hopping mechanism, according to the following point defect equation [7]:

$$2\ Ce_{Ce} + O_O ---> 2\ Ce'_{Ce} + V^{..}_O + 1/2\ O_2(g) \qquad (2)$$

Annealing in air, in fact, completely oxidizes the residual cerous ions to a fully ceric structure, in which the electrical conductivity is confined to the ionic contribution.

TDM revealed a substantial dimensional variation coupled with the redox reaction. Further investigation will be undertaken to establish the readiness and the extent of thermal cycling the material can withstand without major mechanical breakdown.

ACKNOWLEDGMENTS

Financial support from the italian Ministero della Pubblica Istruzione through the University of Trieste is gratefully acknowledged.

REFERENCES

1 A.Magistris and G.Chiodelli, Science and Technology of Zirconia III,
   Advances in Ceramics vol 24, Ed. S.Somiya and M.Yoshimura, Am.Ceram.Soc., Westerville, OH 1988, in press.
2 G.Chiodelli, A.Magistris, E.Lucchini and S.Meriani, Science of Ceramics
   vol. 14, Ed. D.Taylor, The Institute of Ceramics, Stoke-on-Trent, (1988),
   903-908.
3 A.Rouanet, C.R.Hebd, Seances Acad.Sci., Ser.C, 266,(1968), 908-911.
4 S.Meriani, J.Physique, 47, C1, (1986), 485-498.
5 E.Tani, M.Yoshimura and S.Somiya, J.Am.Ceram.Soc., 66, (1983), 506-
   510.
6 S.Meriani and G.Spinolo, Powder Diff., 2 (4), (1987), 485-488.
7 W.D.Kingery, H.K.Bowen and D.R.Uhlman, Introduction to Ceramics, John
   Wiley and Sons, New York, 1976, Chapter 4.

POLYMER PROCESSING EQUIPMENT IN THE PRODUCTION OF ZIRCONIA TOUGHENED ALUMINA COMPOUNDS SUITABLE FOR INJECTION MOULDING OF TECHNICAL CERAMIC COMPONENTS

N. Theilgaard - Dept. of Plastics Technology

Technological Institute, Denmark

Abstract:

A method whereby Yttria stabilized tetragonal zirconia is mechanically dispersed into alumina by means of a high speed fluidizing mixer and a twin-screw compounding extruder will be described. A comparative evaluation of the resulting dispersions will be made with a dispersion produced by a chemical precipitation method using SEM analysis.

X-Ray radiography analysis results, 4-pt bend stregths and weibull modulus test results will be presented and discussed. The possibilities of using this technique in commercial applications will be considered.

Synopsis

The purpose of the work carried out in this contribution was to investigate the possibilities of using mixing and compounding methods commonly used in the plastics industry to produce composite ceramic materials suitable for injection moulding which compare with those composites produced by other well-known methods by the ceramics industry.

The ceramic materials studied were made from ALCOA alumina powder CT 3000 SG and Dynamit Nobel yttria stabilized tetragonal zirconia powder Dizircon F-5Y.

The aim of the study was to initially produce two compounds, one which comprised entirely of the alumina ceramic powder which had been incorporated into a previously optimized organic binder system and the other comprised of a ceramic powder consisting of approx. 14 vol% yttria stabilized zirconia and approx. 86 vol% alumina incorporated into the same binder system as for the alumina compound.

PROCESS FLOW DIAGRAM AND EQUIPMENT

STAGE I	*COMPOUNDING*	1. High speed fluidising mixer - 4l capacity
		(Fig 1)
		2. Laboratory Twin-screw compounding extruder
		(Fig.2)
STAGE II	*INJECTION MOULDING*	Reciprocal Screw Injection moulding machine
STAGE III	*BINDER REMOVAL*	Atmosphere/Vacuum furnace - max temp. 1100 deg C
STAGE IV	*SINTERING*	High Temp Furnace Max Temp. 1600 deg C

Both compounds were injection moulded using mould geometries suitable for 4-point bend tests. Two different moulds having different ejector pin systems were evaluated. The resulting moulded parts of the two compounds were then simultaneously debinderised and sintered.

The mechanical properties of the samples were evaluated by 4-point bend tests and K_{IC} determinations. The homogeneity of the compounds were evaluated by SEM and the resulting micrographs were compared to those micrographs obtained from samples made by a chemical precipitation method(*).

X-ray diffraction was also carried out on the strained surfaces of some of the samples used in the 4-point bend tests.

All the SEM work has been carried out using a CAMBRIDGE INSTRUMENTS Stereoscan 360 by the Chemical Laboratory at the Technological Institute.

The evaluation of the mechanical properties, 4-point bend testing, K_{IC} determinations and X-ray diffraction analysis has been carried out in close co-operation with the Metallurgy department at RISØ National Laboratory, Denmark.

Density and open porosity evaluations were carried out by the Technical University of Denmark (DTH).

Method

The ceramic powders and all the components of the binder system were premixed in the high speed fluidising mixer. The premix was then fed into the twin-screw compounding extruder.

The conditions for compounding were as follows:

Screw configuration	: 48 L/D
Temp. zones 1 to 7	: 220 °C
Temp. zone 8 (feed)	: 22 °C
Temp. zone 9 (melt, measured)	: 230 °C
Pressure at die head	: 5 bar
Current	: 5 amps
Screw speed	: 300 l/min

(*) By courtesy of the Metallurgy Department, RISØ National Laboratory, Denmark.

Fig 1. Principle of operation
of the HENSCHEL
fluidising mixer

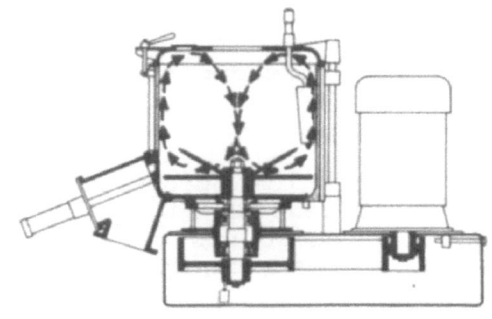

Fig 2. Twin-Screw
Compounding
Extruder

The compound was extruded through a single strand die and air cooled. The compound was subsequently granulated by crushing manually.

Samples were taken at various intervals during the compounding to evaluate the solids content of the compound.

The granulated material was then loaded onto a reciprocal screw injection moulding machine. Two different moulds, with geometries suitable for carrying out 4-point bend tests, having different ejectorpin systems, were used for evaluating the injection mouldability of the compounds. Fig. 3 shows the geometries and the positions of the ejector pins on the samples.

The resulting samples were sintered and tested for 4-point bend strengths, K_{IC} values, density and open porosity evaluations, XRD and SEM analysis.

Conclusion

The results indicate that, by using the compounding method described above, it is possible to produce a well dispersed zirconia/alumina composite ceramic as shown by the SEM micrographs.

The increased 4-point bend strengths, i.e., from a Max 380.8 MPa for alumina to 504 MPa for zirconia/alumina, as well as the increased K_{IC} values from 3.84 MPa·m$^{1/2}$ for alumina to 4.90 MPa·m$^{1/2}$ for the zirconia/alumina indicate that the presence of zirconia incorporated by this method results in a tougher ceramic. These values also compare very favourably with the results obtained from the tests carried out on the samples made by the chemical precipitation method(1), i.e., 4-point bend strength Max.346 MPa and K_{IC} 5.16 MPa·m$^{1/2}$.

The toughening mechanism is further confirmed by the XRD results where low strength failure does not give rise to a monoclinic phase whereas in the case of the higher strengths there is an increase in the monoclinic phase corresponding to an increase in the bend strengths. The causes for the low strengths can, in fact, be traced back to defects introduced during the injection moulding process as is indicated by the SEM fracture micrograph shown in Figure 1.7. These samples break before the phase transformation from tetragonal to monoclinic zirconia can occur.

The above results imply that the method of compounding ceramic materials described in this paper can be seriously considered as an alternate method and is being implemented at present industrially.

314

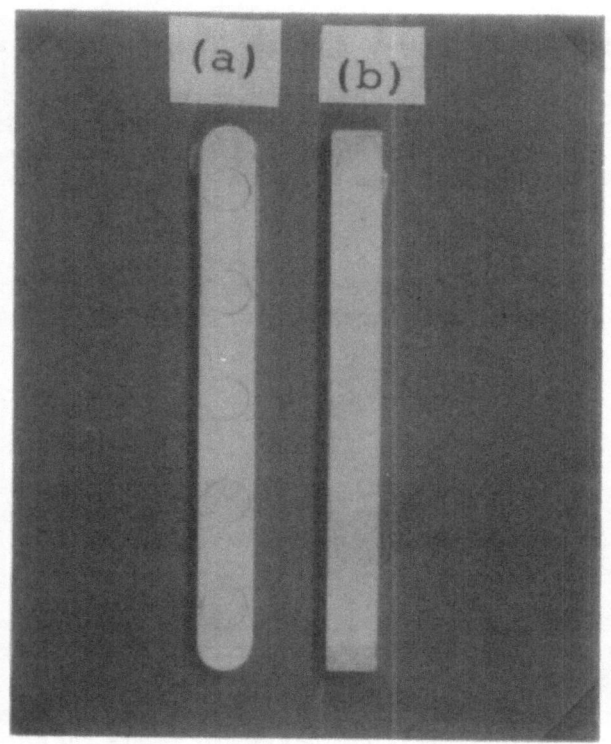

Fig. 3

Samples showing the geometries and positions of the
ejector pins in the injection moulded parts.

a. Sample produced in Trial 1
   NB. 5 ejector pins along the part

b. Sample produced in Trial 2
   NB. Core ejector pin

RESULTS FROM TRIAL 1   Fig. 4

Test	TP IV	TZP IV
Apparent density/open porosity %		
Green body	2.84/0.20	2.98/0.20
After binder removal	3.82/32.67	4.03/34.8
After sintering	3.84/0.06	4.17/0.14
4-point Bend test		
Bend strength 50%, MPa	263MPa	346MPa
Weibul modulus, m	22.5	15.6
No. of samples	6	5

RESULTS FROM TRIAL 2   Fig 5

Test	TP IV	TZP IV
Apparent density after sintering	3.93 +/- 0.06	4.23 +/- 0.11
Open porosity %	0.14 +/- o.14	0.49 +/-0.20
KIC MPa.m$^{1/2}$ (average of 3 measurements)		
sample no. 3 (271MPa)		4.90 +/-0.40
sample no. 8 (453.2MPa)		4.70 +/-0.20
sample no. 11(367.8MPa)		4.20 +/-0.53
sample no. 3(344.5MPa)	3.84 +/-0.39	
Hardness GPa		
sample no. 3		16.27 +/- 0.0
sample no. 8		16.57 +/-0.21
sample no.11		16.13 +/-0.20
sample no. 3	17.01 +/-0.21	

Weibul plots for 4-point bend tests of injection moulded sintered samples from Trial 2. Fig. 6

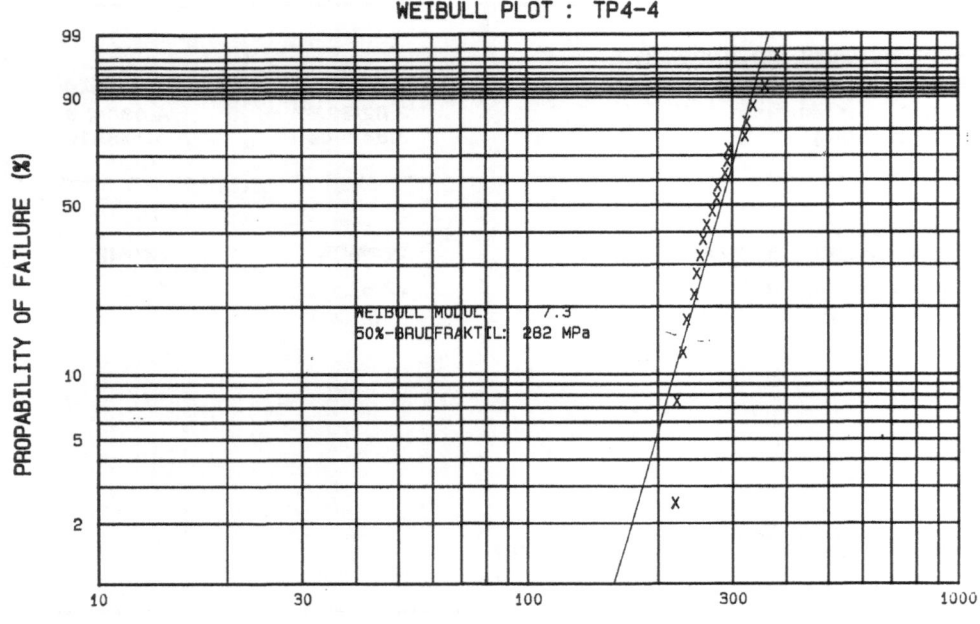

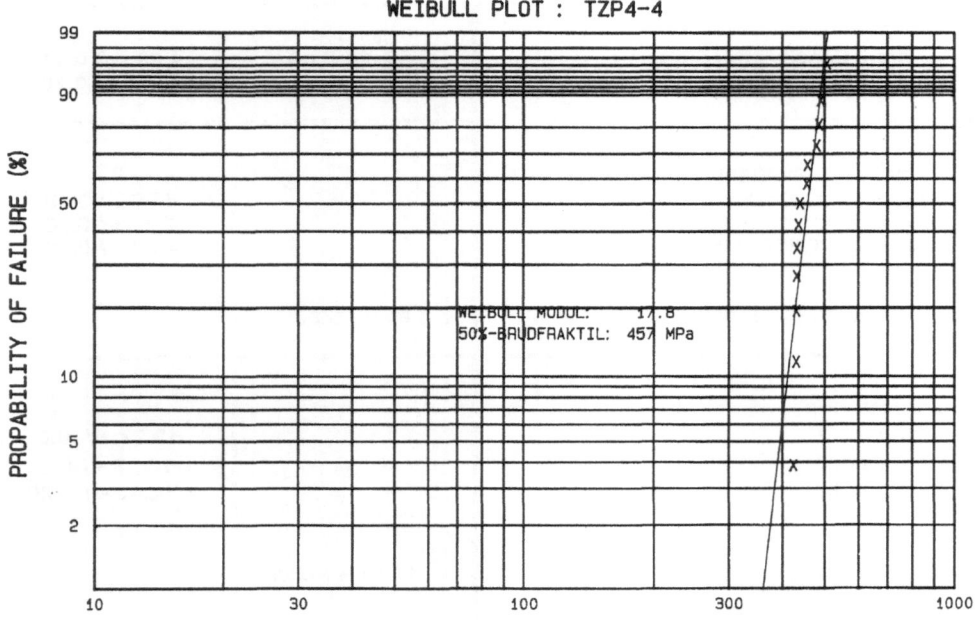

XRAY DIFFRACTION RESULTS

Fig.7

Sample No. 3
Bend strength 271MPa

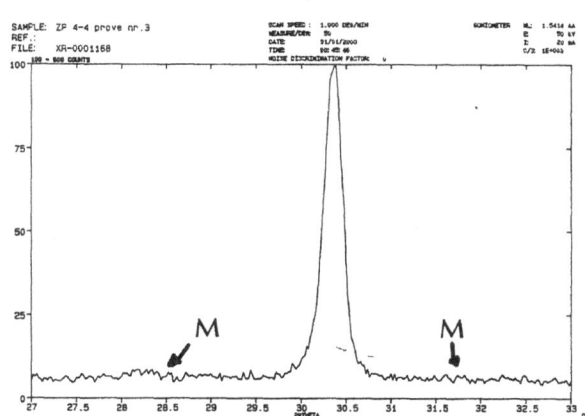

Sample No. 11
Bend strength 367MPa

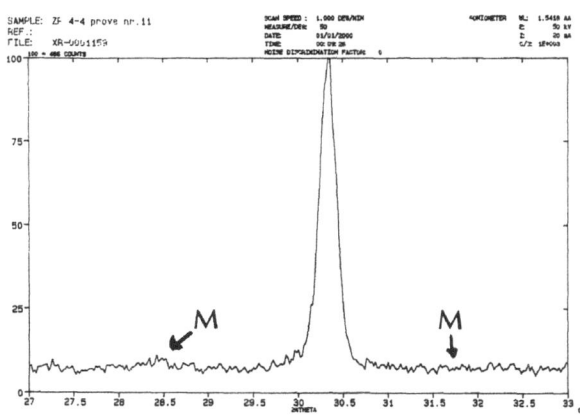

Sample No. 8
Bend strength 453MPa

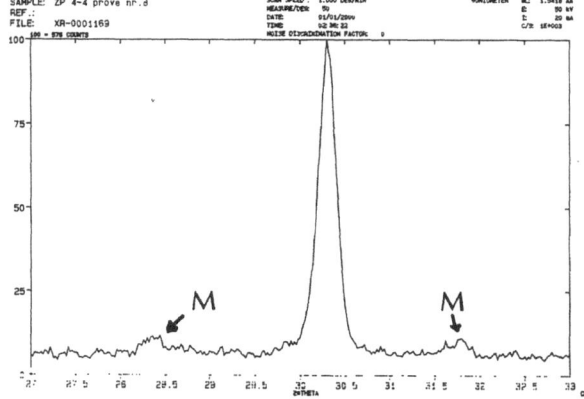

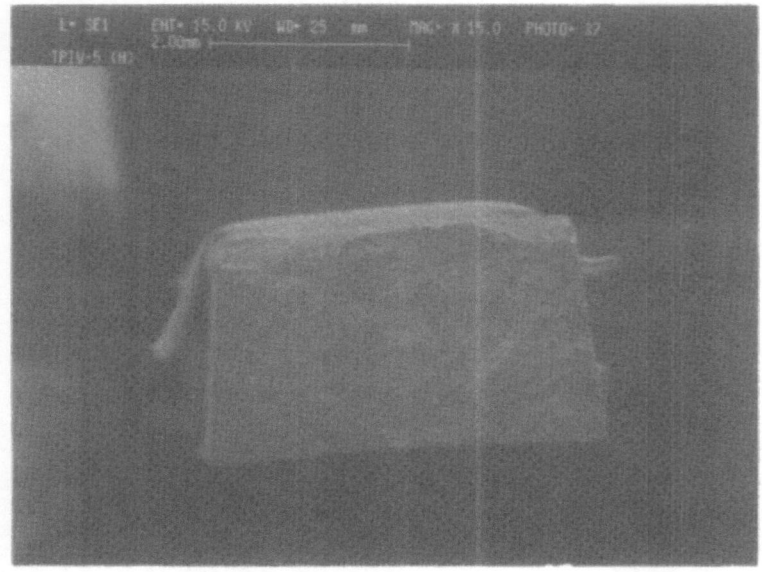

Fig . 8. TP IV-5(322.7MPa) Fracture surface (Mag X15)

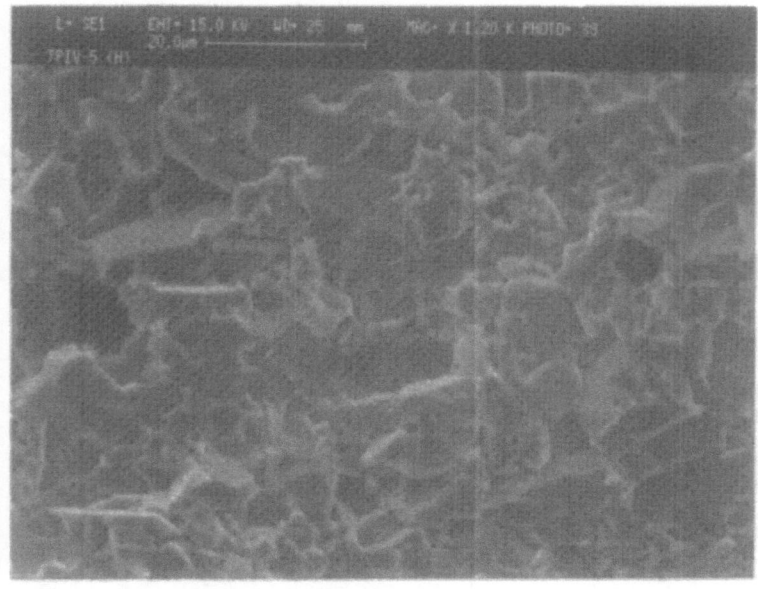

Fig. 9. TP IV-5(322.7MPa) Fracture surface (Mag X1.2K)

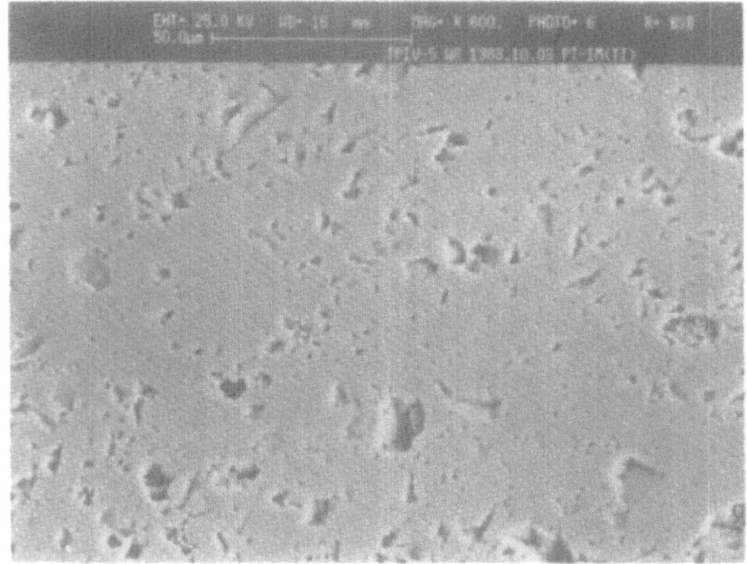

Fig. 10. TP IV-5(322.7MPa) Microstructure after polishing and thermal etching Mag X600

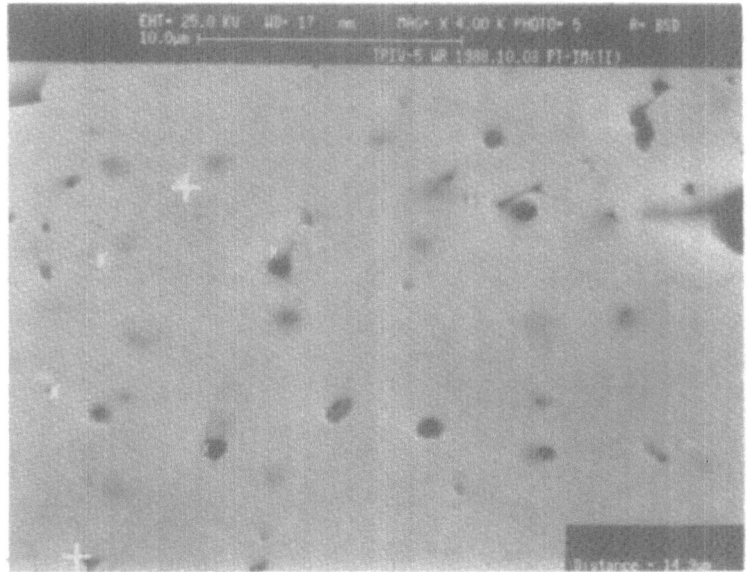

Fig. 11. TP IV-5(322.7MPa) Microstructure after polishing and thermal etching Mag X4K
NB. Extensive secondary grain growth and closed porosity

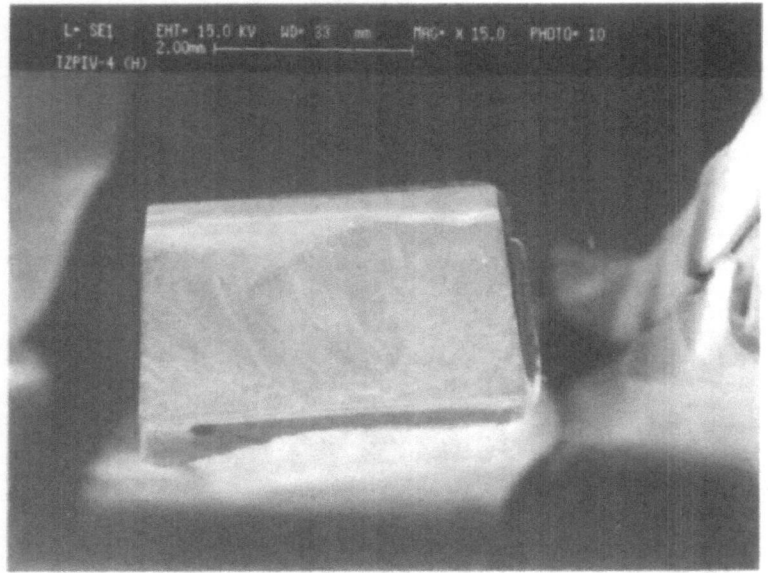

Fig. 12.TZP IV-4 (490 MPa) Fracture surface (Mag X15K)

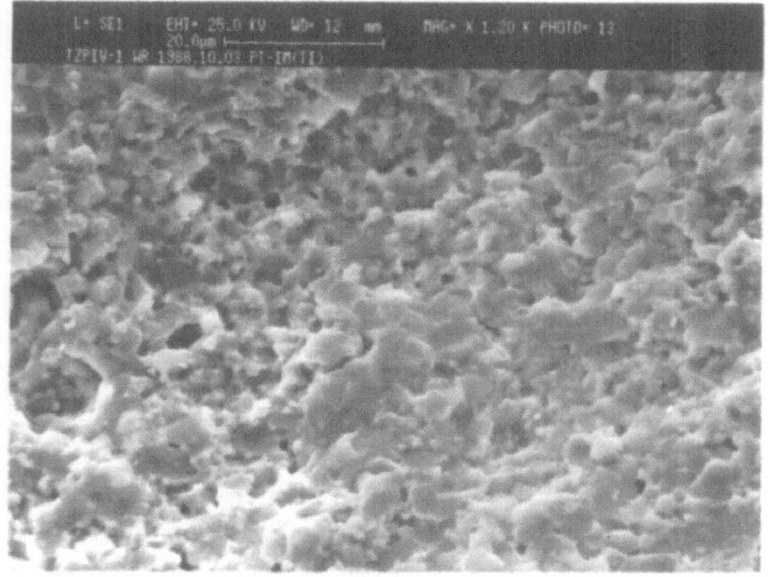

Fig. 13. TZP IV-1 (479 MPa) Fracture surface (Mag. X1.2K)

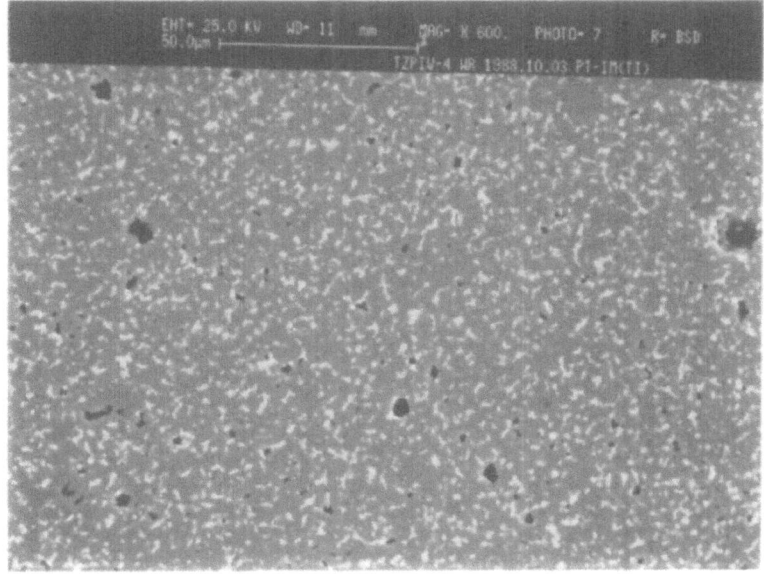

Fig. 14. TZP IV-4(490 MPa) Microstructure after polishing and thermal etching Mag X600
NB. Dispersion of Zirconia(white) in alumina matrix (grey)

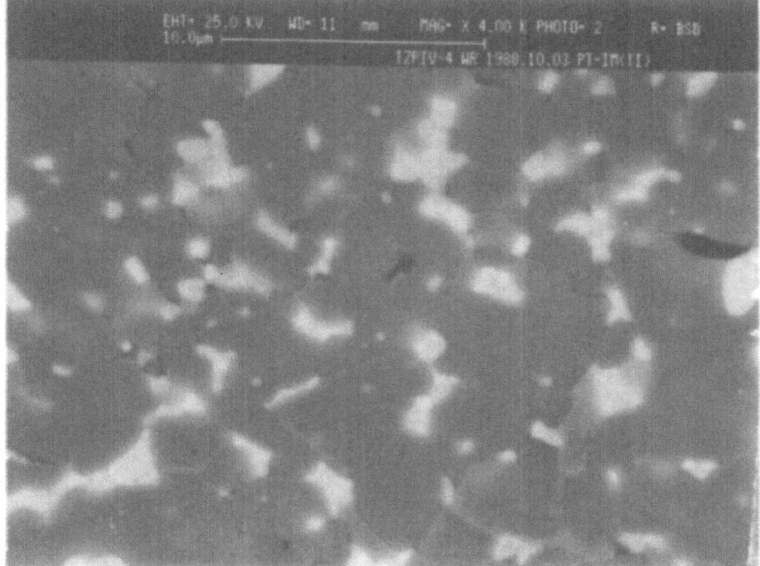

Fig. 15. TZP IV-4 (490MPa) Microstructure after polishing and thermal etching Mag X 4K
NB. Zirconia packed along the boundaries of alumina grains - alumina grain size restricted by the zirconia

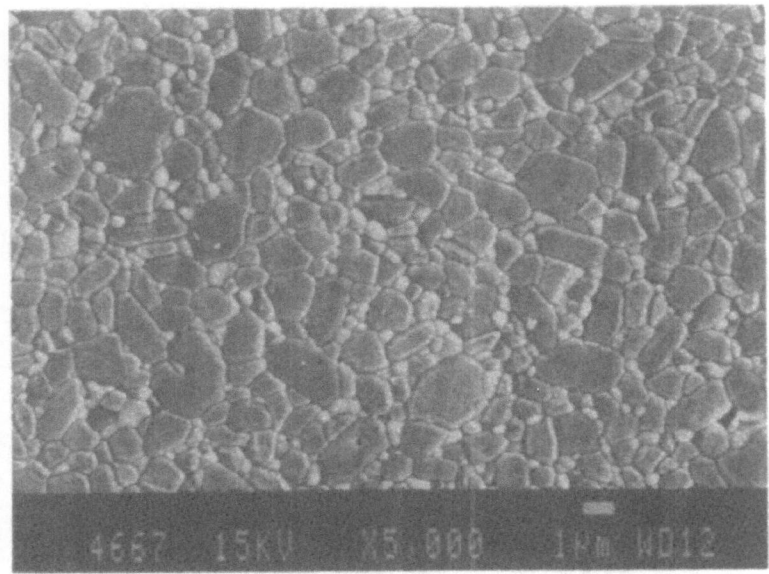

Fig. 16. SEM micrograph of a sintered zirconia toughened alumina made by chemical precipitation*

* By courtesy of the Metallurgy dept., RISØ National Laboratories.

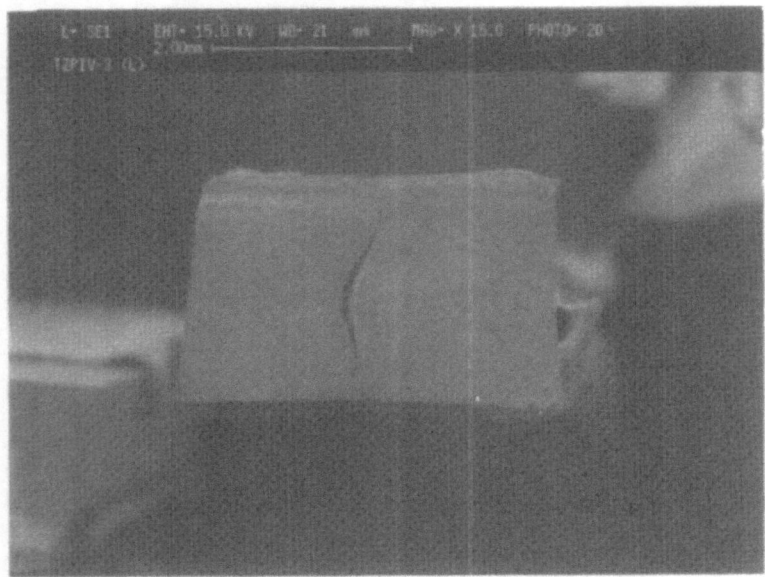

Fig. 17. SEM micrograph of a fracture surface with defect caused by processing conditions during injection moulding. Sample TZP IV-3

BINDER SYSTEM COMPONENTS

POLYSTYRENE
POLYETHYLENE
PARAFFIN WAXES
LUBRICANTS
COMPATIBILISERS/SURFACTANTS
SINTERING AIDS
DEFLOCCULANTS

COMPOSITION OF COMPOUNDS

Compound No.	Al_2O_3 Vol %	ZrO_2 Vol. %	Vol . Fraction of solids in green body
TP IV	100	0	61.84%
TZP IV	approx. 86	approx 14	61.90%

Fig 18

REFERENCES:

1 - S. Maschio and O. Toft Sørensen.
Mechanical Properties and Microstructure of Zirconia Toughened Alumina.
Risø-M2675, Risø National Laboratory, Denmark, 1988

2 - R. Stevens.
Zirconia and Zirconia Ceramics.
Magnesium Electron Publication No. 113, July 1986.

MICROSTRUCTURE DEVELOPMENT DURING SINTERING OF ULTRA FINE GRAINED Y-TZP

G.S.A.M. THEUNISSEN, A.J.A. WINNUBST,
W.F.M. GROOT ZEVERT and A.J. BURGGRAAF.

University of Twente, Faculty of Chemical Technology
Lab. for Inorganic Chemistry, Materials Science and Catalysis
P.O. Box 217, 7500 AE Enschede, The Netherlands.

Abstract: Ultra-fine grained, weakly agglomerated Y_2O_3 doped tetragonal ZrO_2 powders with a primary particle size of about 8 nm were prepared by means of two different hydrous-gel precipitation techniques. These methods are respectively the hydrolysis of a metal alkoxide ("alkoxide" method) or a metal chloride ("chloride" method) solution. The sintering behaviour of these powders is compared with a commercial powder (Tosoh TZ3Y). After isostatic compaction at 400 MPa the chloride, alkoxide and TZ3Y compacts densify to 97% relative density after 10 hours sintering at 1050, 1200 and 1100°C respectively. The change in crystallite size and pore morphology has been studied as function of time. The sintering kinetics are probably determined by the aggregate structure within the green compact which is different for the investigated powders. A nanoscale ceramic can be obtained (grain size 53 nm) by sintering a chloride compact during 6.5 hr at 1044°C.

1. Introduction

Polycrystalline 100% tetragonal ZrO_2-Y_2O_3 ceramics (Y-TZP) are regarded as materials exhibiting high strength and toughness [1]. In order to obtain a fully stabilized tetragonal structure the ceramic grain size must be less than 0.8 μm in case of a 3mol% Y_2O_3 doped ZrO_2 ceramic [2]. However this critical grainsize is reduced drastically when the Y-TZP is exposed to humid atmospheres at elevated temperatures [3,4]. In this case the grain size of 3mol% Y_2O_3-ZrO_2 should be less than 0.25 μm [3]. In the past it has been demonstrated that a ceramic with small grain sizes and nearly full density at relative low temperatures can be obtained when ultra-fine grained, weakly agglomerated powders are used as starting material. The primary crystallite size in this powder amounts 8-10 nm [5,6]. One way to prepare these small crystallites is by means of wet-chemical methods, e.g. gel-precipitation techniques. The resulting fine crystallites however tend to form aggregates and agglomerates. In order to prevent the formation of strong agglomerates the precipitated gels has to be treated in such a way that a minimal degree of cross-linking is formed. This has been discussed by several authors [7-11]. The state of agglomeration influences the pore size distribution within powder compacts during pressing and therefore also the sintering behaviour of the compacts. It is known [12,13] that aggregates and agglomerates densify more rapidly than the surrounding matrix and therefore give rise to cracks and large stresses which can not be removed during sintering. The smaller the agglomerates or aggregates are the smaller the remaining defects will be. The microstructure development during compaction has been studied extensively by Van de Graaf et al. [7], Groot Zevert et al. [8] and Lecloux et al. [14]. From these results it is shown that the studied powders differ in primary crystallite size, in agglomerate density and in the way the agglomerates can be broken down.

In earlier work [7] it has been shown for cubic materials that the aggregates are converted at low temperatures (900°C) to crystallites which are considerably larger than the primary crystallites and that densification and grain growth proceed in several stages.

In the present paper the attention is focussed on the development of the microstructure during sintering of tetragonal ZrO_2-3mol% Y_2O_3 ceramics made by three different methods (including the commercial Tosoh powder). It will be shown that nanoscale ceramics can be produced by controlling all the steps in the synthesis proces.

2. Experimental procedure

Zirconia powders with a composition of $(100-x)ZrO_2.xYO_{1.5}$ (denoted as ZYx) were prepared by two different gel-precipitation techniques. Both implied the hydrolysis of a diluted Zr-Y precursor solution in an excess of hydrolysing agent.
One method is the hydrolysis in water of a metal-alkoxide-benzene solution as described by Van de Graaf et al.[7] (the "alkoxide" method). This was performed in a dispersion turbine reactor as described in [7] to decrease the agglomerate size.
After hydrolysis the gel was washed three times with water and subsequently filtered and wet-milled with ethanol.
The second method is the precipitation of a solution of metal chlorides in an excess of an 25wt% Ammonia solution (the "chloride" method). During hydrolysis the pH was kept at a value of 11 or more. After the precipitation reaction was completed, the gel was thoroughly washed with a water/ammonia mixture in order to remove the chloride [6,8,15]. The gel was subsequently washed 3 times with ethanol to remove free water. Hydrolysis and washing were performed in the same reactor as for the alkoxide method. Both gels were dried in air for 15 hours (120°C) and calcined at 550°C for two hours. These powders were compared with the commercial TZ3Y powder (Tosoh Co. Ltd, Tokyo).

Powder compacts were pressed isostatically at a pressure of 400 MPa. Sintering took place at temperatures between 700 and 1500°C in air (heating rate 2°C), in either a furnace or a Netzsch 402E dilatometer. X-ray fluorescence spectrometry, using a Philips PW 1410 spectrometer, was used for the analysis of the overall composition. Nitrogen adsorption and desorption isotherms were measured at 77K by means of a Carlo Erba Sorptomatic Series 1800 instrument. Pore size distributions and pore volumes were calculated from the desorption branch of the hysteresis loops, according to the method described by Dollimore et al.[16]. Pores with radii larger than about 10 nm were determined by mercury-intrusion porosimetry. These measurements were performed with a Carlo Erba Porosimeter Series 200 at pressures up to 200 MPa. Densities were measured by the Archimedes method in Hg. Crystallite sizes as a function of sintering temperature or time were determined by means of X-ray line broadening measurements [17] in the region < 100 nm, using a Philips X-ray diffractometer PW 1370 with CuK_α-radiation. Corrections for instrumental broadening and K_α-splitting have been applied. A Jeol 200CT transmission electron microscope was used for direct observations of grain sizes in the region < 100 nm. Grain sizes above 100 nm were calculated by the method of Mendelson [18] using a Jeol JSM-35CF scanning electron microscope.

3. Results and Discussion

The composition of the powders used in this study, as determined by X-ray fluorescence, was in all cases ZY5.8. The properties of these powders are listed in table 1. From this table it can be seen that large differences are present between the three powders. These different powder microstructures give rise to different compaction behaviour. This has previously been discussed by Groot Zevert et al. [8].

Synthesis	Agglomerate strength (MPa)	Agglomerate density (%)	SBET $(m^2 \cdot g^{-1})$	d (XRLB) (nm)
Alkoxide	45	22	101	8
Chloride	80	30	125	8
Tosoh	40	36*	21	34

Table 1: Powder properties.

After isostatic pressing at 400 MPa a pore distribution, as indicated in Fig. 1, was obtained. At this point it was concluded by Groot Zevert et al. that the alkoxide powder consisted of small aggregates (18 nm) containing approximately 10 primary crystallites with a size of 8 nm. The chloride powder hardly contained any aggregates. For this powder a stringlike structure is assumed, with areas of higher density at the crossings of the strings. The Tosoh powder contained large dense aggregates [8]. After compaction the compacts were sintered in a (tubular or room) furnace at various temperatures during a sintering time of 10 hours. The results are shown in Fig. 2. From this figure it can be concluded that large differences exist in sintering behaviour between the three compacts. The sintering behaviour of the

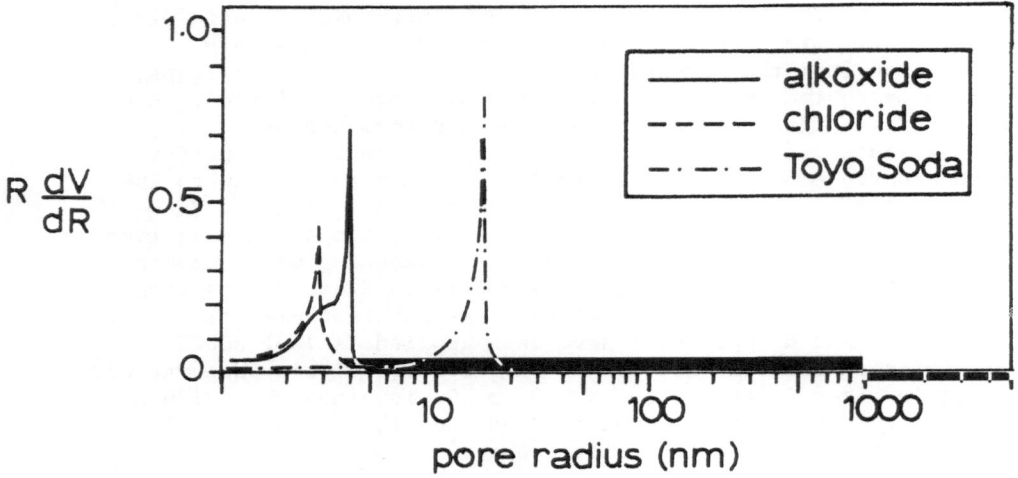

Fig. 1: Pore size distribution data within the powder compacts after isostatic pressing at 400 MPa.

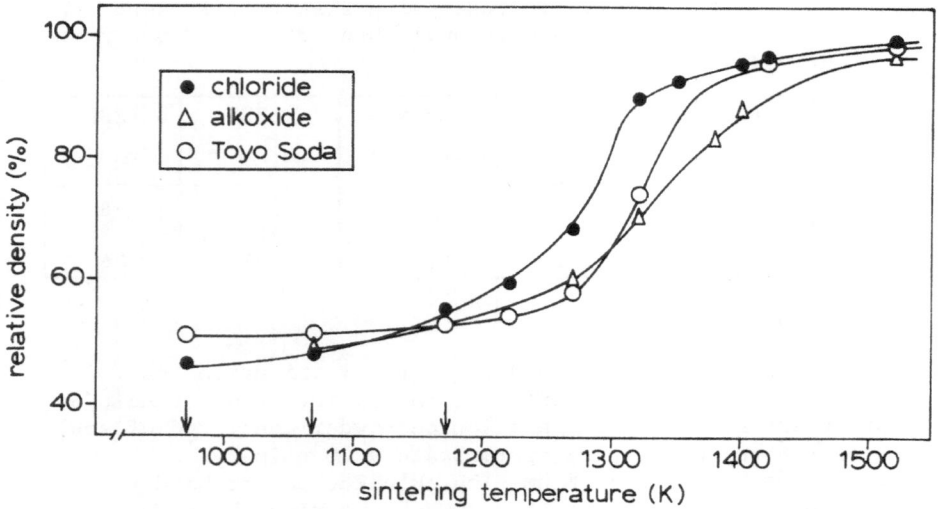

Fig. 2: Densification behaviour of green compacts after stepwise sintering for 10 hours.

chloride powder seems to be the best (>97% at 1150°C). At temperatures of about 1250°C it is possible to densify both the chloride and Tosoh powder to more than 98% of the theoretical density. At this point the relative density of the alkoxide powders is not more than about 95%. The temperature region where densification takes place is for both the chloride and the Tosoh powder limited to about 100°C. For the chloride powder however this region is shifted to a somewhat lower temperature compared to the Tosoh powder (1000-1100°C). In Fig. 3 a dilatometer curve for a chloride and a Tosoh powder are given. From this figure it can also be concluded that the sintering behaviour of the chloride compact is different from the Tosoh compact. A remarkable difference is seen in the low temperature region of 500-900°C. In the Tosoh compact an increase in volume can be seen as is expected to happen because of thermal expansion. In the chloride compact however a slight decrease in volume is observed. This is expected to be caused by a fast recrystallisation of small aggregates consisting of some primary crystallites to a new crystallite. This was also observed by Van de Graaf [7] in case of a cubic nanocrystalline alkoxide powder at about 900°C. This will be discussed furtheron.

At several temperatures, before overall densification starts (as indicated in figure 2 by arrows), pore sizes and pore size distributions are measured. The results of these measurements are presented in Fig. 4. From this it can be seen that the average pore size increases with increasing sintering temperature as has

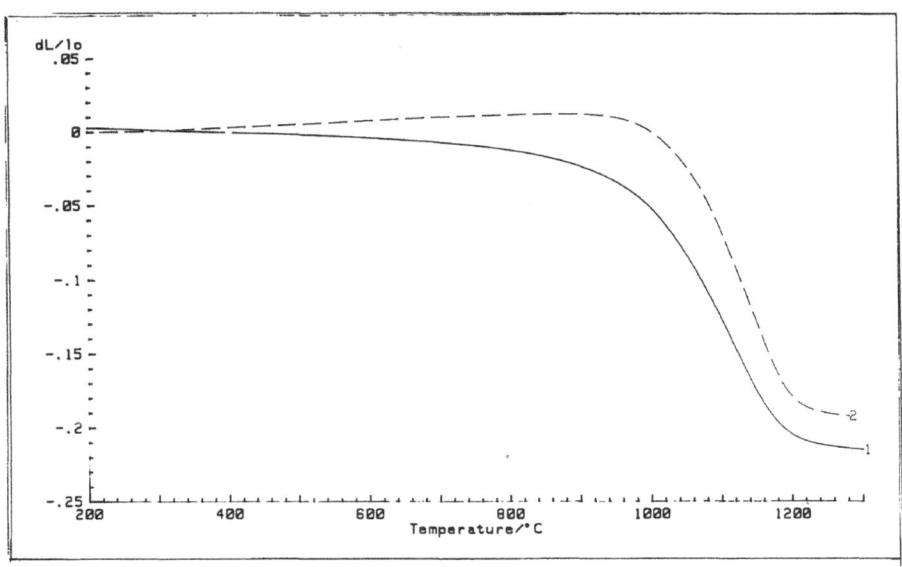

Fig. 3: Dilatometer curve of a "Chloride" compact (1) and a Tosoh compact (2). Heating rate: 2°C/min.

also been observed by Van de Graaf et al. for a cubic alkoxide
powder containing 17 at% Yttrium [7]. By comparing the different
compacts (for example at a temperature of 800°C) it is noticed
that the pore radius of the Tosoh compact (about 18 nm) is much
larger than the pore radius of the chloride and the alkoxide
compact at the same temperature (respectively 8 and 9 nm). This
was expected because it is known that the Tosoh compact contains
large aggregates whereas the primary units in the alkoxide and the
chloride powder are much smaller [8].

As is known from literature [12,13,19], regions with high packing
density will densify more rapidly than regions containing a low
number of contact points per unit volume. Between aggregates the
number of contact points per unit volume is rather low and
therefore the mutual sintering between the aggregates will be
negligible in this temperature region. As a consequence
densification in this temperature range will be restricted to the
internal aggregate structure. In the course of this process, the
pore volume within the aggregates will be transported to the
surface of the aggregates. This results in a complete
densification of the aggregates. Making use of the simple
geometrical relation:

$$\bar{d}_T = \bar{d}_{aggr} * (1 - \epsilon)^{1/3} \qquad (1)$$

the average crystallite size $\bar{d}_T$ of the densified aggregate after
sintering at T°C is calculated to be 16 nm (daggr = 18 nm, ϵ
(porosity) = 0.3 [8]), in case of the alkoxide compacts. In the
chloride compacts however also a fast recrystallisation seems to
take place. In this case however no real aggregates are present.
However if we take the regions where the crystallite strings cross
each other (which involves about 5-6 primary crystallites) and we
assume that full densification of these regions have taken place,
a new grain size of about 13 nm is calculated using equation 1. It
is assumed that in both cases (chloride and alkoxide powder) the
temperature at which the recrystallisation takes place lies
between 600 and 700°C. Preliminary DSC (differential scanning
calorimetry) measurements also point to a recrystallisation
temperature between 600 and 700°C. At 700°C the crystallite size
in the chloride compact is determined to be 14 nm, whereas at
800°C the crystallite size is 16 nm (XRLB).

Assuming that the new formed crystallites are arranged in simple
cubic arrays the radius of a sphere inscribed in the pore (Rpore)
can be calculated using the equation as given by Iler [20]:

$$R_{pore} = 0.732 * R_{cryst} \qquad (2)$$

in which R_{cryst} represents the radius of the new formed
crystallite.

Using equation 2, a pore radius of 5 nm is calculated in case of
the chloride compact and 6 nm in case of an alkoxide compact (when
the crystallites are 13 and 16 nm respectively). For a chloride

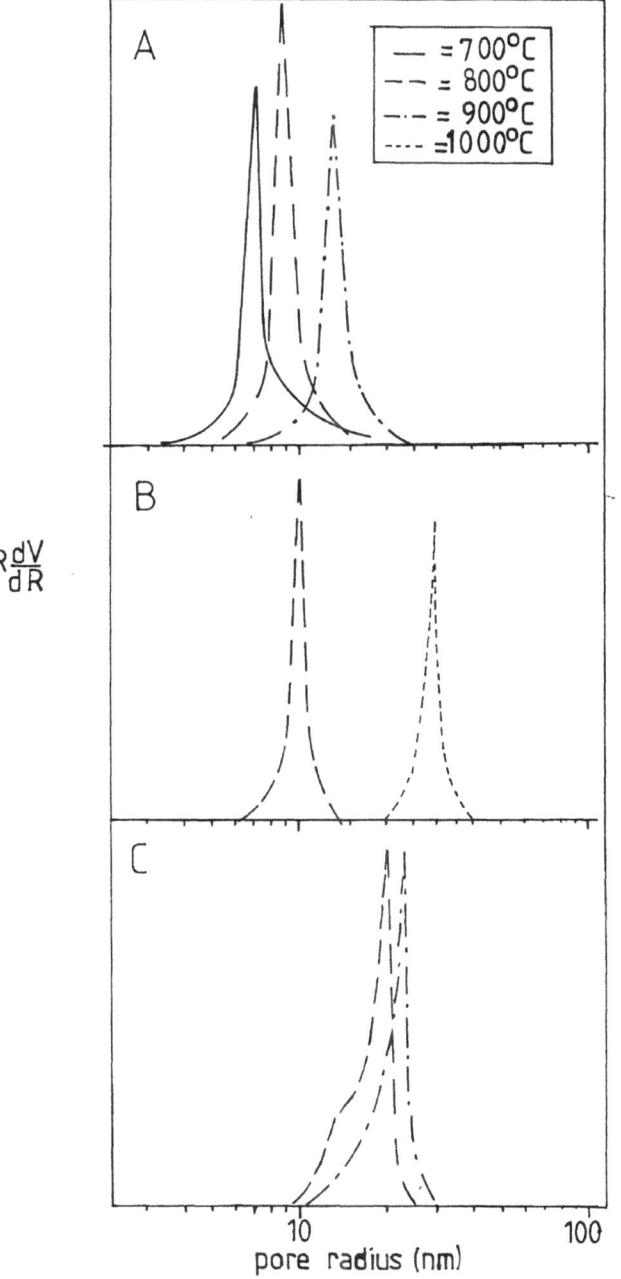

Fig. 4: Pore size distribution data after stepwise sintering for
10 hours.
a) Chloride compact
b) Alkoxide compact
c) Tosoh compact

compact sintered at 700°C (lowest temperature at which the pore radius is measured (Fig. 4)) a pore radius of 5.5 nm is calculated (crystallite size 14 nm). The actually measured value is about 6.5 nm, which seems to fit reasonably well with the calculated value.

As can be seen in Fig. 3 only a slight macroscopic densification takes place in the 600-700°C temperature region, whereas the aggregates densify completely. This means that microscopic (back) stresses can be created. Their effects are studied now. Because of the relative large primary crystallites in the Tosoh compact, a fast recrystallisation is not expected at these temperatures (and therefore no shrinkage at low temperature).

To obtain a small grain size in the ultimate ceramic (as is necessary for obtaining nanoscale ceramics or the tetragonal phase) control of grain growth is very important.

Grain growth behaviour can be interpreted by means of a kinetic theory as is developed by Brook [21] in which growth equations of the general form are used:

$$G^n - G_o^n \cong \frac{D}{T} * t \qquad (3)$$

where n is a constant for a given growth mechanism; D is the diffusion constant of the slowest ion; G_o and G are the starting grain size for the actual mechanism and the grain size resulting from grain growth at a time t at a certain temperature T. In our

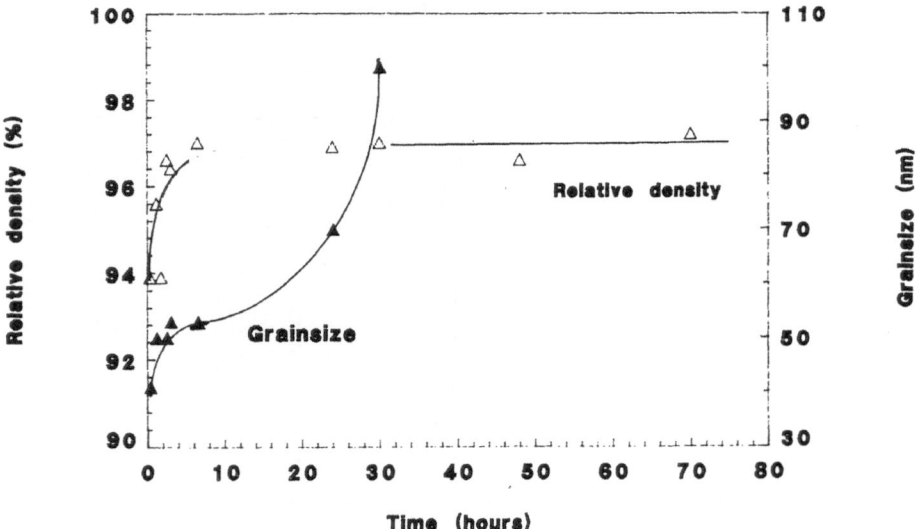

Fig. 5: Relative density and grainsize of a "chloride" compact sintered at 1044°C.

experiments a chloride compact (containing 3 mol% Y_2O_3) was sintered at a temperature of 1044°C for several sintering times.

Relative densities and grain sizes are presented in Fig. 5.
It is clear from the former discussion that the size of a primary
crystallite (in the powder) is not the correct value for G_o in
equation 3. Probably the size of a densified aggregate has to be
taken. However when arriving at 1044°C, with a heating rate of
2°C/min, the grain size was already grown to 41 nm. This means
that equation 3 cannot be used, because 2 parameters are changed
now. The fast grain growth at low temperature in the initial
stages is thought to be caused by lack of an "impurity drag". At
higher temperatures segregation of Yttrium takes place [22]. This
segregation layer is formed during the heating up part of the
proces and probably slows down the grain growth by means of an
"impurity drag" mechanism [21]. Therefore the grain growth as a
function of time at a temperature of 1044°C is much slower
compared to the grain growth during the heating up proces to
1044°C. After about 6.5 h of sintering the proces of grain growth
speeds up again. At this moment it is not clear yet if the data
represent a single sintering mechanism or more than one, because
of lacking information at sintering times between 6.5 and 25 hour.
This is under investigation now. A grain size of 53 nm (XRLB) and
a relative density of 97% are obtained after 6.5 hr of sintering.
In this way it is possible to obtain a nanoscale ceramic.
Even at a lower temperature of 1000°C the same density (>97%) is
obtained after 250 hr of sintering (Fig. 6). At this moment the
grain size of this ceramic is not known yet. In future the grain
size will be measured as a function of sintering temperature and

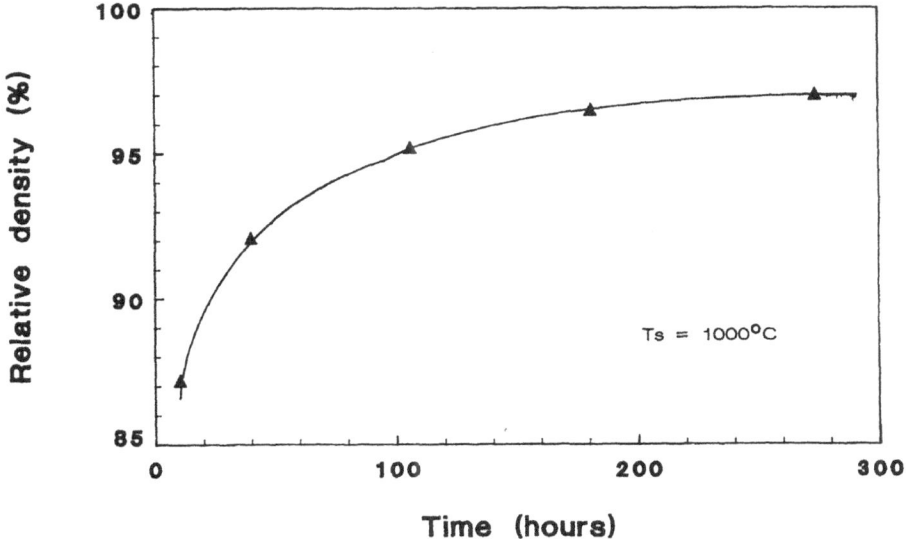

Fig. 6: Relative density of a "chloride" compact sintered at
1000°C.

activation energies will be calculated to obtain further insight in the possible mechanisms.

4. Conclusions

1. It is possible to produce ultra fine grained, weakly agglomerated, tetragonal zirconia powders with a primary crystallite size of about 8 nm.
2. The "chloride" powder shows the best sintering behaviour.
3. In case of "chloride" and "alkoxide" powders a fast recrystallisation step takes place at a temperature between 600 and 700°C.
4. The resulting crystallite size is calculated to be 16 and 13 nm for the "alkoxide" and the "chloride" compact respectively.
5. It is possible to densify a "chloride" compact to almost full density at 1000°C after 250 hours of sintering.
6. The same is possible at 1044°C after 6.5 hours of sintering. The resulting grain size is then only 53 nm and a nanoscale ceramic can be produced.
7. At higher temperatures (about 1000°C) the grain growth is expected to be slowed down by means of an "impurity drag" mechanism.

Acknowledgements
A.M. Bongers is acknowledged for performing the gasadsorption/desorption measurements. J.M. Jacobs is acknowledged for performing the dilatometry experiments. These investigations were partly supported by the Innovative Research Program on Technical Ceramics (IOP-TK) with financial aid of the Dutch Ministry of Economic Affairs.

References
1. N. Claussen, in "Science and Technology of Zirconia II" ed. by N. Claussen, M. Rühle and A.H. Heuer (American Ceramic Society, Columbus, Ohio, 1984) p. 325.
2. F.F. Lange, J.Mater.Sci., 17 (1982) 240.
3. A.J.A. Winnubst, A.J. Burggraaf, to be published in Advances in Ceramics vol. 24, Science and Technology of Zirconia III, The American Ceramic Society, Inc. (Columbus, Ohio) 1988.
4. T. Sato, S. Shimada, J.Am.Ceram.Soc., 68 (1985) 356.
5. W.H. Rhodes, J.Am.Ceram.Soc. 64 (1981) 19.
6. A.J.A. Winnubst, G.S.A.M. Theunissen, W,F.M. Groot Zevert, A.J. Burggraaf in "Science of Ceramics 14" edited by D. Taylor, The Institute of Ceramics, Shelton, Stoke-on-Trent, Staffs., UK (1988) pp. 309-314.
7. M.A.C.G. van de Graaf, J.H.H. ter Maat, A.J. Burggraaf, J.Mat.Sci. 20 (1985) 1407-1418.
8. W.F.M. Groot Zevert, A.J.A. Winnnubst, G.S.A.M. Theunissen, A.J. Burggraaf, submitted for publication in J.Mat.Sci..

9. M.A.C.G. Van De Graaf, A.J. Burggraaf, Adv.Ceram. <u>12</u> (1984) 744.

10. R. Pampuch, K. Haberko, Mat.Sci.Mon. <u>16</u> (1983) 623.

11. T. Kosmac, V. Krasevec, R. Gopalakrishnan, M. Komac, in Advances in Ceramics vol. 24, Science and Technology of Zirconia III, The American Ceramic Society, Inc. (Columbus, Ohio) 1988.

12. B. Kellett, F.F. Lange, J.Am.Ceram.Soc. <u>67</u> (1984) 369.

13. W.F. Dynys, J.W. Halloran, J.Am.Ceram.Soc. <u>67</u> (1984) 596.

14. A.J. Lecloux, P. Verleye, J. Bronckart, F. Noville, P. Marchot, J.P. Pirard, Reactivity of Solids <u>4</u> (1988) 309.

15. P.D.L. Mercera, J.G. van Ommen, E.B.M. Doesburg, A.J. Burggraaf, J.R.H. Ross, Proceedings of Symposium on Advanced Ceramics, 19,20 December 1988, (London), to be published in "Britisch Ceramic Proceedings" the Institute of Ceramics (Stoke-on-Trent).

16. Dollimore, G.R. Heal, J.Appl.Chem. <u>14</u> (1964) 109.

17. K.P. Klug, L.E. Alexander, "X-ray Diffraction Procedures" (Wiley and Sons, New York, 1974).

18. M.I. Mendelson, J.Am.Ceram.Soc. <u>52</u> (1969) 443-446.

19. M.A.C.G. Van de Graaf, T. van Dijk, M.A. de Jongh, A.J. Burggraaf, Sci. Ceram. <u>9</u> (1977) 75.

20. R.K. Iler, The Chemistry of silica: solubility, polymerization, colloid and surface properties, and biochemistry, New York; Chichester [etc.]: Wiley, 1979.

21. R.J. Brook, Ceramic Fabrication Processes in "Treatise of Materials Science and Technology", Vol. 9, edited by F.F.Y. Wang, (Academic Press, New York, 1976) pp. 331-364.

22. G.S.A.M. Theunissen, A.J.A. Winnubst, A.J. Burggraaf, "Segregation aspects in the ZrO2-Y2O3 ceramic system" accepted for publication in J.Mat.Sci.Letters.

The Grain Size Dependence of the Mechanical Properties in TZP Ceramics

J. Wang, M. Rainforth and R. Stevens
School of Materials
University of Leeds

Abstract

TZP ceramics, containing 2, 2.5, and 3 mol% Y_2O_3 have been fabricated to give a grain size range of 0.5 to 2.0 μm and the microstructure and mechanical properties (fracture toughness and fracture strength) determined. The grain size has been demonstrated to be an important parameter in controlling the mechanical properties and can itself be controlled by the sintering temperature and time.

The grain size and the stabilizer addition are related to the size of the transformation zone and the extent to which both these parameters determine the toughening increment is discussed.

1. Introduction

Tetragonal Zirconia Polycrystals(TZP) describes a new generation of toughened ceramics developed in the last few years[1,2]. A fracture toughness of more than 20 $MPam^{0.5}$ and a fracture strength of more than 2 GPa have been obtained on the toughest ceramic materials[3,4]. The current research interest is being focused on their engineering applications, which necessitate optimization of the processing parameters which influence their performance and a full understanding of environmental effects on these materials[5-7].

It is well recognized that the amount of the stabilizing agents (i.e., Y_2O_3, CeO_2) is one of the variables which critically influence the mechanical properties[2,8,9]. A tougher ceramic is achieved with zirconia containing 2 mol% Y_2O_3 than with materials containing either 2.5 or 3 mol% Y_2O_3[8,9]. Materials containing less than 2 mol% Y_2O_3 show decreased toughness and strength as a consequence of the spontaneous t $\longrightarrow$ m transformation, which induces severe microcracking, which if of sufficient density can coalesce to form macrocracks during cooling from the sintering temperature[2,9]. TZPs containing more than 3 mol% Y_2O_3 again show a decrease in the mechanical properties, but as a result of the reduced transformability. The factor can be expressed as a function of the amount of the transfomable tetragonal zirconia in the transformation zone and the actual size of the transformation zone[10,11]. The chemical free energy change associated with the t $\longrightarrow$ m transformation, which is the driving force for the transformation toughening, is reduced by the increased amount of the stabilizing agent[12].

It is well known that TZP ceramics suffer from environmental degradation when aged in a water/humid atmosphere in the temperature range 100 to 400°C[5-7]. The low temperature ageing degradation, especially under hydrostatic pressure and moist conditions, is also strongly influenced by their microstructural parameters such as grain size and the composition[13,32]. Coarse grained TZPs show more severe degradation than do fine grained TZPs. A critical grain size has been established experimentally for a range of TZPs containing various amount of stabilizing agent[14], below which degradation is not observed.

The grain size of most commercially available TZP ceramics ranges from 0.2 to 1.0 μm[19], with the Y_2O_3 content varying from 2 to 3 mol%. The composition dependence of the mechanical properties in these fine grained TZP ceramics has been widely studied. Both the fracture toughness and fracture strength peak at a particular composition in the composition range[4,9]. The fracture toughness peaks at 2 mol% Y_2O_3 and fracture strength at 2.5 or 3 mol% Y_2O_3. It is noticeable that the peak toughness does not coincide with the peak strength in terms of composition.

A recent literature survey on TZP ceramics has indicated that only limited information is available concerning the grain size dependence of the mechanical properties in TZP ceramics[1], although it has been theoretically recognized that the transformability and size of the transformation zone are two paramount important parameters in determining the mechanical performance of the TZP ceramics[10,11]. The transformability and the size of the transformation zone are expected to increase with increasing grain size as a consequence of increased possibility of transformation nucleation[15-17] and autocatalitical transformation[18,19].

The objective of the present investigation is, therefore, to study the grain size dependence of the mechanical properties in TZP ceramics.

2. Experimental Work

Y_2O_3-doped zirconia powders containing 2, 2.5, and 3 mol% Y_2O_3 respectively were obtained from Toyo Soda

Manufacturing Co, Japan. The as-received powders were deagglomerated by wet-milling in proponal for 24 hrs with zirconia as the milling media. The milled powders were then compacted in a steel die of 45 mm diameter at a pressure of 30 MPa, followed by sintering at temperatures ranging from 1400 to 1600 °C for various times, from 2 to 30 hours. The grain sizes which resulted were in range of 0.5 to 2 μm. The sintered specimens were cut into bars of 2.5x3.5x>20 mm, which were then polished to a 6 μm finish. The mechanical property testing included three point bend tests for fracture strength(span 20 mm), single edge notch beam (SENB) (notch 400 μm)[20] and microindentation (load 10 kg) to determine fracture toughness and hardness. The microstructural characterization included examination of the fracture and polished surfaces using SEM, and phase analysis using X-ray diffraction. Grain size measurements were made using the line interception method, when more than 200 grains were counted for each of the measurements.

3. Experimental Results

Table 1 shows the sintering temperatures and the sintering times employed for fabrication of the specimens. Fine grained materials were obtained at the lower sintering temperatures (1400 to 1450 °C). More cubic phase, Figure 1(a,b,c,d), was present in the specimens sintered at the higher temperature (1600 °C) and for longer periods at the low temperatures (1500°C/20 hours, 1550°C/8 hours). Table 2 shows the phase analysis of the as-sintered specimens, compared with the phase analysis given in reference (4) for the same materials. The density of the as-sintered

Table 1. Sintering Temperatures and Sintering Times Used in
Firing of the Specimens

Specimens	Composition mol% yttria	Sintering Temperature / C	Sintering Time / hrs
1	2, 2.5, 3	1400	2
2	2, 2.5, 3	1450	2
3	2, 2.5, 3	1500	2
4	2, 2.4, 3	1500	20
5	2, 2.5, 3	1500	30
6	2, 2.5, 3	1550	2
7	2, 2.5, 3	1550	8
8	2, 2.5, 3	1600	2

Figure 1. Polished surfaces of the sintered TZP ceramics
at 1400°C/2 hrs (a); 1500°C/2 hrs (b); 1600°C/2 hrs (c);
and 1500°C/ 20 hrs. Grain size and composition
distribution of the specimens change with the different
thermal treatments. The large grains in (d) are
cubic phase.

Table 2. Phases of the Sintered Specimens Measured Using XRD
Analysis

2.0Y/1400-1600/ 2	Tetragonal + Cubic (trace)
2.5Y/1400-1600/ 2	Tetragonal + 17 wt% Cubic (4)
3.0Y/1400-1600/ 2	Tetragonal + 28 wt% Cubic (4)
2.0Y/1500/ 20	Tetragonal + 11 wt% Cubic
2.0Y/1550/ 8	Tetragonal + 12 wt% Cubic
2.5Y/1500/ 20	Tetragonal + 33 wt% Cubic
2.5Y/1550/ 8	Tetragonal + 38 wt% Cubic
3.0Y/1500/ 20	Tetragonal + 33 wt% Cubic
3.0Y/1550/ 8	Tetragonal + 38 wt% Cubic
2.0Y/1500/ 30	Monoclinic + 38 wt% Cubic (Cracked
2.5Y/1500/ 30	Monoclinic + 38 wt% Cubic (Cracked
3.0Y/1500/ 30	Monoclinic + 39 wt% Cubic

specimens (>99% theoretical density) did not change to any extent with the different sintering temperatures and the times.

Figure 2 shows the grain size dependence of the three point bend strength. It can be seen that the fracture strength of materials containing 2 mol% Y_2O_3 does not change significantly with increased grain size, only decreasing slightly with increasing grain size. In contrast, the fracture strength of materials containing both 2.5 and 3 mol% Y_2O_3 shows a maximum at a particular grain size. The fracture strength of materials containing 3 mol% Y_2O_3 peaks at a slightly higher grain size (1.44 μm) than that (1.22 μm) of the materials containing 2.5 mol% Y_2O_3. The maximum strengths achieved in materials containing the different amounts of Y_2O_3 are, however similar (1000 to 1100 MPa). The fall in the fracture strength at grain sizes of 1.86, 2.0 and 2.2 μm for TZP-2, 2.5 and 3 mol% Y_2O_3 respectively corresponds to the critical grain size for each of the compositions respectively, at which spontaneous transformation occurs on cooling.

Figures 3 and 4 show the grain size dependence of the fracture toughness of the materials, determined using both the microindentation and single edge notch beam(SENB) methods respectively. It is generally accepted that SENB overestimates the fracture toughness for transformation toughened ceramics.

It is noted from Figure 3 that the indentation toughness shows a maximum at a particular grain size for each of the compositions. The grain size at which the fracture toughness reaches a maximum varies with the amount of Y_2O_3 in the TZPs. The relevant grain sizes for materials containing 2, 2.5 and 3 mol% Y_2O_3 are 0.87, 1.34 and 1.46 μm respectively.

Figure 4 shows the SENB toughness as a function of

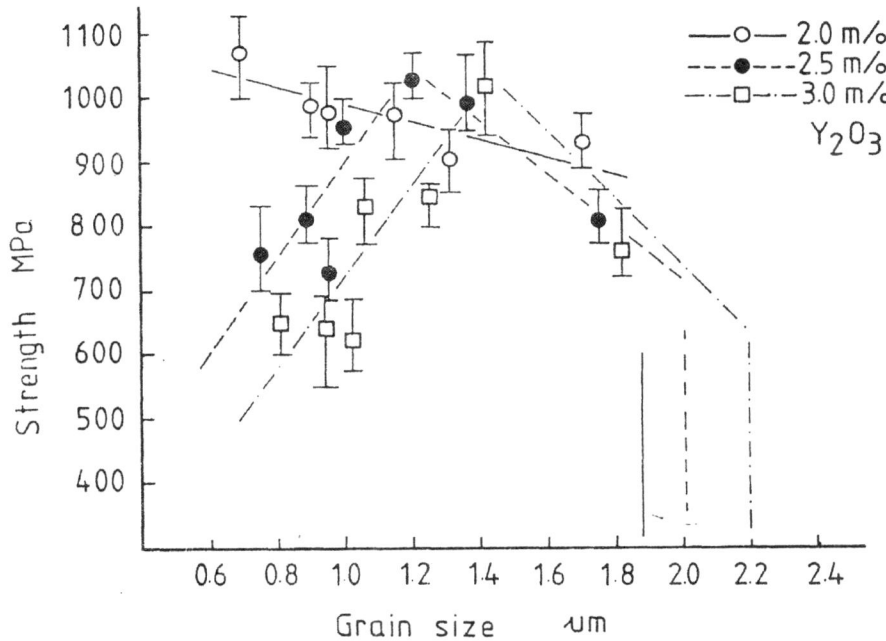

Figure 2. Three point bend strength as a function of grain
size for materials containing 2, 2.5 and 3 mol% Y_2O_3,
respectively. The strength peaks at a particular grain
size for both TZP-2.5 and -3 mol% Y_2O_3 ceramics.

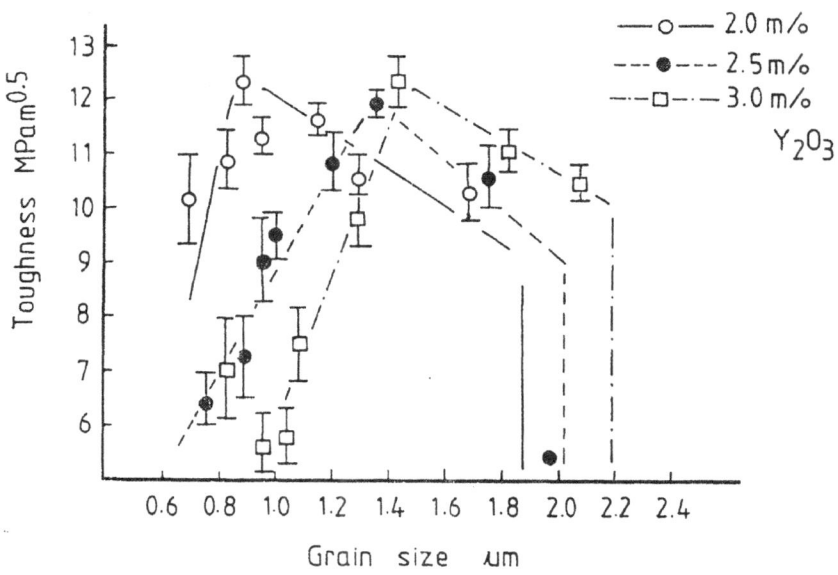

Figure 3. Indentation Toughness (10 kg) as a function of grain
size for the materials containing various amount of Y_2O_3.

grain size, demonstrating a similar trend as in Figure 3, although the SENB technique overestimates the toughness values. The optimum values occur at similar grain sizes for the materials containing 2, 2.5 and 3 mol% Y_2O_3.

Figure 5 shows the dependence of the three point bend strength on the amount of Y_2O_3 for various grain sized materials. The relationship between the amount of Y_2O_3 and the fracture strength is dependent on grain size of the materials. When the grain size is less than 1 μm (i.e., 0.76 and 0.95 μm grain sized materials), the fracture strength decreases significantly with increasing amount of Y_2O_3 from 2 to 3 mol%. The 1.02 μm grain sized material shows a slight decrease in the fracture strength with increasing Y_2O_3 content. In contrast, the strength is a maximum at 2.5 mol% Y_2O_3 for 1.20 μm grain sized materials. The fracture strength increases with increasing Y_2O_3 addition in the composition range investigated when the grain size is 1.37 μm. The overall decrease in the fracture strength with increasing Y_2O_3 addition for 1.77 μm grained material is due to both the presence of considerable amount of cubic phase and the spontaneous t $\longrightarrow$ m transformation of the large tetragonal zirconia grains.

Figures 6 and 7 show the composition dependence of the fracture toughness, determined using both the microindentation and SENB techniques, for the range of materials. It can be seen from Figures 6 and 7 that the relationship between the fracture toughness and composition is also dependent on the grain size of the materials investigated. In Figure 6, the fracture toughness decreases with increasing Y_2O_3 content from 2 to 3 mol% for 0.89, 1.02 and 1.20 μm grain sized

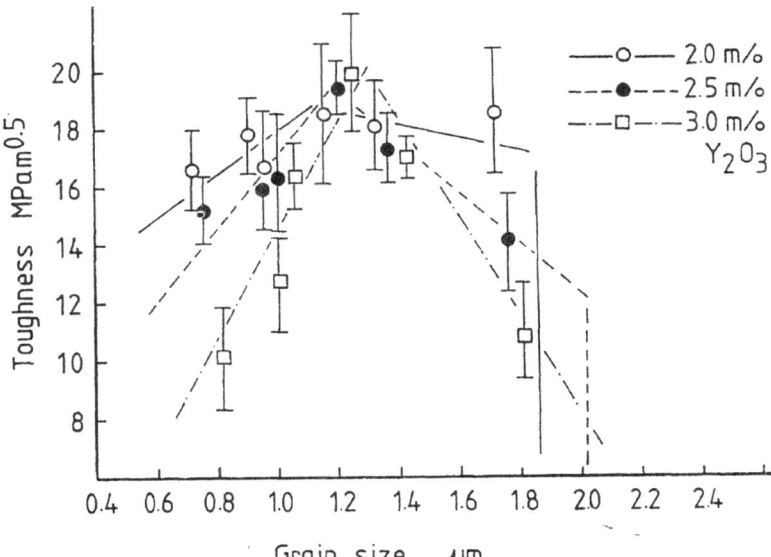

Figure 4. SENB toughness as a function of grain size for the
same materials as in Figure 3. The SENB toughness shows
the same trend as the indentation toughness in Figure 3.
However, SENB overestimates toughness values of the
transformation toughened ceramics.

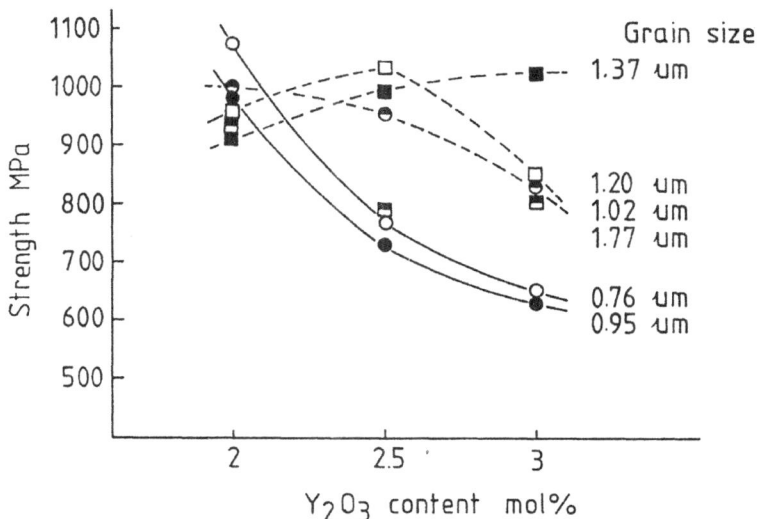

Figure 5. Three point bend strength as a function of
Y_2O_3 additon for materials with various grain sizes.
When the grain size is less than 1.20 μm the strength
decreases with increasing Y_2O_3 addition. However
the strength increases with increasing Y_2O_3 addition
when grain size is 1.37 μm.

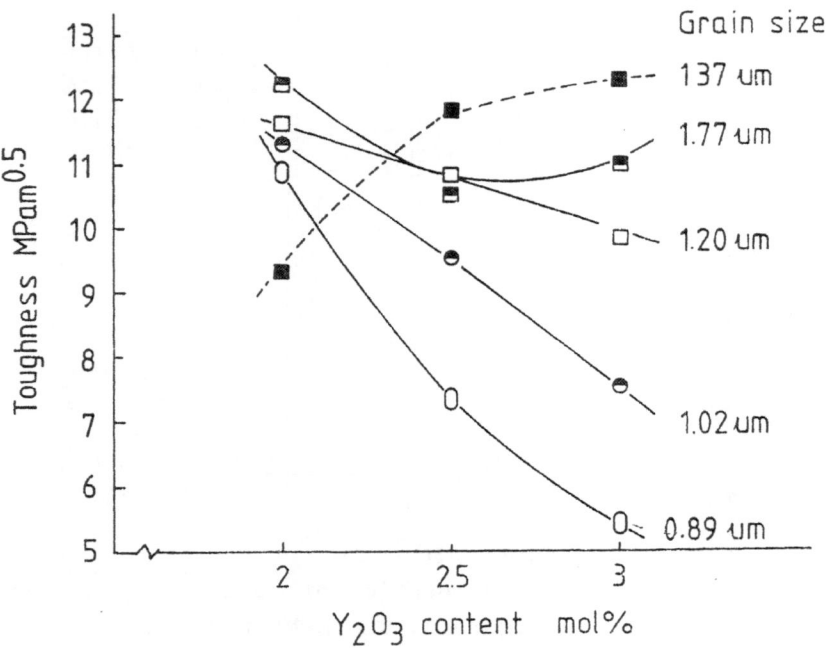

Figure 6. Indentation toughness as a function of Y_2O_3 addition for the various grain sized TZP ceramics. The toughness decreases with increasing Y_2O_3 addition for fine grained materials. In contrast, The toughness increases with increasing Y_2O_3 addition when grain size is 1.37 μm. The decrease in indentation toughness with increasing Y_2O_3 addition for 1.77 um grained material is considered to be due to the presence of cubic phase.

materials. In particular, the 0.89 μm grain sized material shows a more significant decrease with Y_2O_3 content than does either 1.02 or 1.20 μm grain sized materials. The composition dependence of the fracture toughness for these fine grained TZPs is consistent with the experimental results reported in literature[2,8,9]. In contrast, the toughness increases with increasing Y_2O_3 addition when the grain size is 1.37 μm. The overall decrease in the indentation toughness from 2 to 3 mol% Y_2O_3 for 1.77 μm grain sized material is due to the presence of a considerable amount of cubic phase.

In Figure 7, the SENB toughness values show a similar trend, the 0.76 and 0.95 μm grain sized materials decreasing their toughness with increasing Y_2O_3 content(especially from 2.5 to 3 mol%). The fracture toughness for the 1.02 μm grain sized material is almost independent of Y_2O_3 content, whereas the 1.20 μm grain sized material shows a linear increase in the SENB toughness with increasing Y_2O_3 content. The 1.37 and 1.77 μm grain sized materials show a minor and significant decreases in the toughness respectively.

4. Discussion

The effect of grain size on toughness and strength in transformation toughened ceramics is a complex issue. Two operating factors determine the final properties. On one hand, an increase in grain size gives an increase in the transformability (i.e., tetragonal to monoclinic), therefore increasing toughness and strength. The limit of the parameter is the grain size at which spontaneous tetragonal to

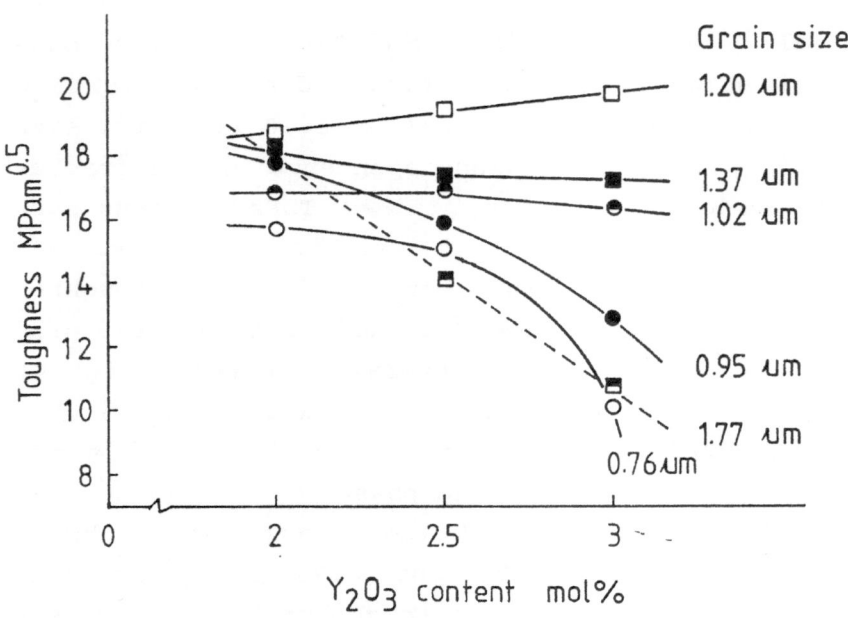

Figure 7. SENB toughness as a function of Y_2O_3 addition for the same specimens as in Figure 6. The SENB toughness shows similar trends as for the indentation toughness in Figure 6.

monoclinic transfromation occurs during cooling from sintering temperature, leading to degradation in the mechanical properties. In contrast, an increase in grain size gives an increase in the Griffith flaw size thereby reducing strength and toughness. In addition, high sintering temperature and prolonged sintering time may introduce a large amount of cubic phase, enhancing thermal mismatch stresses and reducing the amount of tetragonal available for transformation. The relative contribution of each of these factors will be discussed in the following sections.

The exact role of the grain size in the nucleation of the t $\longrightarrow$ m transformation still remains unclear. The relevance of the present results to the available theories will be discussed in Section 4.2.

Finally, a discussion will focus on the influence of test methods on the toughness values measured.

4.1 Effect of Grain Size on Transformability and Properties

As is shown in Figures 2 to 4, both the fracture toughness and fracture strength peak at a particular grain size for each composition. The fall off in the mechanical properties above the peak grain size is a result of high sintering temperature(1600 ^{o}C) or prolonged periods of time at 1500°C(20 hours) and at 1550°C(8 hours). The prolonged sintering time at such temperatures results in a substantial amount of cubic phase, the presence of which gives a negative effect on the mechanical properties as a consequence of reduced amount of transformable tetragonal and thermal mismatch stress. The very rapid fall off in strength and toughness at larger grain sizes is associated with spontaneous tetragonal to monoclinic transformation during cooling from sintering temperatures. For smaller grain sizes, both the fracture toughness and fracture strength(Figures 2 to 4) increase almost linearly with increasing grain size. This region will now be discussed in detail.

The toughness of transformation toughened ceramics is determined primarily by the amount of tetragonal to monoclinic transformation that occurs around a propagating crack. This quantity can be described by two parameters, namely, the fraction of available tetragonal grains which transform(Vf) and the distance from the crack to which transformation occurs(h). These parameters are related to the toughness by the equation developed by Evans et al[10,11]:

$$\Delta K = \eta.E.\Delta V.V_f \sqrt{h}/(1-\nu) \tag{1}$$

in which E and ν are the elastic modulus and Poison's ratio of the material, ΔV the volume dilation associated with the phase transformation, Vf the fraction of the transformable tetragonal zirconia, h the width of the transformation zone and η is a constant determined by the nature of the transformation[10].

Experimental results in literature are not in full agreement as to the applicability of this equation or more specifically, as to whether toughness increases linearly with $Vf\sqrt{h}$. The transformation zone(width) in fine grained(0.3-0.6 μm) TZPs containing 2 to 3 mol% Y_2O_3 was studied by Swain[22] and Masaki and Sinjo[23]. It was shown that the zone size decreased with increase in Y_2O_3 content. Transformation zones(depths) of 4.6, 1.6 and 1.5 μm were estimated for TZPs containing 2, 2.5 and 3 mol% Y_2O_3 respectively. Swain[22] experimentally confirmed that the fracture toughness increased linearly with increasing $V_f\sqrt{h}$, consistant with the theoretical prediction by Evans et al[10,11]. Masaki and Sinjo[23] on the other hand, observed a non-linear relation between fracture toughness and $Vf\sqrt{h}$, an apparent contradiction to the theoretical prediction[10,11]. Masaki and Sinjo also observed the fracture toughness to be independent of the grain size for their fine grained TZP ceramics, ranging from 0.3 to 0.6 µm. This range in grain size is, however, small, and so little variation in mechanical properties would be expected.

The influence of increased grain size on critical flaw size can be described using the Griffith's equation[21],

$$\sigma = K_{1C} / (2 \pi C)^{\frac{1}{2}} \qquad (2)$$

the fracture strength (σ) is proportional to the fracture toughness (K_{IC}) when the critical flaw size (C) is constant, regardless of grain size. It was calculated from the Griffith equation that the critical flaw size of the sintered materials in this work was ~30 µm, which is considerably larger than the grain size range, 0.5 to 2 µm. Thus thermal mismatch cracks developed by increasing the grain size are not large enough to degrade strength and toughness. Therefore, this effect can be discounted and the Evans equation should apply. In other words, both the fracture toughness and the fracture strength of the TZP ceramics should be proportional to the product of the square root of the transformation zone width ($\sqrt{h}$)

and the amount of transformable tetragonal zirconia in the transformation zone (Vf).

Based on the equation derived by Evans et al (equation 1)[10,11] and known elastic constants (η = 0.23, ΔV = 0.05, E = 220 GPa and ν = 0.31, Ko = 3.0 MPam$^{0.5}$), values of Vf$\sqrt{h}$ were calculated, Table 3, and are plotted in Figure 8. These values are larger than those obtained by either Swain[22] or Masaki and Sinjo[23] for their fine grain sized materials (0.3 to 0.6 µm). The increased values in the present work are considered to be a result of the increased grain size. Extrapolation of the experimental results predicts the value of Vf$\sqrt{h}$ for the material of 0.4 µm grain size to be 1.4-1.5x10^{-3}µm, which is consistent with Masaki and Sinjo's experimental result[23]. Thus the Evans' model appears to applicable over a wide range of grain sizes and therefore the toughness and strength is entirely dependent on the value of Vf$\sqrt{h}$.

To understand the individual contributions to the fracture toughness by Vf, the fraction of transformable tetragonal zirconia in the transformation zone, and h, the transformation zone size, it is necessary to determine one of them. Several methods have been developed to estimate the transformation zone (depth) using XRD, including Double Wave Diffraction[24], progressive surface removal and phase identification[25]. It is possible to carry out phase analysis on the fracture surface using XRD. Table 4 shows such results determined on the fracture surface for fine grain sized TZP ceramics(0.5-1.0 µm)[4]. It is seen that fraction of transformed monoclinic phase varies with change in Y_2O_3 additon. One problem with such results is that reverse m $\longrightarrow$ t transformation occurs on the fractured surface[26,27]. Moreover it is also

Table 3 Calculated Transformation Żone Size for Various
Grain Sized TZP Ceramics

Composition	Grain size / μm	Transformation Zone Size $Vf\sqrt{h}(\mu m)^{\frac{1}{2}}$	h (μm)	
2 mol% yttria	0.3	1.36	3.80	(Ref. 22)
	0.4	1.43	4.20	(Ref. 22)
	0.6	1.43	4.20	(Ref. 22)
	0.7	1.60	5.22	
	0.8	1.70	6.29	
	0.9	2.09	8.91	
	1.0	1.84	6.94	
	1.2	1.91	7.45	
	1.3	1.40	4.00	
	1.7	1.35	3.72	
2.5 mol% yttria	0.8	0.76	1.16	
	0.9	0.96	1.86	
	1.0	1.33	3.63	
	1.1	1.45	4.26	
	1.2	1.75	6.13	
	1.4	1.98	7.98	
	1.8	1.66	5.67	
3.0 mol% yttria	0.8	0.87	1.61	
	0.9	0.51	0.54	
	1.0	0.53	0.58	
	1.1	1.00	2.04	
	1.3	1.58	4.66	
	1.4	2.07	8.72	
	1.9	1.78	6.47	
	2.1	1.66	5.67	

Table 4 Fraction of Transformed Monoclinic Zirconia on
the Fractured Surface[22]

Composition/Sintering Temperature C/ Time hrs	Fraction of Monoclinic on Fractured Surface
2.0Y/1400-1600/ 2	72 wt% Monoclinic
2.5Y/1400-1600/ 2	24 wt% Monoclinic
3.0Y/1400-1600/ 2	16 wt% Monoclinic

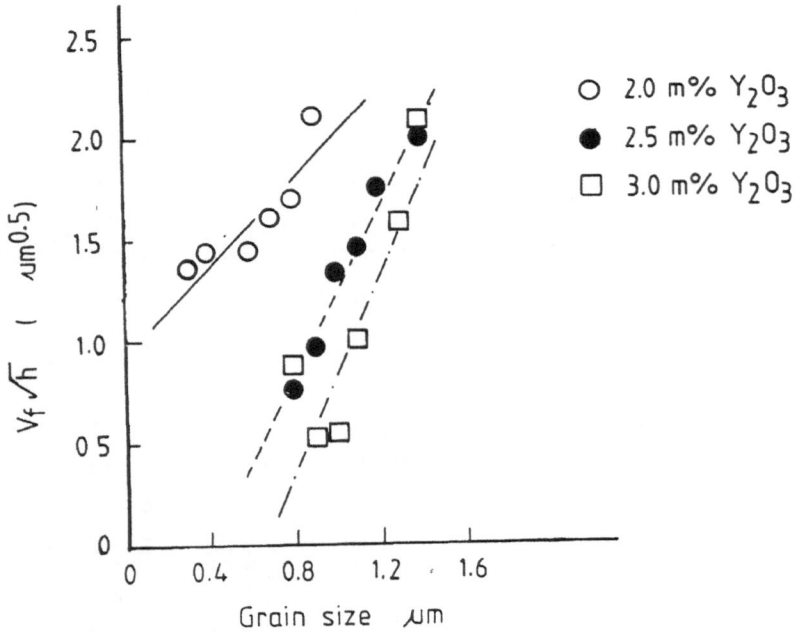

Figure 8. The dependence of transformation zone size on grain size for TZPs containing 2, 2.5 and 3 mol% Y_2O_3, respectively. The values of $Vf\sqrt{h}$ were calculated using Evans' equation and the indentation toughness values.

difficult to attain a large flat fractured surface suitable for XRD. Such results should therefore be treated with caution.

As is shown in Figure 1, the large grained materials contain a substantial amount of cubic phase, a consequence of prolonged sintered time at either 1500 or 1500 °C. The existance of the large cubic grains will obviously reduce the value of Vf, the fraction of transformable tetragonal in transfromation zone. However, the formation of the high yttria cubic grains is associated with a decrease in Y_2O_3 content of resulting tetragonal grains, which will become more transformable. The thermal mismatch stresses associated with co-existance of tetragonal and cubic phases may also result in more tetragonal grains to undergo the t $\longrightarrow$ m transformation. Therefore, a value of Vf would be expected to be 70% for the larger grained TZP ceramics, which is in agreement with the phase analysis results(Table 2). Using this, values of the transformation zone size(h) were calculated, using Equation 1, Table 3. Transformation zone size is seen to increase with grain sizes up to 1.77 μm above which it falls due to the presence of cubic phase. The addition of non-transformable second phase(e.g., Al_2O_3) would similarly reduce the transformation zone size of TZPs.

4.2 Effect of Grain size on the Nucleation of the t $\longrightarrow$ m Transformation

There are two conflicting theories which are used to explain the nucleation and growth of the t $\longrightarrow$ m martensitic transformation, namely the "non-classical" model by Heuer and Ruhle[35,36] and the "classical" model by Chen and Chaio[15,16]. Both the "classical" and the non-classical" theories suggest that the transformation is nucleation controlled although there is a disagreement on how the nucleation is induced. The non-classical theory involves a continous sequence of of states along a reaction path, i.e., there is a locally continous distortion of the parent phase into the martensite product in a small but finite region, until a critically sized nucleus of the martensite phase is reached. Therefore the nucleation is a homogeneous process. On the basis of thermodynamics and in-situ TEM observation, Heuer and Ruhle[35,36] suggested that the nucleation was invariably stress assisted, the stress arising from thermal expansion mismatch or morphology effects[35]. For TZP ceramics, grain boundaries are sites of localized residual stresses arising from the thermal expansion anisotropy of tetragonal zirconia ($\alpha a = 7.1 \times 10^{-6}$/C and $\alpha c = 11.4 \times 10^{-6}$/C). Unfortunately, they did not give a direct relation between grain size and possibility of nucleating the martensitic transformation.

In contrast, the classical theory believes that the nucleation is a hetergenous process, the nucleation actually occurs in the stress field of defects. The nucleus for the transformation is believed to occur by microstructural defects such as microcracks, lattice defects or the areas where the concentration of Y_2O_3 is below the average[30,31]. Chen and Chaio[15,16] demonstrated a clear relation between the grain size and the possibility of nucleating the martensitic transformation. It therefore appears that the "classical" model, which suggests that the nucleation is defect-controlled, is more applicable to the present case of TZP ceramics.

According to the classical theory[16],

$$F = 1 - \exp(-C \, \Delta g_{ex}^m \, D^n) \qquad\qquad (3)$$

in which F is the transformation probability of the particle; Δg_{ex}^m is a parameter referring to the driving force for the transformation; D is the size of the grain, and n=2 or 3 for surface and volume nucleation respectively. There is higher probability for a large tetragonal zirconia grain to nucleate the t $\longrightarrow$ m transformation than for a smaller one.

Another factor which may influence the stability of the tetragonal zirconia polycrystals with respect to the difference in grain size is the difference in surface energy. Garvie[33] and Lange[12] considered that the stability of tetragonal zirconia grains was associated with the grain size effects. Using thermodynamics, they demonstrated the existance of the critical grain size for the transformation. That the metastable tetragonal zirconia grains exist below the transformation temperature is a consequence of grain size effect. It is therefore appears that large grains have a high potential to undergo the transformation in terms of surface area effects.

4.3 Comparison of SENB and Indentation Methods for Measuring Transformation Toughened Ceramics

As is shown in Figures 3 and 4, SENB provided larger values of the fracture toughness than obtained by microindentation and is usually considered to be an overestimate. SENB generates a larger transformation zone, compared to microindentation such that the energy absorbed is much higher. In non-transformation toughened ceramics, the opposite is true (as an example, the present authors observed a SENB toughness of 3.70 MPam$^{0.5}$ for Refel Silicon Carbide, compared with a microindentation toughness of 5.45 MPam$^{0.5}$ for the same material). This is because SENB does not account for crack nucleation, only crack propagation[34].

In highly transformation toughened ceramics, transformation zone sizes change this factor. The notch width of 400 μm introduced an increased transformation zone and therefore caused a high toughness value, compared the case of microindentation. The transformation size increases with increasing notch width. This point can be illustrated by comparing the fracture surfaces for three point bend strength test and an SENB specimen, Figure 9(a,b). In comparison, the SENB specimen shows a smooth fracture surface, Figure 9(b), with no fracture origins visible, indicating that the fracture is a crack propagation process. This implies that a pre-introduced crack exists at the notch tip. The crack is fully embedded by a transformation zone, which resulted in the overestimated SENB toughness compared with microindentation toughness.

5. Conclusions

(1) The mechanical properties of TZP ceramics have been shown to be strongly dependent on the grain size.

(2) The fracture toughness and fracture strength peak at a particular grain size, the value of which is dependent on the Y_2O_3 content.

(3) The toughness is a linear function of $Vf\sqrt{h}$, the product of the volume fraction transformed and the

square root of the transformation zone size adjacent to the propagating crack.

(4) For large grained TZPs, the transformation zone size can extend to~9 µm from the crack.

(5) The relationship between grain size and mechanical properties favours the classical nucleation theory for the t $\longrightarrow$ m transformation.

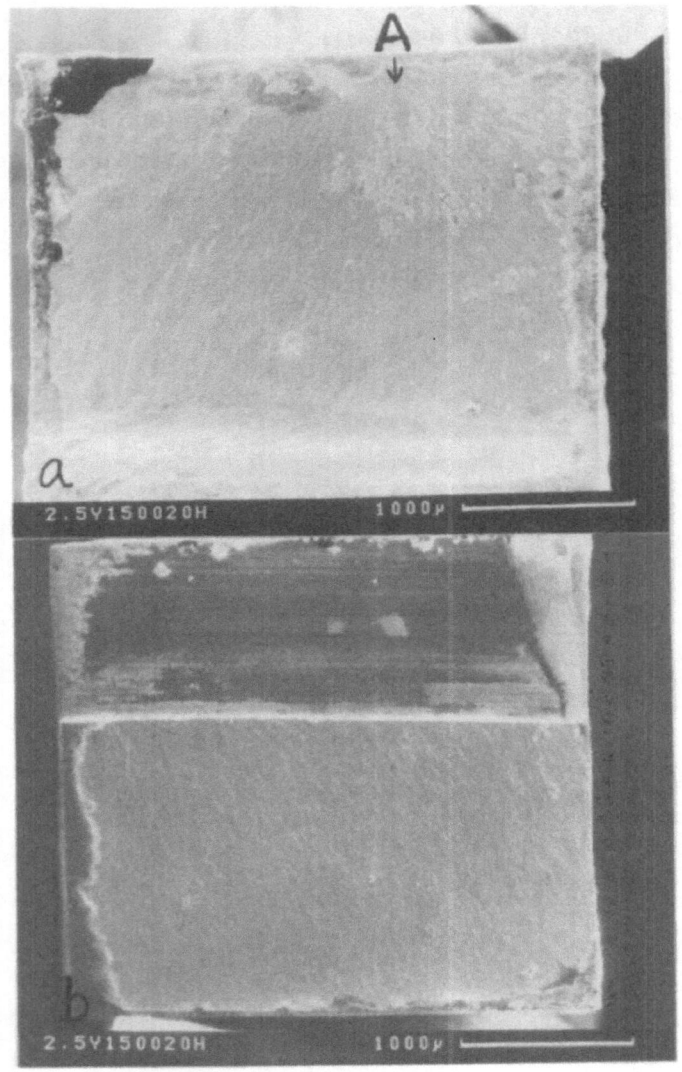

Figure 9(a,b) SEM photographs showing fracture surfaces of a three point bend strength specimen (a) and SENB testing bar (b). It is interesting to note that the fracture started from point A in (a). The SENB specimen shows a much more smooth fracture surface, indicating existance of a transformation zone at the notch tip.

References

1. I. Nettleship, Itn. J. High Tech. Ceram., $\underline{3}$ (1987) 1.

2. K. Tsukuma, and M Shimada, J. Mat. Sci., $\underline{20}$ (1985) 1178.

3. T. Masaki, J. Amer. Ceram. Soc., $\underline{69}$ (1986) 519.

4. K. Tsukuma, Y. Kubota and T. Tsukidate, PP.382 in Advances in Ceramics, Vol. 12, Science and Technology of Zirconia II. Edited by N. Claussen, M. Ruhle and A.H. Heuer. The American Ceramics Society, Columbus, OH (1984).

5. K. Tsukuma, Y. Kubota and K. Nobugai, Yagyo-Kyokai Shi (J. Jpn. Ceram. Soc.), $\underline{92}$ (1984) 133.

6. F. F. Lange, G. L. Dunlop and B.I. Davis, J. Amer. Ceram. Soc., $\underline{69}$ (1986) 237.

7. T. Sato and M. Shimada, J. Amer. Ceram. Soc., $\underline{68}$ (1985) 356.

8. F. F. Lange, J. Mat. Sci., $\underline{17}$ (1982) 240.

9. K. Haberho and P. Pampuch, Ceram. Int., $\underline{9}$ (1983) 8.

10. A.G. Evans and R. M. Cannon, Acta Metall., $\underline{34}$ (1986) 761.

11. R. McMeeking and A.G. Evans, J. Amer. Ceram. Soc., $\underline{65}$ (1982) 242.

12. F. F. Lange, J. Mat. Sci., $\underline{17}$ (1982) 225.

13. T. Sato, S. Ohtaki, T. Endo and M. Shimada, J. Amer. Ceram. Soc., $\underline{68}$ (1985) C-320.

14. M. Watanabe, S. Iio and I. Fukuura, PP.291 in Advances in Ceramics, Vol. 12, Science and Technology of ZrO_2 II. Edited by N. Claussen, M. Ruhle and a.H. Heuer. The American Ceramics Society, Columbus, OH (1984).

15. I-W. Chen and Y-H Chiao, Acta Metall., $\underline{31}$ (1983) 10.

16. I-W Chen and Y-H Chiao, Acta Metall., $\underline{33}$ (1985) 1827.

17. I-W Chen, Y-H Chiao and K. Tsuzaki, Acta Metall., $\underline{33}$ (1985) 1847.

18. M. Matsui, T. Soma and I. Oda, PP.371 in Advances in Ceramics, Vol. 12, Science and Technology of ZrO_2 II. Edited by N. Claussen, M. Ruhle and A.H. Heuer. The American Ceramics Society, Columbus OH (1984).

19. M. Ruhle, N. Claussen and A. H. Heuer, PP.352 in Advances in Ceramics, Vol.12, Science and Technolgy of ZrO_2 II. Edited by N. Claussen, M Ruhle and A.H. Heuer. The American Ceramics Society, Columbus OH (1984).

20. R. W. Davidge, "Mechanical Behaviour of Ceramics", Cambridge Press (1984).

21. A.A. Griffith, Phil. Trans. R. Soc., (London) $\underline{A221}$ (1920) 163.

22. M. V. Swain, Acta Metall., $\underline{33}$ (1985) 2083.

23. T. Masaki and K. Sinjo, Ceram Int., $\underline{13}$(1987) 109.

24. R.C. Garvie, R.H.J. Hannink and M.V. Swain, J. Mat. Sci. Lett., $\underline{1}$ (1985) 437.

25. N. Claussen and M. Ruhle, PP.137 in Advances in Ceramics, vol.3, Science and Technology of ZrO_2. Edited by A.H. Heuer and W. Hobbs. The American Ceramics society, Columbus OH (1981).

26. T. W. Coyle and R.M. Cannon, Amer. Ceram. Soc. Bull., $\underline{60}$ (1981) 3177.

27. D. B. Mmarshall and M.R. James, J. Amer. ceram. Soc., $\underline{69}$(1986) 215.

28. I. Nettleship, PhD Thesis, Department of Ceramics, The University of Leeds (1987).

29. M. V. Swain, J. Amer. Ceram. Soc., $\underline{68}$ (1985) C-97.

30. M. Ruhle, A Strecker, D. Waidelich and B. Kraus, PP.256 in Advances in Ceramics, Vol.12, Science and Technology of Zirconia II. Edited by N. Claussen, M. Ruhle and A.H. Heuer. The American Ceramics Society, Columbus OH (1984).

31. L.H. Schoenlein, M. Ruhle and A.H Heuer, PP. 275 ibid.

32. H.Y. Lu, H.Y. Lin and S.Y. Chen, Ceram. Int., $\underline{13}$ (1987) 207.

33. R.C. Garvie, J. Phys Chem., $\underline{69}$(1965) 1238.

34. M.V. Swain and N. Claussen, J. Amer. Ceram. Soc.,
    66 (1983) C-27.

35. M. Ruhle and A.H. Heuer, PP.14 in Advances in Ceramics,
    Vol. 12, Science and Technology of ZrO_2 II. Edited
    by N. Claussen, M. Ruhle and A.H. Heuer. The American
    Ceramics Society, Columbus OH (1984).

36. A.H. Heuer and M. Ruhle, Acta Metall., 33 (1985) 2101.

STRUCTURE OF STABILIZED ZIRCONIUM DIOXIDE FIBERS.

I.N. Yermolenko*, T.M. Ulyanova*, P.A. Vityaz** and I.L. Fyodorova**.
* Byelorussian Academy of Sciences, Minsk, USSR.
** Byelorussian Powder Metallurgy Association, Minsk, USSR.

ABSTRACT

Forming processes, microstructure and thermal stability of fibrous zirconium dioxide stabilized by magnesium and yttrium oxides are investigated. It is shown that the stabilization of tetragonal or cubic structures can occur at low temperatures during forming of oxide fibers. Due to the high dispersion of the oxide particles and the fibers reactivity, cubic solid solutions in binary systems ZrO_2-MgO, ZrO_2-Y_2O_3 are obtained with stabilizing additives contents equal to 13.3 and 12 mol%, respectively. Binary and ternary solid solutions keep their thermal stability up to 2000°C in air. In nitrogen a partial nitrogenation of ZrO_2 occurs and by interaction with amorphous carbon zirconium carbides are formed. In this case the ternary solid solution decomposes into three solutions of cubic structure and close crystal lattices parameters.

INTRODUCTION

The fibers of ZrO_2 obtained by oxidization of polymer salt-including fibrous materials have a number of specific properties, high thermal stability and reactivity. These properties are defined by a combination of crystallografic and structural factors, by the fiber composition and formation method.
The oxidation of polymer fibers impregnated with salt solutions allows to synthezise oxide fibrous materials with predetermined purity and high particles dispersion. This procedure gives the possibility of obtaining ZrO_2 fibers with different stabilization extents. Differently from ZrO_2 compacts, the porous zirconia fibers have greatly reduced the volume hindrance of polymorphous transformations and their properties are greatly affected by this feature.
The aim of this work was to investigate effect of the composition in binary systems ZrO_2-MgO and ZrO_2-Y_2O_3 on forming of solid solution with cubic structure, their sinterability and thermostability.

EXPERIMENTAL

Zirconia fibers were obtained according to the technique described previously(1). In binary system ZrO_2-MgO the contents of the second component varied in the 5-30 mol% range. In the system ZrO_2-Y_2O_3, in the 5-15 mol% range. Ternary solid solution contained 18 mol% Y_2O_3 and 2 mol% MgO.

Structure variations, sintering kinetics and grain growth were studied in the 400-1600°C temperature range . The stability of the ternary solid solution was explored at 900-2000°C in air, in nitrogen and in a reducing atmosphere in contact with amorphous carbon. Isothermal annealing were carried out for 0.5-10 hours at the aforementioned temperatures. We investigated composition, crystal structure, specific surface and size of oxide grains, . the morphology, the porosity and the strength of the fibers.

RESULT AND DISCUSSION

The results of X-ray phase investigation of zirconium dioxide fibers with and without yttrium oxide additives indicated that at the very begining (400-450°C), at the burn-out of carbon during the oxidation of polymer fibers, the ZrO_2 formed a tetragonal structure. With increasing temperature up to 800-900°C, one could observe on X-ray diffractograms either a monoclinic structure or a mixture of tetragonal and monoclinic phases when the binary system comprises 5-10 mol % Y_2O_3. By increasing the yttrium oxide contents up to 12-15 mol%, a solid solution with cubic structure was obtained (Fig.1).

When to the original hydrocellulose fibers, mixtures of zirconium and magnesium salts were added, one could obtain a solid solution with a cubic structure at 13 mol% MgO and 1550°C. This result is well correlated with the published data (2). Fig.2 represents the diagrams of binary systems ZrO_2-MgO from references (2) and (3) (dashed lines) with our results plotted as solid lines. Solid solutions with cubic structure were obtained at 1550°C and didn't decay by air annealing in the 1300-2000°C temperature range and subsequent cooling. Stabilization of the cubic structure with relatively low magnesium oxide content is perhaps due to the homogeneous distribution of zirconium and magnesium atoms within the fibers. In fact the polymer fibers' impregnation is performed with true salts solutions, which at the organics oxidation, burn-out, stage produce highly dispersed oxide particles which sinter in absence of volume hindrance.

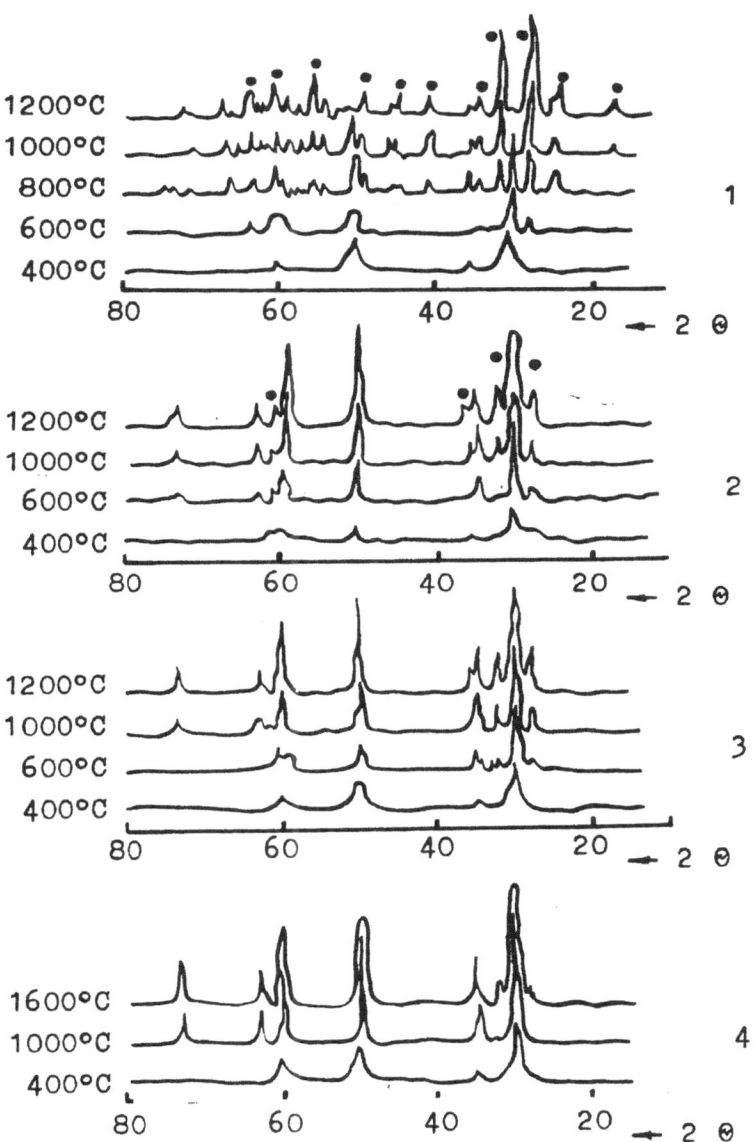

Fig. 1 : X- rays of ZrO_2 -based fibres, annealed at different temperatures:1) ZrO_2 -monoclinic; 2) $ZrO_2+5mol.\%$ Y_2O_3 3) $ZrO_2+10mol.\%$ Y_2O_3; 4) $ZrO_2 15\%$ mol. Y_2O_3

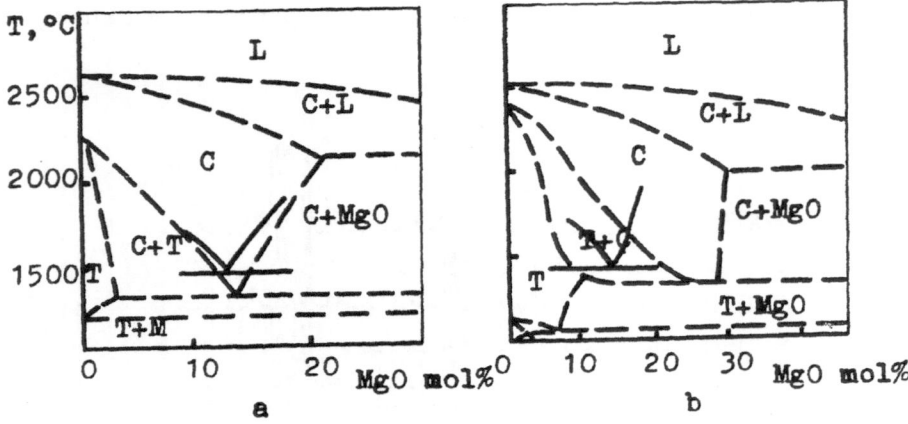

Fig. 2: Diagrams of ZrO_2 -MgO systems from references 2 (a) and 3 (b) (dashed lines) and obtained in this work (solid lines).

As it has been shown by the investigation of

the thermostability of the ternary solid solution comprising 18 mol% Y_2O_3 and 2 mol% MgO, it maintained its structure after annealing in air up to 2000°C.

Together with X-ray analysis technique we used IR-spectroscopy because it is very sensitive to the absorption bands of the individual oxides and monoclinic ZrO_2 modification. The absorbtion spectra of monoclinic ZrO_2 powders and fibers are presented in Fig. 3a together with IR-spectra of the ternary solid solution. The analysis of the spectra allowed to conclude that solid solutions were stable in the mentioned temperature range.

As a result of heating in a nitrogen atmosphere at 2000°C, a small quantity of zirconium nitride was formed on the ZrO_2 fiber's surface. On the X-ray diffractograms the newly formed phase is present as an impurity of the well-defined reflexes of ZrO_2 cubic structure.

As a result of the thermal treatment of the ternary solid solution in contact with amorphous carbon, which was formed during the carbidization of the phenol-formol or epoxy resin, a Zr carbide formation began above 1300°C. At 1700°C and above ternary solid solution decayed into three cubic solid solutions with close C.L. parameters. Solid solution decay was not followed by any segregation of individual metal oxides (Fig. 3b). The phenomenon is due to the fact that some ZrO_2 of the fibrous solid solution reacts with carbon, forming zirconium carbide.

Therefore the components' ratio may changes; the solid solution becomes poorer in ZrO_2 and richer in Y_2O_3. Different solid solutions may form according to the Y_2O_3 rich side of the binary diagram(4).

The highly-dispersed state of the zirconium solid solutions in the fibers promotes sintering during further thermal treatments. For describing sintering kinetics the power functions are used. The use of the generalized kinetic equation was explored:

$$X = 1-\exp(-kT^n) \qquad (1)$$

where X is a fraction of reacted matter,
k is a reaction constant,
n is a factor characterizing the process' step.

Sintering kinetics of porous materials may be characterized by the simple length change vs. annealing times at different temperatures. The experimental results give $\lg(-\lg(1-X))$ plots versus $\lg T$, the slope of which determines of power "n" in equation (1). Relations of "n" with annealing temperature "T" of fibrous ZrO_2-monoclinic and of ZrO_2+ 10mol% Y_2O_3 are shown on Fig. 4c. These curves show that the sintering of high-dispersed ZrO_2 fibers is caused by different mechanisms. In the 600-800°C temperature range surface diffusion prevails, between 800-1000°C - surface and volume diffusion occur, above 1000°C - the volume diffusion prevails. It may be considered that the variation of ZrO_2-fiber specific surface is connected with surface diffusion, which in temperature region between 1200 and 1600°C changes slightly. Zirconia grains growth is explained to a better extent by a volume diffusion process. In experiment the sintering processes are considered going on simultaneously when elementary acts of vacancies

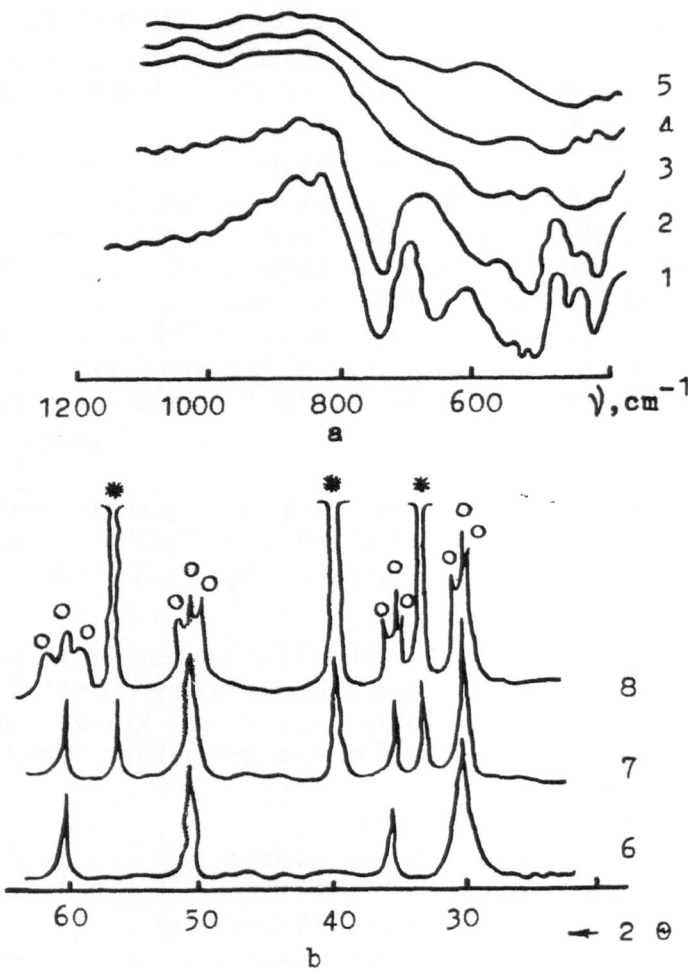

Fig. 3: IR- spectra (a) of powder (1) and fibrous (2) monoclinic ZrO_2 ; fibrous tetragonal ZrO_2 annealed at 900 °C (3) , 1600°C (4), 2000 °C (5); X-rays (b) of triple solid solution ZrO_2+18mol. % Y_2O_3+ 2mol.% MgO, annealed in contact with amorphous carbon at 1150 °C (6), 1400 °C (7), 1800 °C (8)

In order to define the activation energy of the sintering process in fibrous zirconium dioxide and its solid solution under

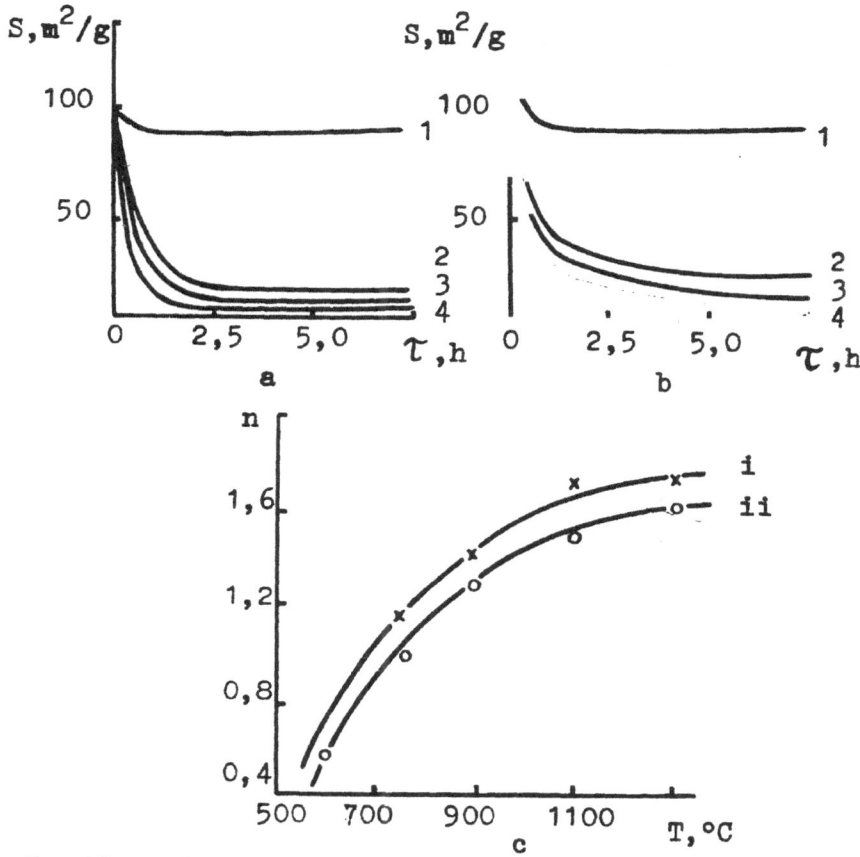

Fig .4: Specific surface of fibrous ZrO_2 (a) and ZrO_2+10mol.% Y_2O_3
 (b) versus annealing time at temperatures 600 °C (1), 800 °C (2),
1000 °C (3), 1200°C (4);
Factor "n" versus annealing temperature (c) for fibrous ZrO_2 (i) and
ZrO_2+10mol.% Y_2O_3 (ii).

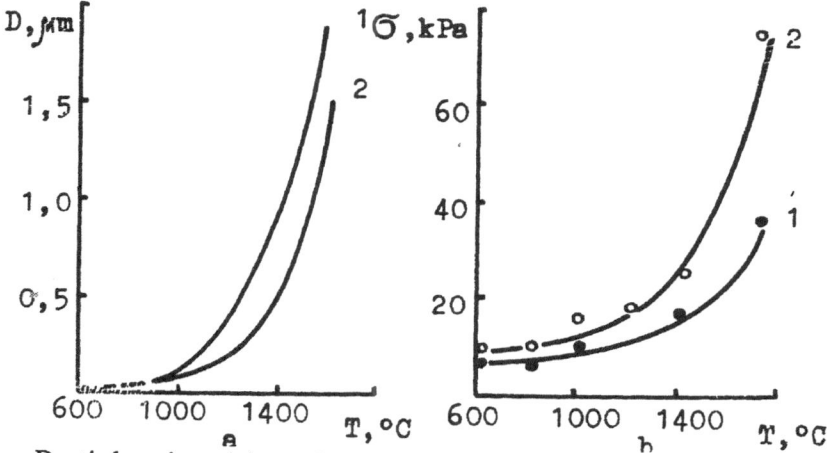

Fig. 5 :   Particle size (a) and strength versus annealing temperature

a constant heating rate, the Arrenius equation was used. From two experiments with heating rates V_1 and V_2 and temperatures T_1 and T_2, the activation energy can be determined by equation (5):

$$\ln(V_2 \cdot T_1 / V_1 \cdot T_2) = (E_a/R)\ (1/T_1 - 1/T_2) \qquad (2)$$

where E_a is activation energy,
R is universal gas constant.

In the case of sintering of monoclinic ZrO_2 fibres the activation energy was calculated to be 85,5 and for the case of solid solution of ZrO_2 with 10 mol% Y_2O_3 - 92,5 KJ/mol. These values are well correlated with other experimental data that show that introducing yttrium oxide in ZrO_2 fibers the grain growth and sinterability is slowed down.

Sintering of fibrous ZrO_2 is accomplished by reducing specific surface and changing of oxides' grain sizes in fibers. At temperatures between 600 and 1000°C grain size was not more than 0.01-0.03 μm. With increasing the annealing temperature up to 1600-1800°C grain size increased greatly up to 1 μm (Fig. 5 and 6). At the same time strengthening of zirconia fibers took place (Fig. 5b).

It should be noted that grain growth explained by a volume diffusion mechanism doesn't cause mechanical degradation. Furthermore, in Fig. 6 one can see that individual oxide particles at temperatures 1400-1600°C join to one another bridging fibers and creating a part of a ceramic frame.

Fig. 6: Structure of zirconia fibres in system ZrO_2-Y_2O_3-MgO at 800 °C (a), 1200 °C (b), 1400 °C (c), 1600 °C (c).

CONCLUSIONS

Synthesis of ZrO_2 fibers by oxidization of salt-containing polymer fibrous materials permits to obtain zirconia solid solutions with different extent of stabilization and homogeneous component distribution. Due to high dispersion of ZrO_2 particles it is possible to stabilize the cubic structure by introducing 12 and 13,3 mol% of yttrium and magnesium oxides respectively. Solid solutions of zirconium dioxide are thermally stable at heating in air and in nitrogen in 1300-2000°C temperature range. However by heating in contact with amorphous carbon the formation of carbide occurs and solid solution decays into three other ones with cubic structure.

However, synthesized zirconia fibers have a good sinterability and thermostability which make it possible to use them for different purposes.

REFERENCES

1. P.A.Vityaz, I.L. Fyodorova, I.N. Yermolenko and T.M. Ulyanova, Ceramics International, 9, (2), 46-47, (1985).
2. D.L. Porter and A.H. Heuer, J.Am.Ceram.Soc., 60, 543, (1977).
3. D.S. Rutman, Yu.S. Toropov, C.Yu. Pliner et al., High-temperature materials from zirconium dioxide. Moscow, Metallurgy, (1985).
4. P.J. Duwez, Electrochem. Soc., 98, 360, (1951).
5. P.S. Kislyi and M.A. Kuzenkova, Sintering of refractory materials. Kiev, Naukova dumka, (1980).

AUTHORS' INDEX

Page